Advances in Mathematics Education

Advances in Mathematics Education is a forward looking monograph series originating from the journal ZDM – Mathematics Education. The book series is the first to synthesize important research advances in mathematics education, and welcomes proposals from the community on topics of interest to the field. Each book contains original work complemented by commentary papers.

Researchers interested in guest editing a monograph should contact the series editors Gabriele Kaiser (gabriele.kaiser@uni-hamburg.de) and Bharath Sriraman (sriramanb@mso.umt.edu) for details on how to submit a proposal.

Gail F. Burrill • Leandro de Oliveria Souza
Enriqueta Reston
Editors

Research on Reasoning with Data and Statistical Thinking: International Perspectives

Springer

Editors
Gail F. Burrill
Michigan State University
East Lansing, MI, USA

Leandro de Oliveria Souza
Federal University of Uberlândia
Ituiutaba, Brazil

Enriqueta Reston
University of San Carlos
Cebu City, Philippines

ISSN 1869-4918 ISSN 1869-4926 (electronic)
Advances in Mathematics Education
ISBN 978-3-031-29458-7 ISBN 978-3-031-29459-4 (eBook)
https://doi.org/10.1007/978-3-031-29459-4

This Springer imprint is published by the registered company Springer Nature Switzerland AG
The registered company address is: Gewerbestrasse 11, 6330 Cham, Switzerland

Preface

Background

Today, the world is facing a challenge in educating its citizens. New technologies have enhanced many aspects of the quality of life, health, education, access to information, and communication. However, the improvement in the speed of communication, with the fast-growing capacity of technological devices, software, apps, social media, and so on, is changing lives, interactions, and the way in which discourse is structured and takes place. The concept of democracy is being remodeling by the impact of these changes, and statistics can be an important tool to investigate, argue, inform, convince, and guide rules of social behavior.

Many initiatives for improving statistics teaching and learning started in the 2000s. According to Pfannkuch (2018), this can be seen in the reform of statistics curricula in many countries, which recommended moving from just a focus on descriptive statistics to active exploration of data and probability modeling. Such curricular reforms build on research that uncovers and develops ways to enhance the study of how students reasoned from data (Frischemeier, chapter "Reading and Interpreting Distributions of Numerical Data in Primary School," this volume; Smucker and Manouchehri, chapter "Elementary Students' Responses to Quantitative Data," this volume). Since then, we have also encountered new technologies that support the learning of statistics through simulations (Burrill, chapter "An International Look at the Status of Statistics Education," this volume; Estrella, chapter "The Mystery of the Black Box: An Experience of Informal Inferential Reasoning," this volume). Globalization and big data (Monteiro and Carvalho, chapter "Interdisciplinary Data Workshops: Combining Statistical Consultancy Training with Practitioner Data Literacy," this volume) have forced us to reconsider the contributions of statistical thinking and the need for interdisciplinary pedagogical approaches as well as the need to confront the challenge of preparing teachers to maximize the potential of technology in their classrooms.

Studies (e.g., Burrill & Ben-Zvi, 2019; Martínez-Castro et al. chapter "Critical Citizenship in Statistics Teacher Education," this volume) and many curricular document (e.g., Bargagliotti et al., 2020) have highlighted that being able to provide sound evidence-based arguments and critically evaluate data-based claims are important skills that all citizens should develop to cope with the challenges of the twenty-first century workplace. Contemporary professional skills demand that people know more than just techniques to analyze and critique data; they also need to build strategies for collecting data, communicating results, and disseminating their analysis for critical discussion. Investigating with data is a change of paradigm that requires skills that integrate statistical thinking, reasoning, communication, and technology.

In today's world, rapid technological advances have facilitated the production and management of large data sets in diverse forms. Being able to create value with data by converting them to meaningful information, to critically evaluate and effectively utilize the information for decision-making, and to understand social and natural phenomena are important twenty-first century skills for all citizens. Thus, the importance of teaching and learning statistics as the "science of learning from data" is increasingly gaining recognition at all educational levels. The study of statistics provides students with tools, skills, ideas, and dispositions to use in order to react intelligently to information in the world around them. Because of the focus on improving students' ability to think statistically, statistical literacy and reasoning is becoming part of the mainstream school and university curriculum in many countries, although not as many as would be desirable (see Part I, this volume). Emerging issues in statistics education relate to dealing with "big data" as "data scientists," the use of statistics in thinking about social changes and policy decisions, and the impact these have on the school and the university curricula. Considering these trends, statistics education is a growing and exciting field of research and development that will enable us to build from the knowledge we have accumulated in the past about teaching and learning statistics to move forward in productive ways. This book brings up recent research on how statistics educators have been investigating teaching and learning statistics given the changing nature of the contemporary scene and the diverse places and countries in which researchers work.

ICME-14 Topic Study Group 12: Teaching and Learning Statistics

At the 14th International Congress in Mathematics Education (14th ICME) held on July 11–18, 2021, academic work on major issues in statistics education research was presented and discussed in Topic Study Group (TSG)-12 on Teaching and Learning Statistics. TSG 12 provided the venue for statistics educators, teachers, and researchers to present research on and to discuss issues in statistics education related to the teaching and learning of statistics in school and at the tertiary level including global trends; development and assessment of statistical literacy; connection of statistics education to social and political issues; preparation and professional development of statistics teachers and of statistics teacher educators; the impact of "big data" and technology-rich learning environments in statistics education; and the connection between learning statistics and learning data science.

Due to the global mobility restrictions brought about by the COVID-19 pandemic, the 14th ICME was a hybrid conference with an on-site component held at East China Normal University in Shanghai, China and an online conference platform via Zoom. TSG-12 had four online sessions with a total of 21 oral paper presentations and 7 poster presentations representing 23 countries. Except for 1 on-site presentation by a Chinese national, all papers were presented virtually with around 20 to 30 participants per session.

After TSG-12 at the 14th ICME, the post-conference discussion focused on submission of selected papers into a monograph to be published by Springer. The Call for Papers among the TSG-12 presenters was initiated by TSG-12 team members, Gail Burrill, Enriqueta Reston, and Leandro Souza. This volume of papers from TSG-12 of ICME-14 is a collection of peer-reviewed papers, each of which shares some common ground with the original paper presented during TSG 12. In addition, short papers on the perspective of statistics education in seven countries were included to provide more context in presenting an international perspective of statistics education. Thus, except for the opening section, individual papers of this book were based on the papers the authors presented at ICME-14 for TSG 12 and the discussion that ensued. These peer-reviewed research reports on teaching and learning statistics in this book have the potential to make the findings of statistics education research accessible to teachers, practitioners, and researchers and can contribute to the literature and a potential impact on teaching practice. To this end, this book is divided into five sections, each dealing with an aspect of research on teaching and learning statistics.

Part I: Statistics Education Across the World

This introductory section presents a survey of statistics education from a global perspective beginning with "An International Look at the Status of Statistics Education" by Gail Burrill and followed by seven short country reviews of statistics education in South Africa, Germany, Turkey, the United States, the Philippines, Brazil, and New Zealand from the perspective of statistics educators from these countries. Through examination of curricular documents and correspondence with statistics educations in Australia, Brazil, Canada (three provinces), Columbia, England, Finland Germany, Japan, Korea, New Zealand, Spain (Catalonia), South Africa, Turkey, and the United States, Burrill categorized curriculum intention and implementation in these countries into three levels; namely: (1) where the curricular objectives relate to introductory data analysis and the computation of statistical summary measures or formulaic probabilities; (2) where the curricular objectives go beyond data analysis to include some inference, but the ideas are primarily mathematical in the treatment of statistics; and (3) where the ideas extend beyond formulas for procedures, the curriculum is data based, and simulation-based procedures are a precursor for formal inference.

Part II: Data and Young Learners

This section presents three chapters that focus on the role of data in teaching statistics for young learners at school level in the United States, Germany, and Denmark. All three chapters argue for early development of young learners' statistical thinking and reasoning through data and data representations appropriate to their level.

In chapter "Elementary Students' Responses to Quantitative Data," Smucker and Manouchehri examine the development of statistical thinking among a cohort of elementary students in the United States surrounding the creation of graphical displays. Students showed flexibility in adapting techniques from categorical data to create displays that took into account various wingspans of class members, although for some class members it was difficult to separate graphing from the "rules" they had encountered in prior instruction.

In chapter "Reading and Interpreting Distributions of Numerical Data in Primary School," Frischemeier argues that the ability to read and interpret distributions is a cornerstone when analyzing data, but in primary school, this ability is often limited to reading and interpreting categorical data displayed in tallies, pie charts, and bar graphs. He reports on a teaching-learning arrangement that aimed to support primary school students (aged 10–11) in extracting information from graphs displaying numerical data.

In chapter "Young Learners Experiencing the World Through Data Modelling," Johansen reports how young learners can make experiences of *Allgemeinbildung* through data modelling. With roots in German philosophy, *Allgemeinbildung* aims towards students' self-development and the development of autonomy. As a contribution of statistics education towards this goal, she describes a teaching experiment conducted in a Danish third grade class (aged 9–10) where students performed all parts of the data modelling process. Using video recordings, the presentation of empirical data shows examples of essential statistical reasoning including posing questions relevant for data modelling, reasoning about how to structure data, and reasoning about how the data modelling sheds new light on the chosen topic. The chapter discusses how these students' experiences can be a potential resource for their *Allgemeinbildung*.

Part III: Data and Simulation to Support Understanding

This section presents four research reports on the use of data to explore and support understanding of certain statistical concepts at different levels of the educational system. Two of these chapters focus on conceptions of mathematics teacher educators and preservice teachers, that is, use of data-based tasks to explore conceptions on the informal line of best fit among mathematics teacher educators in Turkey, and conceptions on the margin of error among preservice teachers in the United States.

The other two chapters focus on students' conception of data in histograms among grades 10–12 students in the Netherlands and the development of informal statistical reasoning among seventh graders (12–13 years old) in Chile.

In chapter "Investigating Mathematics Teacher Educators' Conceptions and Criteria for an Informal Line of Best Fit," Gunbak-Hatil and Akar examine 11 mathematics teacher educators' conceptions for an informal line of best fit using a qualitative research design. Through intensive task-based interviews designed to understand teacher educators' conceptualizations of lines representing the relationship between two variables, their results showed that five of the teacher educators had "representer," four had "predictor," and two had "signal" as their dominant conceptions for an informal line of best fit.

In chapter "Introducing Density Histograms to Grades 10 and 12 Students: Design and Tryout of an Intervention Inspired by Embodied Instrumentation," Boels and Shvarts present the design and tryout of an intervention inspired by embodied instrumentation to explore student learning of density histograms. Through a sequence of tasks based on students' notions of area, students re-invented unequal bin widths and density in histograms. The results indicated that while students had no difficulty choosing bin widths or using area in a histogram, the reinvention of the vertical density scale required intense teacher intervention. This study contributes to a new genre of tasks in statistics education based on the design heuristics of embodied instrumentation.

In chapter "Margin of Error: Connecting Chance to Plausible," Burrill describes a simulation-based formula-light approach to answering the question "How it is possible to know about a population using only information from a sample?" in a course for elementary preservice teachers. Applet-like dynamically linked documents allowed preservice teachers to build "movie clips" of the features of key statistical ideas to support their developing understanding. The live visualizations of simulated sampling distributions of "chance" behavior enabled students to see that patterns of variation in the aggregate are quite predictable and can be used to reason about what population might be likely or "plausible" for a given sample statistic. The findings, using an adaptation of the Solo taxonomy, suggest that the approach leads to a relatively high level of student understanding.

In chapter "The Mystery of the Black Box: An Experience of Informal Inferential Reasoning," Estrella, Mendez-Reina, Rojas, and Salinas present the results of the implementation of a learning sequence designed by a Lesson Study Group, with the purpose of promoting informal inferential reasoning of seventh graders (12- to 13-year-olds). The results show that students' responses indicate they are able to coordinate their informal knowledge about the context and data in the problem to outline statements beyond the data they possess. The authors conclude that the learning sequence implemented in online teaching enabled the progressive development of some components of informal statistical inference.

Part IV: Data and Society

This section connects statistics education with data and society to assert the social relevance of teaching and learning statistics in all levels of education. With different areas of focus, the chapters in this section present research on developing critical citizenship among mathematics teacher educators in Colombia, statistical literacy and big data literacy to support the needs of constructive, concerned, and reflective citizens in Brazil, and the use of interdisciplinary data-based workshops in three African countries to gain data awareness, literacy, and skills in dealing with data arising from social issues.

In chapter "Critical Citizenship in Statistics Teacher Education," Martínez-Castro, Cardona-Zapata, and Jones argue that statistical investigation have the potential to contribute much to the development of critical citizenship among prospective mathematics teachers. Their research focused on 10 prospective teachers pursuing a degree in mathematics education in a well-known public university in northwestern Colombia. The teachers voluntarily participated in the study and carried out four classroom statistical investigations related to social crises. From the work and discussions in eight lessons where prospective teachers empirically studied various crises in society (global warming, minimum wage, malnutrition and obesity, and gender inequality in Colombia) using statistical tools and supplementary information from ideograms, autobiographies, and narratives, their findings support the stance that statistical investigations can promote statistical thinking and encourage the development of a sense of agency in preservice teachers.

In chapter "Toward Statistical Literacy to Critically Approach Big Data in Mathematics Education," Monteiro and Carvalho problematize the conceptualizations of mathematics and statistical literacy and highlight the need to integrate the new challenges and critical issues from data-science associated with Big Data in statistics education. They emphasize the need to develop school and teacher education curricula that address learning and teaching processes associated with a literacy that will enable learners to understand big data in daily life.

In chapter "Interdisciplinary Data Workshops: Combining Statistical Consultancy Training with Practitioner Data Literacy," Parsons, Stern, Szendroi, and David-Barrett report an approach to interdisciplinary data workshops they developed to bring together mathematical science students and practitioners to work on problems with real data in the practitioners' area of expertise. The approach was conceived through collaborative research projects aimed at tackling development challenges in Africa using data-based approaches to corruption in public procurement and farmer experimentation in agriculture. Their implementation of the approach in four workshops in three African countries provided important learning outcomes both for students, experiencing a problem-solving approach to working with real data in a genuine context, and for practitioners to gain data awareness, literacy, and skills.

Part V: Statistical Learning, Reasoning, and Attitudes

This section presents research that connects to the learning of statistics from authors from the United States, Sweden, Germany, and Spain. The chapters in this section focus on student learning related to the role of reasoning and argumentation, characteristics of sustainable learning, students' beliefs and attitudes, and the nature of statistical tables in textbooks and how these connect to statistical learning.

In chapter "Distinctive Aspects of Reasoning in Statistics and Mathematics: Implications for Classroom Arguments," Connor and Peters discuss distinctions between reasoning in statistics and mathematics and suggest that teachers need to support their students in making arguments that use appropriate reasoning for the subject in which they are engaged. Using diagrams of argumentation to illustrate differences in probabilistic and contextual reasoning in statistics compared to other areas of mathematics that rely on deductive reasoning and inductive or abductive reasoning to examine patterns, they present illustrations of classroom arguments illustrating different kinds of reasoning. They introduce the Teacher Support for Collective Argumentation (TSCA) framework as an analytical tool for researchers to investigate teacher education initiatives to facilitate teachers' construction of arguments that promote students' development of appropriate reasoning in mathematics and statistics.

In chapter "Sustainable Learning of Statistics," Innabi, Marton, and Emanuelsson bring attention to sustainable learning as an important idea related to teaching statistics. Using Variation Theory as a framework for the chapter, the authors claim that teaching statistics by considering variation in data and distribution might help students to strengthen and sustain their generic learning. Sustainable learning is used in the same sense as generative learning that continues and prepares students for the future.

In chapter "How Students' Statistics Beliefs Influence Their Attitudes: A Quantitative and a Qualitative Approach," Berens and Hobert focus on students' beliefs and attitudes in learning statistics. They present results of a survey on beliefs about statistics among 471 students of social sciences in a German university and establish connections to students' attitudes towards statistics. They found that rules-based beliefs about statistics have negative impacts on the students' attitudes, while investigative perspectives on statistics have positive impact on all dimensions of attitudes.

In chapter "Algebraization Levels of Activities Linked to Statistical Tables in Spanish Secondary Textbooks," Gea, Palluta, Artega, and Batanero analyze the algebraization levels involved in the mathematical activity linked to statistical tables in a sample of 18 Spanish secondary school textbooks (12- to 15-year-old students) from three different publishers. They performed a content analysis based on the classification of the statistical tables and the levels of algebraic reasoning described by earlier researchers. The results show an increase in algebraization levels required to apply both statistical and arithmetic knowledge and algebraic reasoning at the school level.

East Lansing, MI, USA — Gail Burrill

Ituiutaba, Brazil — Leandro de Oliveira Souza

Cebu City, Philippines — Enriqueta Reston

References

Bargagliotti, A., Franklin, C., Arnold, P., Gould, R., Johnson, S., Perez, L., & Spangler, D.A. (2020). *Pre-K-12 guidelines for assessment and instruction in statistics education II (GAISE II)*. American Statistical Association and National Council of Teachers of Mathematics.

Burrill, G., & Ben-Zvi, D. (2019). *Topics and trends in current statistics education research: International perspectives*. Springer International Publishing. https://doi.org/10.1007/978-3-030-03472-6

Pfannkuch, M. (2018). Reimagining curriculum approaches. In Ben-Zvi, D., Makar, K., & Garfield, J. (Eds.). *International handbook of research in statistics education*. Springer International Handbooks of Education. https://doi.org/10.1007/978-3-319-66195-7_12

Contents

Introduction

Andee Rubin and Robert Gould

Abstract As a way to frame the papers in this book, this introduction describes some of the developments in statistics education over the past several decades that inform the current focus on data science education. Progress in curriculum development, software development and the establishment of scholarly communities are discussed. The concept of "context" as an underlying component of both statistics and data science is examined, with an example that suggests how new methods of collecting and sharing data may expand our view of "context" and raise new questions for how students should encounter data.

Keywords Data science education · Data preparation · Visualization software

The papers in this book are part of a growing movement in today's educational landscape, born of the realization that data plays an increasingly central role in our lives – and that a basic understanding of data is becoming an essential life skill. There is no question about the trajectory of data science education; for the foreseeable future, it will be the focus of considerable development and research. But what of its past? As much as it might seem "new," there is actually a rich history of scholarship in mathematics and statistics education that preceded the current moment. Here are a few highlights of the last thirty years.

A. Rubin (✉)
TERC, Cambridge, MA, USA
e-mail: andee_rubin@terc.edu

R. Gould
UCLA, Department of Statistics, Los Angeles, CA, USA
e-mail: rgould@stat.ucla.edu

G. F. Burrill et al. (eds.), *Research on Reasoning with Data and Statistical Thinking: International Perspectives*, Advances in Mathematics Education,
https://doi.org/10.1007/978-3-031-29459-4_1

1 Curriculum Development

In the early 1990's, a curriculum called *Used Numbers* (1989) was developed by a team centered at TERC, an educational non-profit in Cambridge, Mass. The materials comprised a data-focused "replacement unit" for each grade level, introducing students to foundational concepts such as case, attribute, sorting and classifying, the shape of the data, modal clump, mean, and group comparison. Several units culminated in a data collection and analysis project; the most popular was a fourth-grade activity in which students collected data about injuries on their school playground and made safety recommendations to their principal based on the results. These units were eventually integrated into the *Investigations in Number, Data and Space* K-5 math curriculum (2017), also developed by TERC. (Unfortunately, over the course of the next versions of the curriculum, data activities began to disappear from Investigations, as the Common Core de-emphasized working with data in elementary school.) A line from the preface to *Used Numbers* still inspires us:

> We introduce students to good literature in their early years. We do not reserve great literature until they are older – on the contrary, we encourage them to read it or we read it to them. Similarly, we can give young students experience with real mathematical processes rather than save the good mathematics for later. Through collecting and analyzing real data, students encounter the uncertainty and intrigue of real mathematics…They cope with the real-world "messiness" of the data they encounter, which often do not lead to a single, clear answer. (Russell & Corwin, 1989, p. 1)

2 Software Development

Starting in the 1980's, software developers began to create programs specifically for students learning how to work with data, in contrast to tools for data analysis professionals. Designing these tools required considerations of how people learn about data, what basic concepts needed to be explained and/or illustrated, and what conceptual difficulties students often experienced. One of the earliest programs, called "Stretchy Histograms," let students manipulate the height of bars in a histogram to see how the changes impacted the mean and median of the distribution. Other early pieces of software were TableTop and TableTop Jr., specifically designed to introduce elementary school students to basic data visualization and analysis moves. In the 1990's, a family of tools whose developers knew one another and built on one another's work began to appear in the United States; these include TinkerPlots (2022), Fathom (2022), and CODAP (2022). Their developers shared a commitment to providing tools for students to construct their own graphs using a set of direct manipulation primitives, rather than choosing from a menu of graph types. They also shared a desire to connect work with data and experience with probability by incorporating probability simulation tools into their data analysis tools and regarding the output of simulations as data, subject to the same visualization and analysis moves as any data. Innovative educational technology also arose in 2014 in

New Zealand with the iNZight project (Wild and iNZight Team, 2023) which, among other features, provides interactive animations to help students visualize the effects of sampling variation on statistics and statistical graphics.

The important commonality among all of these tools is their focus on helping students learn about data through research-informed software design. A large number of research projects have used these tools to probe students' developing understanding of data and chance, and in many cases the software has been modified or extended based on the results of the research.

3 Scholarship Communities

During these decades, even as an emphasis on data waned in the mathematics curriculum, several intellectual communities interested in statistics and data education flourished. The International Conference on Teaching Statistics (ICOTS) has been meeting every four years, beginning in 1982, and the United States Conference on Teaching Statistics (USCOTS), which focuses more on undergraduate education, has been meeting every other year since 2005. The sponsoring organizations' websites (The International Association for Statistics Education (IASE, 2022) for ICOTS and the Consortium for the Advancement of Undergraduate (CAUSE, 2022) for USCOTS) have links to many of the proceedings. The *Statistics Education Research Journal* (SERJ, 2022), also sponsored by IASE, has been publishing peer-reviewed articles since 1987, all of which are available online.

Granted, most of these efforts were carried out under the umbrella of "statistics education" because "data science" hadn't been coined yet. But we maintain that many if not all of the basic concepts of data addressed in the past 30 years are just as relevant in this era of data science as they have ever been. Distributions still characterize data, signal and noise are still important concepts, covariation is still an important part of analysis, and regression analysis, long a part of statistics, is the backbone of machine learning. So, what does this shift from "statistics education" to "data science education" entail? We offer below some thoughts and an example of one difference that has stood out to us over the past few years of transition.

4 The Changing Role of "Context"

The increasingly prominent role of data in all aspects of modern life has changed what we have to consider as "context." Until recently, the data around which classroom statistical investigations revolved were assumed to be collected by the students themselves or from formally designed data collection protocols such as randomized assignment and random sampling. However, such an education fails to prepare students to engage in problem-solving in an age in which secondary data are ubiquitous, messy, and likely did not come from random samples. The increasing availability of secondary data means that interesting and important questions can be potentially addressed with data that were not collected for the purpose of answering them. In fact, one way to view data science education is as an extension of statistics education that emphasizes building bridges between data and question-posing (Wise, 2020).

"Context" is not a new term in statistics. Moore (1988) famously defined data as "numbers in context." Cobb and Moore (1997) elaborated on what Moore meant by providing an example of data taken from the Salem witch trials and illustrating how the context informs the analysis and interpretation. Schaeffer (2006) emphasized the role of context as a way of contrasting statistics with mathematics. In mathematics, he wrote, the goal is to remove context through abstraction. In statistics, on the other hand, the goal is to add context back in order to understand where the statistical model succeeds and where it fails and to use the results to solve a very particular problem in a very particular context.

This change in what data comprise calls into question our understanding of context. For one thing, the definition of data as "numbers" feels limiting when data can be words, images, sounds, and more. If students are to use secondary data to answer their own questions, they must interrogate the data and become familiar with how the data came to their desktop (virtual or physical) in order to evaluate its suitability for the intended task. This is one role of context: to establish the parameters of the problem. The other role of context is to consider how the data themselves were organized and presented for analysis or for classroom use. It's important to ask: How did this data file come to sit on my desktop, and what human or computer actions were involved in getting it there? Once a data set is judged to be appropriate for a given problem, the technical aspects of the data themselves must be considered. Are there missing values or unusual observations? Are the data structured appropriately for the intended analysis? These aspects of the data should be considered part of the context of the data.

5 An Illuminating Example

To spell out these ideas better, we provide an example of what is often regarded as a mundane task: data cleaning. We demonstrate the role of context even in data cleaning as an illustration of the increased importance overall of context in data science education.

Data cleaning, like other types of janitorial work, is often seen as a chore best deferred to others. But there are many good reasons for including it in the curriculum. The best justification is that real data are messy, and an academic preparation that contains only clean data will short-change students, who will find that they cannot use any of their data science education outside of the classroom. Another reason is that cleaning data, that is, preparing it for analysis, is far from a chore and can require deep contextual knowledge and sharp deductive and inductive reasoning skills.

Census at Schools (CAS) is an international program in which teens and preteens from around the world complete a survey that provides data for both local and international use. Teachers can request a random sample of responses and use the resulting data in their classroom. Although the program is international, participating countries maintain their own data, agreeing to share some common questions. In the United States, CAS is administered by the American Statistical Association (2022), which shared the data below.

The data contain the self-reported height and arm span (the distance from finger-tip to finger-tip when arms are extended horizontal to the ground) for a large number of students. These are part of a larger collection of body measurements, but these two variables, height and arm span, are useful in classrooms because they are linearly associated and can be used to test the hypothesis (proposed by ancient Roman architect Vitruvius around the first century BC) that arm span is equal to height.

While that may be the case in an ideal world, these data in their raw form are not useful. Not surprisingly, twelve- and thirteen-year-olds are not the most diligent measurers, and the anticipated structure fails to appear when examining the raw data (Fig. 1).

We can see more by "zooming" in on those who are shorter than 300 cm (10 feet) (Fig. 2). At this level of detail, some intriguing structures appear. Understanding these requires students to have an understanding of the context in which the data were generated. First, they need to understand something about the limits of human physiology. (What's the tallest possible person? The shortest?) Second, they need to understand mistakes students might make when measuring and recording data.

The "shadow" below the main cloud, for example, consists of people who are roughly in the middle percentiles of height, but whose arm spans are, on average, half of what they should be. Possibly, these students mis-interpreted arm span to be the distance from one hand to the middle of the back.

The smaller cloud centered around (60,60) could very well be from students who used imperial measurements (inches) rather than metric. The cloud to the left of the main cloud might be students who, for reasons we will never know, entered their height in inches and their arm span in centimeters.

Each of these hypotheses suggests ways of correcting the data that go beyond simply deleting erroneous data. Each of these corrections come with a new set of questions: is it valid to adjust the data in this way? How can we communicate which values are untouched and which are "corrected"? Carrying out the corrections themselves requires some knowledge of mathematical statistics (how is the mean of a collection of points affected by adding a constant value?) and computational know-how.

There are still mysteries in these data. For example, students with advanced data visualization skills might notice a suspiciously large number of points for which height and arm span are exactly equal. Should these observations be "corrected" and if so, how? What is the harm in leaving them alone?

In practice, the classrooms who participate in CAS do not see the data that produced the above figures. Instead, they see a cleaned version with no (or few) anomalies. We are not advocating that all students should see the original, messy data.

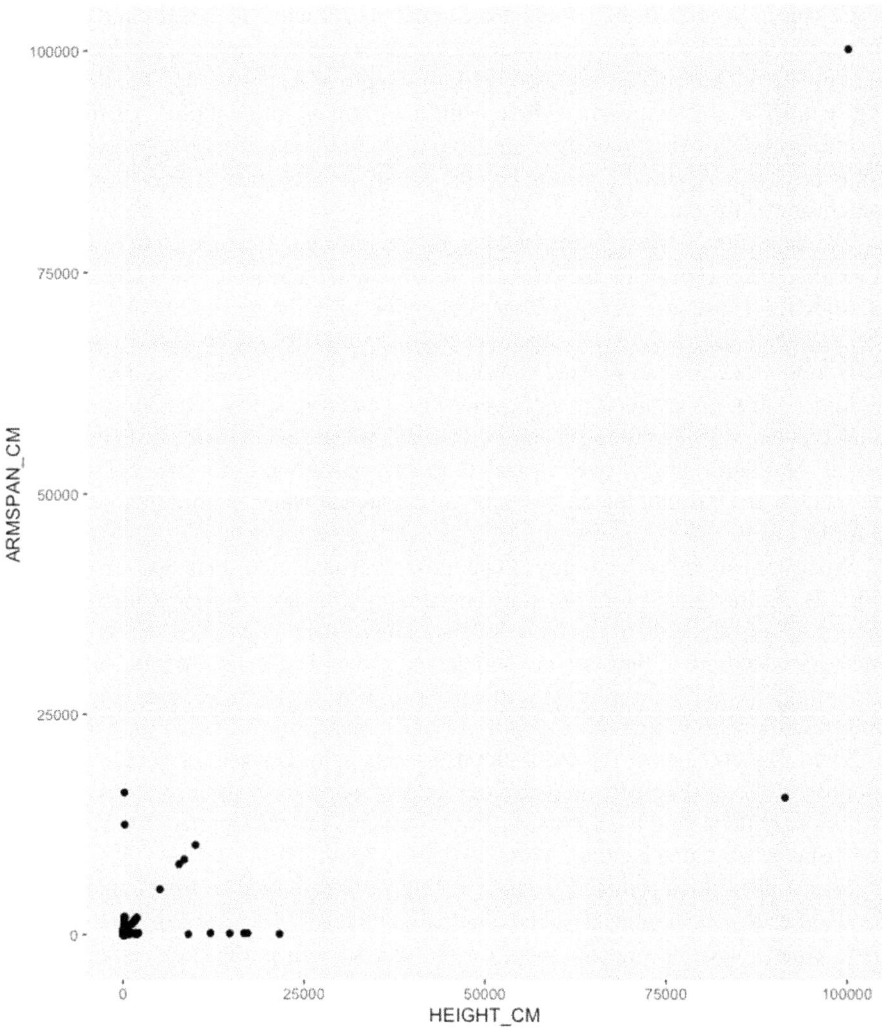

Fig. 1 Self-reported and recorded heights and arm spans (cm) for roughly 9300 middle-school students

Different lessons may require different levels of cleanliness. But there is an important reason for providing students with messy data sets at some point: it forces them to confront the context, both in the substantive sense and also in the sense of the data generating process. Not only does this better prepare students for handling real data beyond the classroom, but it enhances their ability to critically evaluate conclusions based on data sets that might not have been collected as carefully as possible.

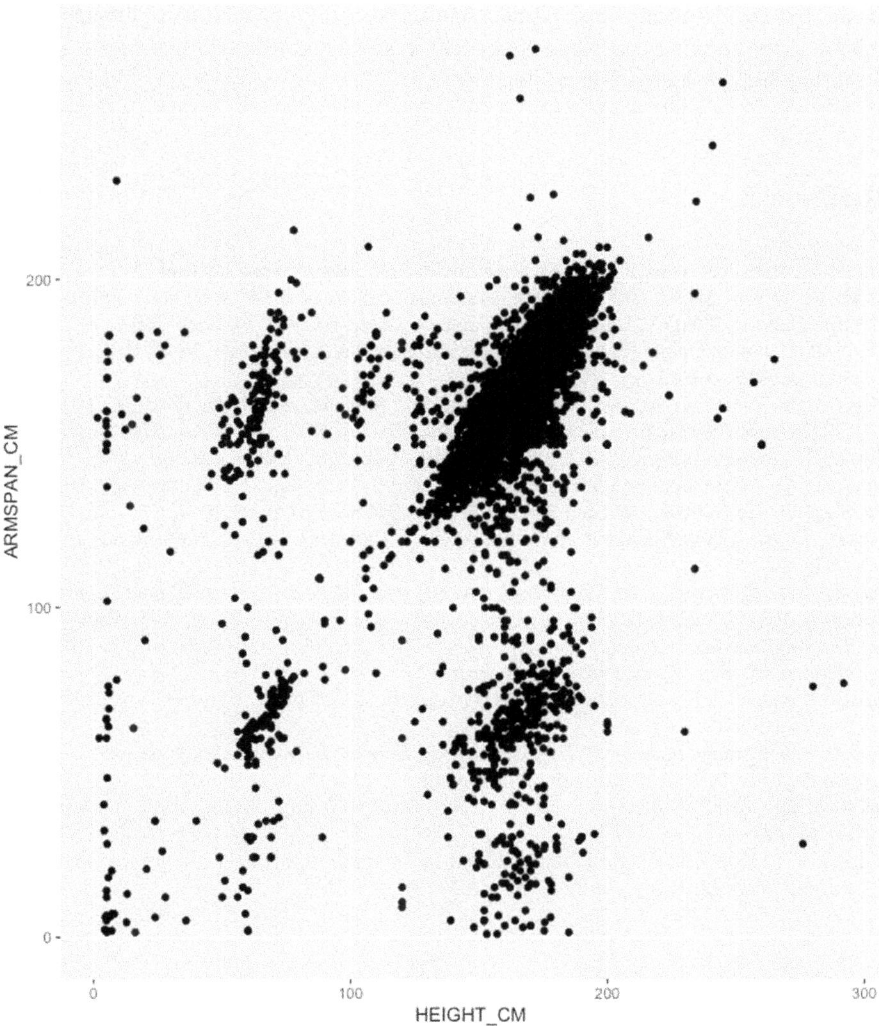

Fig. 2 The same graph "zoomed in" to focus on those with heights and arm spans less than 300 cm (sample size is now approximately 8400)

6 Conclusion

As we enter an age where data will be used more and more frequently for wide-reaching decisions, educating youth about the importance and complexity of context will only grow in importance. We urge statistics educators, data science education researchers and anyone else who wants to understand how to help students reason with data to keep context in mind, especially when algorithms become more “independent” and hard to understand. Hats off to the authors of the papers in this book for thinking hard about statistical learning and for making progress in this important field at such a critical juncture.

References

American Statistical Association (ASA) [website]. (2022). https://www.amstat.org/

Cobb, G. W., & Moore, D. S. (1997). Mathematics, statistics, and teaching. *The American Mathematical Monthly, 104*(9), 801–823. https://doi.org/10.1080/00029890.1997.11990723

CODAP (Common Online Data Analysis Platform) [Computer software]. (2022). Retrieved from https://codap.concord.org

Consortium for the Advancement of Undergraduate Statistics Education (CAUSE) [website]. (2022). https://www.causeweb.org/cause/

Fathom [Computer software]. (2022). Retrieved from https://fathom.concord.org/

International Association for Statistical Education (IASE) [website]. (2022). https://iase-web.org/

Investigations in Number, Data and Space®. (2017). (3rd edition). Pearson.

Moore, D. S. (1988). Should mathematicians teach statistics? *College Mathematics Journal., 19*(1), 3–7.

Russell, S. J., & Corwin, R. B. (1989). *Statistics: The shape of the data.* Dale Seymour Publications.

Scheaffer, R. (2006). Statistics and mathematics: On making a happy marriage. In G. Burrill (Ed.), *Thinking and reasoning with data and chance, 68th NCTM yearbook (2006)* (pp. 309–321). National Council of Teachers of Mathematics.

Statistics Education Research Journal (SERJ) [website]. (2022). https://iase-web.org/Publications.php?p=SERJ

TinkerPlots [Computer software]. (2022). Retrieved from https://www.tinkerplots.com/

Used Numbers. (1989). Dale Seymour Publications.

Wild, C and the iNZight Team. (2023). The iNZightVIT project, University of Auckland, inzight.nz.

Wise, A. F. (2020). Educating data scientists and data literate citizens for a new generation of data. *Journal of the Learning Sciences, 29*(1), 165–181.

Part I
Statistics Education Across the World

An International Look at the Status of Statistics Education

Gail Burrill

Abstract This paper summarizes the status of statistics education in fourteen countries across the world based on curricular documents and interviews with statistics educators from the countries. The discussion includes implementation barriers and moves towards future trends.

Keywords Data science · Implementation barriers · International statistics education · Statistical literacy

1 Country Survey Results

This study looked at the status of statistics education in the following countries: Australia, Brazil, Canada (the provinces of Nova Scotia, Ontario, and British Columbia), Columbia, England, Finland, Germany, Japan, Korea, New Zealand, Spain (Catalonia), South Africa, Turkey, United States. More details on the statistics curricula in Brazil, Germany, New Zealand, Turkey, South Africa, and the United States are given in ensuing chapters in this volume. The investigation included examining curricular documents, reviewing papers/articles describing statistics education in each country, and personal correspondence with statistics educators from the countries. Because countries differ in the number of years of required mathematics, and consequently opportunities to learn statistics vary greatly, and the diverse options available for some students beyond the required curriculum, comparisons among countries are subject to argument. However, overall, it would seem that the content of statistical education at the secondary level falls into three categories.

G. Burrill (✉)
Program in Mathematics Education, Michigan State University, East Lansing, MI, USA
e-mail: burrill@msu.edu

G. F. Burrill et al. (eds.), *Research on Reasoning with Data and Statistical Thinking: International Perspectives*, Advances in Mathematics Education,
https://doi.org/10.1007/978-3-031-29459-4_2

1. The curricular objectives relate to introductory data analysis, often restricted to univariate data with respect to typical graphical representations (box plots, dot plots, histograms, scatter plots) and the computation of statistical summary measures or formulaic probabilities. This seems to be the case in countries such as Brazil, Columbia, South Africa, and Turkey, where the problems often involve simple or non-contextual data and the focus is on calculations. Efforts are underway in Turkey and Brazil to change this, but the reality is that system barriers (e.g., teacher preparation, lack of resources, professional development opportunities) make this difficult.
2. The curricular objectives go beyond data analysis to include some inference, but the ideas in the treatment of statistics are primarily mathematical often with a relatively heavy emphasis on probability. This seems to be the case in Germany, the United Kingdom, Finland, Korea, and the Canadian provinces of Nova Scotia and British Columbia. The International Baccalaureate courses would seem to be largely in this category with a sound but formal mathematical approach to statistics and probability.
3. The ideas extend beyond formulas for procedures, the curriculum is data based, and simulation-based procedures are a precursor for formal inference as in New Zealand and the United States or in the new guidelines for Math B in Japan, which call for simulation to introduce concepts such as hypothesis testing. This could also describe the College Board Advanced Placement Statistics course, although the course has a strong focus on traditional topics such as hypothesis tests in order to be comparable to many introductory college statistics courses.

2 Electives

In some countries such as Korea and some of the Canadian provinces, most statistical concepts beyond numerical summaries for center, variability, and basic graphs are taught in a high school elective course rather than in a course required for all students. The following description is from the 2015 Korean National Mathematics Curriculum (p. 131) and is relatively typical:

> <Probability and Statistics> is a general elective course that students with a firm grasp of the content of <Mathematics> (a general common course) may select in order to pursue higher levels of mathematics. The contents for <Probability and Statistics> consist of three areas; 'Number of Cases','Probability', and 'Statistics'. 'Number of Cases' includes circular permutations, repeated permutations, repeated combinations, and the binomial theorem; 'Probability' includes statistical probability, mathematical probability, properties of probability and its applications, and conditional probability; and 'Statistics' includes random variables, probability distributions, binomial distributions, normal distributions, and statistical estimation.

3 The Reality

For most countries, with the exception of New Zealand, the gap between the written goals or standards and the reality is large. The reasons for this difference consistently cited by those surveyed point to the lack of time in the curriculum for statistics (e.g., South Africa, Catalonia in Spain) and the reluctance on the part of teachers to engage with statistical content (Japan and the United States). And in almost all of the countries, assessments establish barriers such as those in the UK where only 18% of the marks at A level are on statistics or in the US and Australia where assessments primarily focus on other mathematical content and those items that deal with statistics often are not about statistical understanding but rather about performing mathematical operations (e.g., find a missing value so the mean will be 50; if 25 in 100 prefer A, how many will prefer A in a sample of 400?).

New Zealand stands out because the vision of what statistics should look like in K-12 schools and what takes place in practice seems to be aligned (Pfannkuch and Wild 2013). A coherent and systemic effort led by the New Zealand Statistical Association (NZSA) education committee involving statisticians, statistics education researchers, professional developers, and teachers along with the Ministry of Education supported the implementation of the statistics strand of the 2007 New Zealand Mathematics and Statistics Curriculum. The group collaborated to produce curriculum resources that reflected modern and future statistical practice and that incorporated statistics education research findings about student learning. They also worked to ensure key stakeholders in education, including those involved in national assessment, knew about and supported the changes. The results have given New Zealand a leadership role in making statistics education a relevant and vital part of what students learn in schools. Two research projects were started to support the 2007 curricular changes. Their research papers documenting the developing understanding of how teaching and learning can be improved as students engage in statistical activities have provided fodder for other countries as they work towards this goal (e.g., Arnold et al., 2011). This collaborative work has continued; in 2022 New Zealand statistics educators re-examined what and how statistics is taught in schools, particularly in light of the advent of data science, and worked together with the Ministry of Education to refine and advance the national statistics curriculum.

4 Statistical Literacy

Statistical literacy is recognized as one of the three main focal points in the statistics curriculum in New Zealand, identified as a strand in the essential concepts in the National Council of Teachers of Mathematics *Catalyzing Change in High School Mathematics* (2018) and recognized in the new Japanese National Course of Study(personal correspondence). A data-based approach to teaching statistics shapes the curriculum in New Zealand and in some states in the United States. The Erasmus+ project, Strategic Partnership for Innovative in Data Analytics in Schools (SPIDAS) (Fujita et al., 2018) is analyzing practices relating to the teaching of data analytics in school education in Turkey, the United Kingdom, and Catalonia (Spain) with the aim to extend best practice in the teaching of data analytics through student-centered, problem-based learning, focusing on the impacts of extreme weather events.

5 Data Science

Across the world, data science is an emerging field. Many countries (e.g., Israel, Brazil, Germany, South Korea) have a plethora of programs for preparing data scientists at the tertiary level. Reflecting this trend, data science in school movements are becoming increasingly common. Data science is a component of the new curriculum of the Canadian Province of Ontario. Elements of data science such as computational thinking are in the draft revision of the Australian mathematics standards and appear in the Catalonia standards. Data science electives are being created in many states in the US for example, Ohio and Alabama (Drozda et al., 2022). The International Data Science in Schools Project (IDSSP) (2019), focused on teaching students in the last two years of high school how to learn from data, is an international collaborative project that involves educators from Australia, Canada, England, Germany, the Netherlands, New Zealand and the United States. The Education and Mathematics Education Innovation in the UK has created a self-study Introduction to Data Science Short Course (see the report on the pilot phase at https://mei.org.uk/data-science). In Germany an interdisciplinary joint project of didactics of mathematics and didactics of computer science at the University of Paderborn, ProDaBi project (www.prodabi.de), aims to develop and implement a data science curriculum, initially in the upper secondary level, and then for the lower secondary level in the form of a system of stand-alone modules. In Scotland, Data-Driven Innovation, part of the Edinburgh and South East Scotland City Region Deal funded by the Scottish government, has a DDI Skills Gateway programme project, to develop curricular resources on data science for Scotland school students and a set of real-world data science resources for primary and secondary teachers. Some data science initiatives focus on out of school initiatives for secondary students, such as the Belguim Data Camp, University of Pennsylvania's Warton Global Youth Program summer Data Science Academy or TERC Data Clubs for middle school students.

Bibliography

Australian Curriculum, Assessment and Reporting Authority. (2015). *Mathematics curriculum*. https://www.australiancurriculum.edu.au/f-10-curriculum/mathematics/

Belgium. https://www.datacamp.com/blog/datacamp-for-classrooms-is-now-free-to-belgian-secondary-school-teachers-and-students

Brazilian Ministry of Education. National Curriculum Parameters. Secondary Education. http://portal.mec.gov.br/seb/arquivos/pdf/pcning.pdf

British Columbia Ministry of Education. BC's New Curriculum. (2015). https://libguides.uvic.ca/teachingmath/bccurriculum

Framework Curriculum 1-10 compact. An overview of the subjects and content taught in Berlin. (2018). Landesinstitut für Schule und Medien Berlin-Brandenburg (LISUM – State Institute for School and Media Berlin-Brandenburg). https://bildungsserver.berlin-brandenburg.de/fileadmin/bbb/unterricht/rahmenlehrplaene/rlp1-10/RLP_kompakt_1-10_Englisch.pdf

International Commission on Mathematical Instruction Database Project Turkey. https://www.mathunion.org/icmi/activities/database-project/database-project-turkey

National Governors Association Center for Best Practices & Council of Chief State School Officers. (2010). *Common Core state standards for mathematics*. Authors.

Newfoundland & Labrador Department of Education. Mathematics. (2022). https://www.gov.nl.ca/education/k12/curriculum/descriptions/mathematics/#high

New Zealand Ministry of Education. (2014). *The New Zealand Curriculum Mathematics and Statistics*. https://nzcurriculum.tki.org.nz/The-New-Zealand-Curriculum/Mathematics-and-statistics/Achievement-objectives

Nova Scotia. (2014, updated 2022). Government of Nova Scotia Department of Education and Early Childhood Development. High School Full Course List. https://curriculum.novascotia.ca/english-programs/high-school/full-course-list

Ontario Ministry of Education. (2021). *Mathematics Elementary and Secondary Curriculum*. https://www.dcp.edu.gov.on.ca/en/curriculum/secondary-mathematics https://www.dcp.edu.gov.on.ca/en/curriculum/elementary-mathematics/grades/g8-math

Republic of South Africa Department of Education. (2011).*Curriculum and Assessment Policy Statement Grades 10–12 Mathematics*. http://dsj.co.za/wp-content/uploads/PDF/FET-_-MATHEMATICS-_-GR-10-12-_-Web1133.pdf

STEM parts of the National Core Curriculum in Finland. (2019, November). *European Schoolnet. Scientix*. http://storage.eun.org/resources/upload/482/20191112_140830068_482_Finnish_curriculum_Scientix_English-Nov2019.pdf

TERC Data Clubs. https://www.terc.edu/dataclubs/

TIMSS. (2015). *Encyclopedia. The mathematics curriculum, in primary and lower secondary grades*. TIMSS & PIRLS International Study Center. Boston College. https://timssandpirls.bc.edu/timss2015/encyclopedia/countries/germany/the-mathematics-curriculum-in-primary-and-lower-secondary-grades/

United Kingdom Department of Education. (2021). *National Curriculum in England: Mathematics programmes of study*. https://www.gov.uk/government/publications/national-curriculum-in-england-mathematics-programmes-of-study/national-curriculum-in-england-mathematics-programmes-of-study#key-stage-3

Wharton Global Youth Program Data Science Academy. University of Pennsylvania. https://globalyouth.wharton.upenn.edu/programs-courses/data-science-academy/

References

Arnold, P., Pfannkuch, M., Wild, C. J., Regan, M., & Budgett, S. (2011). Enhancing students' inferential reasoning: From hands-on to "movies". *Journal of Statistics Education, 19*(2), 1–32. https://doi.org/10.1023/A:1009854103737

Drozda, Z., Johnstone, D., & Van Horne, B. (2022). Previewing the national landscape of K-12 data science implementation. Paper commissioned for the *Workshop on foundations of data science for students in grades K-12*. Board on science education, the board on mathematical sciences and analytics, the computer science and telecommunications board, and the National Academy of Sciences, engineering, and medicine.

Fujita, T., Kazak, S., Turmo, M., & Mansour, N. (2018). *Strategic partnership for innovative in data analytics in schools (SPIDAS)*. State of the Art Review.

International Data Science in Schools Project (IDSSP). (2019). *Curriculum frameworks for introductory data science*. http://idssp.org/files/IDSSP_Frameworks_1.0.pdf

National Council of Teachers of Mathematics. (2018). *Catalyzing change in high school mathematics: Initiating critical conversations*. The Council.

Pfannkuch, M., & Wild, C. J. (2013). Working together to improve statistics education: A research collaboration case study. *Proceedings of the 59th International Statistical Institute World Statistical Congress, 25–30 August 2013, Hong Kong, China*. The Hague, The Netherlands: ISI. http://2013.isiproceedings.org/Files/IPS069-P3-S.pdf

The Brazilian National Curricular Guidance and Statistics Education

Leandro de Oliveira Souza

Abstract This report describes the changes called for in the of teaching statistics in a recent curriculum document intended to guide the teaching of mathematics in Brazilian schools. The discussion includes concerns about how to achieve the desired implementation and, in particular, the challenges teachers will face in implementing the recommendations.

Keywords Base Nacional Comum curricular · Statistical competencies · Teacher preparation · Technology

The current curriculum guidance document in Brazil is the Base Nacional Comum Curricular (Brazil, 2018) (BNCC), which was recently implemented. Due to this recent implementation, many of the teaching systems of Brazilian schools (public and private) are still preparing their curricula in accordance with this document. One of the reasons is that the guidelines were promulgated shortly before the beginning of the pandemic caused by the Covid-19 disease. During this period, most Brazilian schools were closed and face-to-face teaching did not take place, so teachers had to adapt the materials to other realities and meet the demands and teaching possibilities in a timely manner. In 2022, the education systems effectively started implementing the new regulations in schools.

Until then, before the BNCC, the official document that guided Brazilian education was the National Curricular Parameters (Brazil, 1998) (PCN), a document that was published in the late 1990s. Unlike the PCNs, which were the guiding documents of Brazilian Basic Education by which teachers had more freedom to choose how and what to taught, the BNCC assumes a normative and prescriptive character (Cazorla & Giordano, 2021), mainly in relation to the content to be taught in schools,

L. de Oliveira Souza (✉)
Instituto de Ciências Exatas e Naturais do Pontal, Universidade Federal de Uberlândia, Ituiutaba, Brazil
e-mail: leandrosouza@ufu.br

G. F. Burrill et al. (eds.), *Research on Reasoning with Data and Statistical Thinking: International Perspectives*, Advances in Mathematics Education,
https://doi.org/10.1007/978-3-031-29459-4_3

which becomes obligatory and is described grade by grade. Basic Education in this text refers to students at the right age, that is, without delays in school life, aged between 6 and 17 years old [grades 1–9 and grades 1–3].

Some of the changes that took place in the BNCC in relation to mathematics education were the inclusion of an algebra unit beginning in the first years of schooling, and the block of content that was called Information Treatment is now called Statistics and Probability. In the specific case of Statistics and Probability, the biggest changes were that the guidelines in the PCNs were more focused on reading and interpreting representations and/or statistical data, and in the BNCC norms are prescribed in the form of skills/content. The focus in the BNCC is on the construction of tools (thinking, planning, executing, implementing and communicating a project) by the students that help them in the whole process of investigation and statistical communication. In addition, the mandatory use of technological resources for the teaching of statistics was regulated. Other changes are expectations about what students should learn throughout Basic Education. In the specific case of statistics education, these expectations pervade all the years of schooling and are aligned with the Investigative Cycle proposal (Wild & Pfannkuch, 1999) from the perspective that, in the school environment, student participation in classes should be active and participatory.

BNCC has received criticism in recent academic publications, much of which refers to the excess of skills and competences described in the document. The concern is that too many skills are listed to be developed by the students, and there is not enough time to cover them all; furthermore, there are no proposed didactic and pedagogical guidelines (Cyrino & Grando, 2022). Although the document guides the pedagogical work using Transversal Contemporary Themes (e.g., Economy, Environment, Multiculturalism, Health, among others), the recommendations do not bring didactic guidelines that could help the teachers in their work of teaching. This becomes a problem since the initial training of mathematics teachers does not have a multidisciplinary approach, and the undergraduate statistics courses maintain a procedural perspective on learning the content. Few teachers' training courses have didactic proposals for the teaching of statistics as a part of basic education.

Another of the criticisms raised about the BNCC is based on the fact that the skills, which supposedly students should develop, are described in tables and divided by grade/school year with progressive levels of difficulty. When these skills are analyzed in isolation, these frameworks seem to prioritize a series of mathematical contents to the detriment of a more interdisciplinary learning perspective. This is controversial because BNCC suggests an inter-relational pedagogical approach. In order to understand the BNCC policy, it is necessary for the teacher to read the document from a structural perspective. The general competences are described in the introductory chapter with a focus on the broader formation of students across all school subjects. Skills are less broad and are immediately supported by the mathematics content, presented one by one, that should be conveyed throughout schooling. However, with a deeper understanding of the document, if the intention of teachers is aimed at achieving competences throughout schooling, then the use of active methodologies and interdisciplinary pedagogical approaches will be fundamental requirements in the teaching process and are encouraged by the document.

Ten general competencies are described in the BNCC (Brazil, 2018 p. 9), and at least three of them are directly related to the skills that are proposed by the document for the teaching of statistics:

- Valuing and using knowledge historically building on the physical, social, cultural and digital world to understand and explain reality; continuing to learn and collaborate to build a fair, democratic and inclusive society.
- Exercising intellectual curiosity and using the scientific approach, including research, reflection, critical analysis, imagination and creativity, to investigate causes, develop and test hypotheses, formulate and solve problems and create solutions (including technological ones) based on knowledge of different areas.
- Arguing, based on facts, data and reliable information, to formulate, negotiate and defend ideas, points of view and common decisions that respect and promote human rights, socio-environmental awareness and responsible consumption at the local, regional and global levels, with an ethical positioning in relation to the care of oneself, others and the planet.

These competences must be achieved through an effort by all professionals, jointly, from the different areas involved in the schooling process. Competencies are not achievable in the short term; achieving them requires establishing long-term objectives and didactic-pedagogical planning. The competences focus on pillars that refer to the strengthening of democracy. In this sense statistical education is of fundamental importance in the Brazilian curriculum. Skills and content related to statistical education are prescribed in all years of schooling. Six-year-olds, for example, must "carry out research, involving up to two categorical variables of interest and a universe of up to 30 elements, and organize data through personal representations." In addition, they are expected to "report in verbal or non-verbal language a sequence of events relating to a day, using, when possible, the times of events". Regarding probability they should "Classify events involving chance, such as "it will happen for sure", "maybe it will happen" and "it is impossible to happen", in everyday situations." (Brazil , 2018 p. 281).

The PPDAC (Problem, Plan, Data, Analysis, Conclusion) investigation cycle (Wild & Pfannkuch, 1999) permeates all the work with statistical education proposed in the BNCC and must be approached with an increasing degree of depth, as students acquire maturity. For example, around eight years of age (grade three), the research universe should involve up to 50 elements, from which students, in addition to collecting data, must organize them using lists, single or double-entry tables and represent them in single column charts, with and without the use of digital technologies. During this period, the use of electronic spreadsheets is suggested. Around 10 years of age, the use of these worksheets becomes a mandatory requirement. At this age, the planning and data collection of the research carried out must be focused on social practices and the themes chosen by the students. They are expected to interpret and investigate situations that involve data such as those on environmental contexts, sustainability, traffic, or responsible consumption and also to analyze data presented by the media in tables and in different types of graphics. Students should prepare written texts on their thoughts to communicate and synthesize their conclusions. Regarding probability, at this age, students should calculate the experimental probability of a random event, expressing it as a rational number (fractional, decimal and percentage form) and compare the expectations calculated mathematically with the experimental results.

Around the age of 14 (grade nine), when the elementary school cycle ends, students should be able to "analyze and identify, in graphics published by the media, the elements that can induce, sometimes on purpose, reading errors, such as inappropriate scales, subtitles not explained correctly, omission of important information (sources and dates), among others." And with regard to the investigative cycle, it is expected that they have learned to "Plan and execute sample research involving the theme of social reality and communicate the results through a report containing evaluation of measures of central tendency and amplitude (range), appropriate tables and graphs, built with the support of electronic spreadsheets." (Brasil , 2018, p. 319). In secondary school, the content should be expanded.

The BNCC's orientation in relation to statistics and probability is that all citizens need to develop skills to plan an investigation, to collect, organize, represent, interpret and analyze data in various contexts (from the social or scientific sphere), to enable them to make informed judgments and decisions based on their analyses. The document highlights the need to use technologies such as calculators and electronic spreadsheets to evaluate and compare results. By the end of the school cycle, students are expected to be able to plan research and build reports using descriptive statistics, including measures of central tendency and the construction of tables and various types of graphs. Planning includes the construction of relevant questions, the definition of the population to be surveyed, the decision on whether to use a sample and, when applicable, using an appropriate technique of sampling.

The thematic unit that deals with statistics and probability proposes to approach concepts, facts, and procedures present in situations and problems of everyday life, science, and technology from a perspective of active learning on the part of students. The curricula and pedagogical proposals by the education systems should connect the skills prescribed with those from other areas of knowledge, between the thematic units of mathematics and within the skills proposed for statistical education itself. The biggest challenge for teachers is to build their pedagogical approach using a perspective for which they were not prepared in their education. Discussing and making connections between mathematical content, among themes of social relevance or based on experimentation is not common in the preparation of teachers. There are no resources available for teaching the new curriculum; teachers have to prepare by themselves supported by textbooks that are free of charge in public schools. Lopes and Souza (Lopes & de Oliveira Souza, 2016) reflect that implementing curricular innovations is much more complex than making curricular policy prescriptions, because teachers' interpretation and implementation of a reform strongly depends on their personal interpretative framework and, consequently, of their training.

References

Brasil. Ministério da Educação e do Desporto. (1998). *Parâmetros curriculares nacionais.* MEC/SEF.

Brasil. Ministério da Educação. (2018). *Base Nacional Comum Curricular – BNCC.* MEC/SEF.

Cyrino, M., & Grando, R. (2022). (Des)construção curricular necessária: Resistir, (re)existir, possibilidades insubordinadas criativamente. [Necessary curricular (de)construction: Resist, (re) exist, creatively insubordinate possibilities]. *Revista de Educação Matemática, 19* (edição especial), 1–25. https://www.revistasbemsp.com.br/index.php/REMat-SP/article/view/728

Carzola, I., & Giordano, C. (2021). O papel do letramento estatístico na implementação dos temas contemporâneos transversais da BNCC [the role of statistical literacy in the implementation of contemporary transversal themes of BNCC]. In C. Moteiro & L. Carvalho. (Eds.), *Temas emergentes em letramento estatístico [recurso eletrônico]* [emerging themes in statistical literacy] (pp. 88–111). Ed. UFPE, 2021.

Lopes, C. E., & de Oliveira Souza, L. (2016). Aspectos filosóficos, psicológicos e políticos no estudo da probabilidade e da estatística na Educação básica. [Philosophical, psychological and political features while studying probability and statistics in basic education]. Educação Matemática Pesquisa: *Revista do Programa de Estudos Pós-Graduados em Educação Matemática, 18*(3). https://revistas.pucsp.br/emp/article/view/31494

Wild, C. J., & Pfannkuch, M. (1999). Statistical thinking in empirical enquiry. *International Statistical Review, 67*(3), 223–265. https://doi.org/10.1111/j.1751-5823.1999.tb00442.x

Statistics and Probability Education in Germany

Susanne Podworny

Abstract This paper describes the intended and implemented view of statistics and probability education in Germany, where the actual focus in classrooms is often on probability and formal statistics. The discussion includes several projects related to statistical literacy including how work with big data is being developed and supported by the statistics education community.

Keywords Data analysis · Data literacy · Stochastics · Technology

In Germany there are sixteen federal states with sixteen own curricula for schools. In addition, there are national educational standards for mathematics (German "Bildungsstandards"): For elementary school (approximate age 6–10), middle school (approximate age 11–16) and high school (approximate age 17–19) (e.g. Kultusministerkonferenz, 2005, 2012). These educational standards were introduced in 2003 by the German Conference of Ministers of Education as a result of the Programme for International Assessment (PISA) study. They specify which subject-related competencies students should have developed by a certain point in their school career. The goals of the educational standards are to achieve greater transparency in school requirements, to develop competency-based teaching, and to provide a basis for reviewing the results achieved.

With the educational standards, the central idea of data and chance was introduced across all age levels into all curricula of the sixteen German states and became mandatory content of mathematics lessons. The German word "Stochastik" (stochastics) is a (historical) connection between statistics and probability, but in the curricula and in the classroom, the two contents are often treated as unrelated.

S. Podworny (✉)
Universität Paderborn, Paderborn, Germany
e-mail: podworny@math.upb.de

G. F. Burrill et al. (eds.), *Research on Reasoning with Data and Statistical Thinking: International Perspectives*, Advances in Mathematics Education,
https://doi.org/10.1007/978-3-031-29459-4_4

Despite the standards, the data aspect plays a minor role in German mathematics classrooms. There is often a focus on probability. This may go back to the famous contribution of Heitele (1975), who describes fundamental ideas for probability. Only since about 2007, the concept of data is considered equal to chance, also due to the National Council of Teachers of Mathematics (2000) standards and the international discussion.

However, teaching is often reduced to content that is tested in central assessments (grade 3, 8, 13). This means that the broad scope of the subject is not represented (Sill & Kurtzmann, 2019). The standards give the freedom to do more in class, but in reality this is often reduced to a minimum or relegated to voluntary work for groups or elective classes (Krüger et al., 2015). Especially in elementary school stochastics is often not taught at all (Bender et al., 1999), and until the high school it is mostly postponed to the end of a school year 'if there is still time at the end'. In terms of content, single and multi-stage random experiments are introduced in the middle school, and some Bayes is treated. Data analysis is treated rather superficially, although there are suggestions in the sense of the educational standards (Biehler, 2006; Biehler & Hartung, 2006; Eichler & Vogel, 2013).

This changes only in the high school, where stochastics is taught obligatory for one semester to all students. Here, mainly random experiments, the binomial distribution, estimating and testing-mainly confidence intervals-are taught. The entire material is strongly oriented towards inferential reasoning. Nearly no data analysis happens in high school. Stochastics has been a mandatory part of the final exams of high school since 2012.

1 Tool Use in Germany

The typical statistics education tools like Fathom, TinkerPlots and CODAP are available in translated versions for Germany but are used mainly by statistics education researchers. Unlike other countries, Fathom is available in German free of charge (https://www.stochastik-interaktiv.de/fathom/). There are also numerous scientific publications in which the implementation is studied (e.g. for TinkerPlots: Frischemeier, 2017; Podworny, 2019). In the curricula, the use of software is suggested for the representation of data up to grade 10. Simulations are to be used to investigate stochastic situations in high school. For both purposes there are optional textbooks, for example for the use of Fathom (Biehler et al., 2011).

Nevertheless, the use of digital tools in statistics classes is rare because the curriculum's content is only a little data-oriented (Tietze et al., 2002). Therefore there is a recommendation of the Working Group Stochastics of the German Society for Didactics of Mathematics concerning the aims and the design of stochastics lessons for all grades (Arbeitskreis Stochastik der Gesellschaft für Didaktik der Mathematik, 2003).

2 Community and New Developments

In Germany, a national community is active for didactics of stochastics. The peer-reviewed journal "Stochastik in der Schule" (stochastics at school; https://www.stochastik-in-der-schule.de/) publishes articles in German language, which combine theory and practice and give suggestions for teaching. Through the underlying association, teachers and researchers are connected and come into exchange with each other. This also happens at annual national conferences.

Several projects attempt to promote topics like data literacy and statistics and probability education or data science in school. Only recently, the Data Literacy Charter (https://www.stifterverband.org/sites/default/files/data-literacy-charter.pdf) was published by the federal government and is signed by many German ministers, researchers and teachers. The charter focuses on "data literacy and its overall importance in educational processes" (p. 1).

Some projects ensure that ideas for teaching in schools are developed further. One example is the Erasmus+ project "Promoting Civic Engagement via explorations of evidence" (ProCivicStat; http://iase-web.org/islp/pcs/). In this project, teaching materials were developed and researched supporting statistics teaching that enables students to engage with current social issues (Biehler et al., 2018). In the project "Data Science and Big Data at School" (ProDaBi; https://www.prodabi.de/en/), which is funded by the Deutsche Telekom Stiftung, teaching materials are developed in which the new topics of data science and artificial intelligence via data become accessible for school teaching (Biehler & Fleischer, 2021; Frischemeier et al., 2021).

3 Outlook

The educational standards apply to the whole of Germany and prescribe the subject of data and chance for all school levels. However, there seems to be a lack of willingness to implement these topics at various school levels, not least at the political level.

The Data Literacy Charter has triggered movement at the university level, which will hopefully have an impact at the school level as well. Developments such as those in the ProDaBi project are being implemented and are finding their way into the computer science curriculum and hopefully (again) into mathematics curricula in the future.

References

Arbeitskreis Stochastik der Gesellschaft für Didaktik der Mathematik. (2003). Empfehlungen zu Zielen und zur Gestaltung des Stochastikunterrichts [Recommendations on objectives and the design of stochastics lessons]. *Stochastik in der Schule, 23*(3), 21–26.

Bender, P., Beyer, D., Brück-Binninger, U., Kowallek, R., Schmidt, S., Sorger, P., et al. (1999). Überlegungen zur fachmathematischen Ausbildung der angehenden Grundschullehrerinnen und -lehrer [reflections on the mathematics education of prospective elementary school teachers]. *Journal für Mathematikdidaktik, 20*, 301–310. https://doi.org/10.1007/BF03338903

Biehler, R. (2006). Leitidee "Daten und Zufall" in der didaktischen Konzeption und im Unterrichtsexperiment [guiding principle "data and chance" in the didactic conception and in the teaching experiment]. In J. Meyer (Ed.), *Anregungen zum Stochastikunterricht* (Vol. 3). Franzbecker.

Biehler, R., & Fleischer, Y. (2021). Introducing students to machine learning with decision trees using CODAP and Jupyter notebooks. *Teaching Statistics, 43*(S1). https://doi.org/10.1111/test.12279

Biehler, R., & Hartung, R. (2006). Leitidee Daten und Zufall [guiding principle data and chance]. In W. Blum, C. Drüke-Noe, R. Hartung, & O. Köller (Eds.), *Bildungsstandards Mathematik: konkret. Sekundarstufe I: Aufgabenbeispiele, Unterrichtsanregungen, Fortbildungsideen* (pp. 51–80). Cornelson Scriptor.

Biehler, R., Hofmann, T., Maxara, C., & Prömmel, A. (2011). *Daten und Zufall mit fathom - Unterrichtsideen für die SI und SII mit software-Einführung [data and chance with fathom - teaching ideas for SI and SII with software introduction]*. Schroedel.

Biehler, R., Frischemeier, D., & Podworny, S. (2018). Elementary preservice teachers' reasoning about statistical modeling in a civic statistics context. *ZDM, 50*(7), 1237–1251. https://doi.org/10.1007/s11858-018-1001-x

Eichler, A., & Vogel, M. (2013). *Leitidee Daten und Zufall. Von konkreten Beispielen zur Didaktik der Stochastik [guiding principle data and chance. From concrete examples to the didactics of stochastics]*. Springer Spektrum.

Frischemeier, D. (2017). *Statistisch denken und forschen lernen mit der software TinkerPlots [learning to think and research statistically with the software TinkerPlots]*. Springer Spektrum.

Frischemeier, D., Biehler, R., Podworny, S., & Budde, L. (2021). A first introduction to data science education in secondary schools: Teaching and learning about data exploration with CODAP using survey data. *Teaching Statistics, 43*(S1). https://doi.org/10.1111/test.12283

Heitele, D. (1975). An epistomological view on fundamental stochastic ideas. *Educational Studies in Mathematics, 6*(2), 187–205.

Krüger, K., Sill, H. D., & Sikora, C. (2015). *Didaktik der Stochastik in der Sekundarstufe I [didactics of stochastics in secondary school]*. Springer.

Kultusministerkonferenz. (2005). *Bildungsstandards im Fach Mathematik für den Primarbereich [educational standards in mathematics for the primary level]*. Luchterhand.

Kultusministerkonferenz. (2012). *Bildungsstandards im Fach Mathematik für die Allgemeine Hochschulreife [educational standards in mathematics for the general qualification for university entrance]*. Wolters Kluwer.

National Council of Teachers of Mathematics (NCTM). (2000). *Principles and standards for school mathematics*. NCTM.

Podworny, S. (2019). *Simulationen und Randomisierungstests mit der software TinkerPlots [simulations and randomization tests with the software TinkerPlots]*. Springer Spektrum.

Sill, H.-D., & Kurtzmann, G. (2019). *Didaktik der Stochastik in der Primarstufe [didactics of stochastics for primary level]*. Springer.

Tietze, U.-P., Klika, M., & Wolpers, H. (2002). *Mathematikunterricht in der Sekundarstufe II. Band 3. Didaktik der Stochastik [teaching mathematics in secondary school. Volume 3. Didactics of stochastics]*. Vieweg.

New Zealand Statistics Curriculum

Maxine Pfannkuch and Pip Arnold

Abstract This paper describes the statistics strand in the New Zealand school curriculum, one of three mandatory content strands, which all schools are expected to teach. The emphasis is on how, across the years, the curriculum builds knowledge of statistical practice and concepts and promotes thinking statistically in a range of contexts, including a focus on statistical literacy and data consumers. The discussion describes resources available for teachers and the challenges moving forward in teaching statistics in the schools.

Keywords Assessment · Challenges · Resources for teaching statistics · Statistical literacy · Technology

1 Overview

The statistics curriculum in New Zealand is part of a national mathematics and statistics curriculum (Ministry of Education, 1992, 2007) for Years 1 to 13 (5–18-year-olds), which all schools are expected to teach. The curriculum is divided into eight levels: Years 1 to 10 (5–14-year-olds) cover levels 1 to 5, Years 11 to 13 (15–18-year-olds) cover levels 6 to 8. The curriculum has three strands: number and algebra, geometry and measurement, and statistics. Within each strand are a list of achievement objectives for each level. Statistics is divided into three sub-strands: statistical investigation, statistical literacy, and probability. Statistical investigation is based around the PPDAC (Problem, Plan, Data, Analysis, Conclusion) enquiry

M. Pfannkuch (✉)
Department of Statistics, The University of Auckland, Auckland, New Zealand
e-mail: m.pfannkuch@auckland.ac.nz

P. Arnold
Karekare Education, Auckland, New Zealand
e-mail: pip@karekareeducation.co.nz

G. F. Burrill et al. (eds.), *Research on Reasoning with Data and Statistical Thinking: International Perspectives*, Advances in Mathematics Education,
https://doi.org/10.1007/978-3-031-29459-4_5

cycle (Wild & Pfannkuch, 1999). Statistical literacy draws on the work of Gal (2002), with attention to interpreting and evaluating data-based evidence produced by others including the media. Similar to statistical investigation, probability is experientially based whereby students investigate chance situations. Real world data and contextual knowledge are fundamental to the implementation of the curriculum. As Cobb and Moore (1997, p. 801) stated: "statistics requires a different kind of thinking because data are not just numbers, they are numbers with a context." Across the years the curriculum builds knowledge of statistical practice and concepts and promotes thinking statistically in a range of contexts. For example, for the statistical investigation sub-strand in Year 1, with teacher support, students pose and answer investigative questions that are relevant and meaningful to them, gather, sort, count and display category data and discuss their findings (Arnold, 2022). At Year 13, students pose investigative questions after researching and gaining contextual knowledge about the situation, and then conduct their own investigations by using data they have gathered or sourced. Depending on the data they have, they use analytical techniques such as the empirical method of bootstrapping to generate confidence intervals for sample-to-population inferences, the randomization test for experiment-to-causation inferences, Holt-Winters for predictions for times series data or descriptive methods for bivariate data. In a report they communicate their findings.

Alongside the expectation that students will explore data, is the expectation that technology (e.g., CODAP, iNZight) is an integral part of the statistics curriculum. Rather than focusing on constructing data displays, there is an emphasis on using multivariate datasets, posing investigative questions and interpreting and reasoning from data displays and, in the spirit of a data detective, gleaning and telling the stories within the data. From Year 1 (5–6-year-olds) students are enculturated into the practice of statistics and being data producers to learn more in the context sphere about the world they live in. As students are gradually inducted into the world of statistical practice, they are also scaffolded to build their conceptual infrastructure across the years. For instance, in Year 10 (14–15-year-olds) students are introduced to informal inferential ideas such as sample, population, sampling variability, sample distribution, population distribution, comparing two groups, and determining whether they can make a claim about whether one group tends to have higher values than another group (Wild et al., 2011). By Year 13 (17–18-year-olds), students develop their inferential concepts further when they grapple with the inferential ideas underpinning bootstrapping and the randomization test. Hands on activities, interactive visual imagery, and software tools specific to conceptual development are encouraged as learning approaches (e.g., Arnold et al., 2018).

Statistical literacy is focused on the data consumer and is explicitly taught. Initially intertwined with student statistical investigations and probability activities where students evaluate and interpret statements made by their peers, statistical literacy builds to evaluating data collection methods, choices of measures and querying the validity of findings by Year 10. From Year 11 (15–16-year-olds) the focus turns to evaluation of statistical reports in the media and the claims made, moving to interpreting risk, relative risk, and sampling and non-sampling errors in surveys in Year 12 (16–17-year-olds). Culminating in Year 13 (17–18-year-olds), students evaluate a wide range of statistically based reports including surveys and polls, experiments, and observational studies. The Year 13 statistical literacy curriculum drew on the work of Budgett and Pfannkuch (2010). (See https://www.nzqa.govt.nz/ncea/assessment/view-detailed.do?standardNumber=91584 for exam example.) Probability follows a traditional path for chance situations including games of chance, spinners, dice and urns, sample spaces, simulations, distinguishing between theoretical and experimental probability and gradually building strategies for interpreting and calculating probabilities. Year 1 starts with students anticipating possible outcomes in a game of chance and ends in Year 13 with situations involving Poisson, binomial and normal distributions, conditional events and expected values.

2 Resources and Assessment

A major resource for curricular materials is CensusAtSchool (https://new.censusatschool.org.nz/), run by Auckland University Statistics Department since 2003 with some funding provided by the Ministry of Education and Statistics New Zealand. CensusAtSchool conducts a survey of Year 3 to 13 (7–18-year-olds) students biennially, from which samples of data can be used in classrooms. On average, 20,000 students participate in these surveys, which provide rich, meaningful multivariate datasets for students to explore. The site also serves as a repository of curriculum materials produced by teachers. The annual Auckland Statistics Teachers Day, where practicing teachers, professional development facilitators and researchers present workshops and resources, is a major source of curricular materials that are made available to New Zealand teachers. Research papers, applicable to the statistics curriculum, inform teachers of New Zealand research underpinning new learning approaches and new content. The Ministry of Education also hosts other websites (e.g., https://nzmaths.co.nz/) to provide resources, curricular materials and guidance to teachers and publishes workbooks and resources. Independent publishers, as well, provide online resources, workbooks and traditional textbooks, which schools can buy. The state does not mandate specific resources and software for schools, rather schools choose them for their students.

Assessment plays an important role in the implementation of a curriculum, as what is assessed and how it is assessed influences the taught curriculum. In Years 1 to 10 (5–14-year-olds), schools assess their students with teacher-developed assessments and curriculum-based standardized tests available from the Ministry of Education (e.g., e-asTTle Mathematics) in order to benchmark student achievement with other schools. For Years 11 to 13 (15–18-year-olds), there are national qualifications which are set up as achievement standards that are aligned to the curriculum. Generally, at Years 11 and 12 students do six statistics and mathematics standards, whereas at Year 13 statistics can be taken as a full subject. Some standards are externally assessed in a traditional examination, whereas other standards are internally assessed in the schools, which are subject to moderation. This system allows alignment of how students learn to how students are assessed. In Year 13 for example, students conduct statistical investigations employing the software that they use in class for data analysis (e.g., iNZight, https://inzight.nz/) and inference (VIT, https://www.stat.auckland.ac.nz/~wild/VITonline/) in two of the internally assessed standards.

3 Challenges

There are three main challenges to the teaching of statistics in New Zealand. The first challenge is that the curriculum must be future focused for at least the next 15 years and must respond to the rapid changes in the statistics discipline and in technology. Data science is on the horizon bringing new types of data, new methods of inference, classification and predictive modelling and coding. Currently a curriculum refresh is underway, and the question remains about how the developers will integrate data science possibilities. The second challenge is teacher professional development not only to learn and understand new content but also new pedagogies and technologies. The reality of the implemented statistics curriculum is lack of any centrally planned sustained professional development for all teachers, including recent arrivals to New Zealand, mathematics-oriented teachers, examiners, moderators, and primary teachers lacking confidence in teaching statistics (Royal Society Te Apārangi, 2021). By leaving professional development as voluntary and to volunteers interested in promoting statistics education, teacher understanding and delivery of the curriculum has ranged from those producing highly creative ways of teaching statistics to students, in the spirit of the envisaged curriculum, to those reducing statistics to a series of procedures to follow. The third challenge is the assessed curriculum. The benchmark tests provided by the Ministry for primary students do not capture the exploratory nature of statistical investigations and that statistical thinking is different from mathematical thinking, rather they reflect traditional statistics assessments. Even though national assessment at the secondary school level has internal achievement standards, the moderators have influenced how statistics is taught by insisting on key words and phrases resulting in some teachers resorting to rote learning. On the other hand, some examiners for the external standards have propelled the teaching of statistics forward by incrementally over the years steering teachers to focus on new ideas and new understandings of statistics. Despite these challenges, there has been a remarkable shift in statistics teaching from the 1992 to the 2007 curriculum (Pfannkuch et al., 2020). The current challenge is to produce a new curriculum that builds on previous work and pays attention to recent statistics education research and discipline developments in probability and data science.

References

Arnold, P. (2022). *Statistical investigations | Te Tūhuratanga Tauanga: Understanding progressions in The New Zealand curriculum and Te Marautanga o Aotearoa*. NZCER Press.

Arnold, P., Confrey, J., Jones, R. S., Lee, H., & Pfannkuch, M. (2018). Statistics learning trajectories. In D. Ben-Zvi, K. Makar, & J. Garfield (Eds.), *International handbook of research in statistics education* (pp. 295–326). Springer International Publishing AG.

Budgett, S., & Pfannkuch, M. (2010). Assessing students' statistical literacy. In P. Bidgood, N. Hunt, & F. Jolliffe (Eds.), *Assessment methods in statistical education: An international perspective* (pp. 103–121). John Wiley & Sons Ltd..

Cobb, G., & Moore, D. (1997). Mathematics, statistics, and teaching. *American Mathematical Monthly, 104*(9), 801–823.

Gal, I. (2002). Adults' statistical literacy: Meanings, components, and responsibilities. *International Statistical Review, 70*(1), 1–25.

Ministry of Education. (1992). *Mathematics in the New Zealand curriculum*. Learning Media.

Ministry of Education. (2007). *The New Zealand curriculum*. Learning Media.

Pfannkuch, M., Wild, C., Arnold, P., & Budgett, S. (2020). Reflections on a 20-year research journey, 1999–2019. *SET, 1*, 27–33.

Royal Society Te Apārangi. (2021). *Pāngarau mathematics and tauanga statistics in Aotearoa New Zealand: Advice on refreshing the English-medium mathematics and statistics learning area of the New Zealand curriculum*. Author.

Wild, C. J., & Pfannkuch, M. (1999). Statistical thinking in empirical enquiry (with discussion). *International Statistical Review, 67*(3), 223–265.

Wild, C. J., Pfannkuch, M., Regan, M., & Horton, N. J. (2011). Towards more accessible conceptions of statistical inference. *Journal of the Royal Statistical Society: Series A (Statistics in Society), 174*(2), 247–295.

Statistics Education in the Philippines: Curricular Context and Challenges of Implementation

Enriqueta Reston

Abstract Statistics and probability is one of five learning domains in the Philippine's K to 12 Basic Education Curriculum for Mathematics. With emphasis on the spiral progression of topics starting on simple data representations for the statistics component and outcomes of familiar events and experiments for the probability component at the elementary level, this paper describes the intended statistics and probability curriculum and the challenges of implementation, including a lack of resources and the need for teacher professional development, and presents an example of a five-year project intended to meet these challenges.

Keywords Implementation challenges · Professional development · Statistical pedagogy · Spiral curriculum

The Philippines, an archipelago in Southeast Asia with a population of over 100 million as of 2015 Census, has a school system that is considered one of the largest in the region in terms of student enrolment. In 2012, the Department of Education (DepEd) launched the *K to 12 Basic Education* program, a major educational reform that expanded the basic education cycle from 10 to 12 years and enhanced the quality of educational outcomes through reforms at the curriculum level (Department of Education, 2013). Currently, Basic Education now comprises six years of elementary education (between 7 and 12 years old), four years of junior high school (between 13 and 16 years old) and two years of senior high school (between 17 and 18 years old).

E. Reston (✉)
School of Education, University of San Carlos, Cebu City, Philippines
e-mail: edreston@usc.edu.ph

G. F. Burrill et al. (eds.), *Research on Reasoning with Data and Statistical Thinking: International Perspectives*, Advances in Mathematics Education,
https://doi.org/10.1007/978-3-031-29459-4_6

1 The Intended *Statistics and Probability* Curriculum at School Level

Statistics and Probability is one of five learning domains in the *K to 12 Mathematics Curriculum*, along with Numbers and Number Sense, Measurement, Algebra and Patterns, and Geometry. With the twin goals of developing students' critical thinking and problem-solving skills, the *K to 12 Mathematics Curriculum* adopts a spiral progression approach where these five learning domains cut across the grade levels from Grade 1 through 10 with increasing depth and complexity. Unpacking the elementary *K to 12 Mathematics Curriculum* from Grades 1 to 6 for the Statistics and Probability learning domain shows a spiral progression of topics on simple data representations for the statistics component and outcomes of familiar events and experiments for the probability component. Along with these topics is the spiraling of learning competencies in representing data from pictographs to bar graphs, tables and pie graphs for the statistics component and learning to use the language of uncertainty in the likelihood of occurrence of events for the probability component (DepEd, 2013). Figure 1 shows the spiral progression of topics in the curriculum for Statistics and Probability in the elementary school mathematics curriculum.

In junior high school, the curriculum for the Statistics and Probability learning domain for Grades 7 to 10 shows that there is a break in the spiral progression of topics in Grade 9 where Statistics and Probability is not articulated in the curriculum guide (DepEd, 2013) (Fig. 2).

In senior high school, Statistics and Probability is a stand-alone core course in Grade 11, where it is recognized as a discipline in its own right apart from mathematics. It is comprised of six units starting with the topics on random variables, probability and sampling distributions that lay down the foundations of inference and then to the formal inferential methods of parameter estimation and hypothesis testing, correlation and regression analysis.

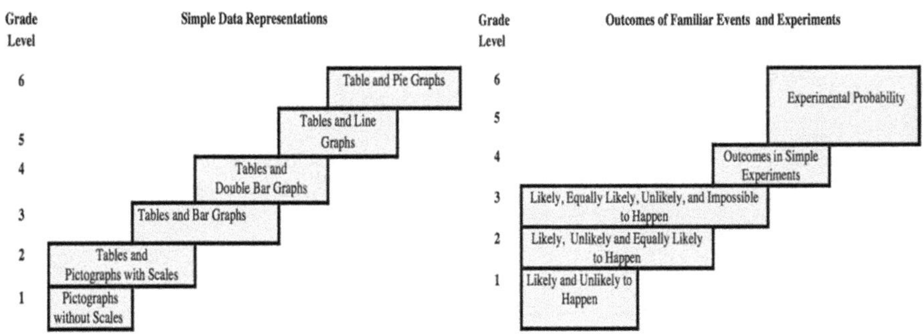

Fig. 1 The Statistics and Probability component in the K to 12 mathematics curriculum for elementary level

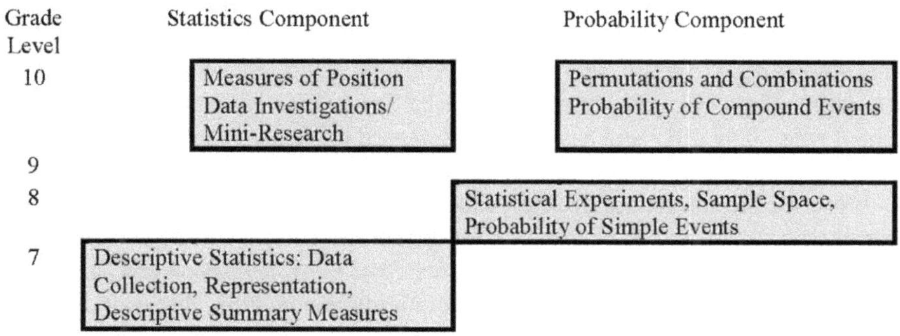

Fig. 2 The Statistics and Probability component in the K to 12 mathematics curriculum for junior high school

2 Challenges in Curriculum Implementation

Prior to the implementation of the *K to 12 Basic Education* program, the teaching of statistics in public schools was limited to the last quarter of third year high school and as an elective course in private schools. In general, problems such as the lack of qualified teachers; lack of locally-produced statistics books and educational materials; inadequate facilities such as computer laboratories, software and other teaching aids; and mechanistic teaching methods that do not enhance the teaching of statistics were identified (David & Maligalig, 2006; Reston & Bersales, 2011).

With the implementation of the *K to 12 Mathematics Curriculum,* the Statistics and Probability learning domain is taught in the fourth quarter of every grade level from Grade 1 to Grade 10 for about 9 to 10 weeks (DepEd, 2013). In senior high school, it is taught as a stand-alone course for one semester term in Grade 11 which spans about five months. This posed further implementation challenges, particularly in the lack of teaching-learning resources and in the preparation of mathematics teachers to teach this learning domain. Teachers who taught statistics classes are typically mathematics teachers, who may not necessarily have adequate training in teaching statistics and probability. Moreover, the Philippine government through the DepEd has provided teaching resources in the teaching of Statistics and Probability at school level. In particular, during the COVID-19 pandemic, DepEd has provided self-learning modules for teaching and learning the different subject areas including the Statistics and Probability domain with alternative learning deliveries in line with its Basic Education Learning Continuity Plan (DepEd, 2020).

With the challenges in the preparation of teaching-learning resources for statistics and probability in senior high school during its first few years of implementation, collaborative efforts were made by DepEd and the Commission on Higher Education to pool together higher education content experts and school teachers. In particular, the *Teaching Guide for Senior High School Statistics and Probability* was prepared by a team of professional statisticians and university statistics educators. The teaching guide was designed to assist senior high school teachers in teaching the course with a mix of lectures and learning activities using active learning strategies and real data (Commission on Higher Education, 2016). This was a vital step needed to teach Statistics and Probability with greater impact with over 1.4 million Grade 11 students enrolled during its first year of implementation in the school year 2016–2017 and has steadily increased to over 1. seven million students by the school year 2020–21 (DepEd, 2021).

While the Department of Education has invested in the development of teaching-learning resources and professional development of teachers for effective curriculum implementation, there is scarce evidence from published reports on the impact of these resources and programs in curriculum implementation, that is, in teaching Statistics and Probability in the K to 12 basic education program. Further, there is a dearth of published studies in the Philippines on statistics teacher development and curriculum implementation at school level. Using the keyword" teaching statistics," a search of articles at the Philippine e-Journal platform (https://ejournals.ph/index.php) revealed some articles on teaching statistics in higher education but none at school level particularly during the implementation of the K to 12 curriculum. In a comparative analysis of statistics education research in Malaysia and the Philippines by Reston & Krishnan (2014), there was only one published article cited on teaching and learning of statistics at school level in the Philippines, that is, the paper by Prado and Gravoso (2011) which explored the integration of technology using video recordings in applying anchored instruction in high school statistics. However, this study was conducted before the implementation of the K to 12 basic education reform.

3 Improving Statistical Pedagogy among K to 12 Mathematics Teachers in the Philippines

To shed light on the implementation of the Statistics and Probability curriculum, this section presents the case of a five-year teacher development project entitled *Improving Statistical Pedagogy among K to 12 Mathematics Teachers,* which was launched in 2015 in response to a needs assessment survey in 2014 where probability and statistics was ranked first as the area where professional development is most needed by the majority (66.3%) of 92 mathematics teachers from selected schools in Metro Cebu, Philippines (Reston & Cañizares, 2019). The project was implemented by the Science and Mathematics Education Department of the University of San Carlos (USC) in Cebu City, Philippines with expert support from Academics without Borders (AWB), a non-governmental organization based in Montreal, Canada. The project included the training of workshop facilitators, development of activity resource books, and two-week parallel workshops for elementary, junior high school and senior high school teachers to enhance their pedagogical content knowledge for teaching statistics and probability across the curriculum. The follow-up with some teacher-participants through email and focus group interviews provided insights on the actual classroom implementation of the curriculum before the

COVID-19 pandemic. When asked about the most valuable ideas they learned in the workshops that they applied in their statistics classes, some of the responses are as follows:

- ICT integration and collaboration
- Starting a lesson with an interesting activity that would connect to the lesson
- Preparing activities that would call for maximum students' involvement
- Using data from real life situations is more applicable for pupils to retain their learnings.
- Teamwork and bridging learning opportunities

4 Recommendations for Future Directions

The challenges for implementing the statistics and probability curriculum include sustained teacher development, contextualized and data-based teaching resources, and the integration of technology to cope with the demands of twenty-first century workplace. With the increased recognition on the importance of statistics education in this data-driven fast-paced age characterized by digitalization and technological innovations, there is the need for statistics education in the Philippines to cope with these challenges and close gaps between the intended and the implemented curriculum so that desired learning outcomes may be attained. From my perspective, future directions in statistics education will include reform efforts towards more responsive and relevant statistics education that integrate big data, technology and data science approaches to meet the demand of statistically literate citizenry and data competence in research and other data-based approaches.

References

Commission on Higher Education. (2016). *Teaching Guide for Senior High School Statistics and Probability*. Quezon City: CHED

David, I., & Maligalig, D. (2006). Are we teaching statistics correctly to our youth? *The Philippine Statistician, 55*(3–4), 1–28.

Department of Education. (2013). *The K to 12 curriculum guide: Mathematics*. Department of Education. Retrieved from http://www.deped.gov.ph/

Department of Education. (2020, July 3). *Basic education learning continuity plan*. Retrieved from Department of Education. https://www.deped.gov.ph/wp-content/uploads/2020/07/DepEd_LCP_July3.pdf

Department of Education. (2021). *Department of Education Education Management Information System Division*. Retrieved from Number of enrollment in all sector senior high school, from AY 2017–18 to 2020–21. https://www.deped.gov.ph/alternative-learning-system/resources/facts-and-figures/datasets/

Prado, M. M., & Gravoso, R. S. (2011). Improving high school students' statistical reasoning skills: A case of applying anchored instruction. *The Asia-Pacific Education Researcher, 20*(1), 61–72.

Reston, E., & Bersales, L. G. (2011). Reform efforts in training mathematics teachers to teach statistics: Challenges and prospects. In C. Batanero, G. Burrill, & C. Reading (Eds.), *Teaching statistics in school mathematics - challenges for teaching and teacher education: A joint study ICMI/IASE study book*. Springer.

Reston, E., & Cañizares, M. (2019). Needs assessment of teachers' knowledge bases, pedagogical approaches and self-efficacy in implementing the K to 12 science and mathematics curriculum. *International Journal of Research Studies in Education, 8*(2), 29–45. Retrieved from https://www.academia.edu/37702617/

Reston, E., & Krishnan, S. (2014). Statistics education research in Malaysia and The Philippines: A comparative analysis for future directions. *Statistics Education Research Journal, 13*(2), 218–231.

Statistics and Probability in the Curriculum in South Africa

Sarah Bansilal

Abstract This paper describes the evolution of statistics in the South African school curriculum and the nature of the statistics curriculum as intended by the most recent curricular documents. A fundamental challenge in realizing the intended curriculum is the need to provide teachers with the knowledge and pedagogical strategies for teaching and assessing statistical understanding, and in particular to ensure that all teachers including those in economically deprived areas have the same access to resources.

Keywords Counting principles · Data handling · Instructional methods · Statistical literacy

Statistics is a relatively new addition to the mathematics curriculum in South Africa and only formed part of the Grade 12 core mathematics assessment from 2008 onwards. Prior to the advent of democracy in South Africa, there were about 18 different education departments each with their own curriculum. In trying to constitute a common curriculum, the Interim Core Syllabus was designed in the early 1990's, which consisted of a list of common topics to be covered in each year of schooling. It is interesting to note that there were no statistics or probability concepts that were assessed in the Grade 12 mathematics examinations (WCED, 1998) at that time.

The next revision was the Curriculum 2005 (C2005) policy which referred indirectly to statistical reasoning when stating that students must be able to collect, summarise, display, and analyse data (Department of Education, 1997). The curriculum was understated, however, and did not include details of specific content and the depth to which these should be covered. It was only in the next stage in the

S. Bansilal (✉)
University of KwaZulu-Natal, Malvern, South Africa
e-mail: Bansilals@ukzn.ac.za

G. F. Burrill et al. (eds.), *Research on Reasoning with Data and Statistical Thinking: International Perspectives*, Advances in Mathematics Education,
https://doi.org/10.1007/978-3-031-29459-4_7

Revised National Curriculum Statements (RNCS) that Data Handling was stipulated as one of four outcomes (DoE, 2002) of the curriculum for grades R to 9. For the Further Education and Training (FET) band, that is, Grades 10 to 12, the RNCS provided an increased emphasis on statistics (Data Handling was one of the strands that was to be assessed in the Grade 12 core mathematics examinations), starting from 2008. With the latest curriculum revisions (DoBE, 2011), referred to as CAPS, the status of statistics and probability has grown further as detailed below:

> From grades R to 9, all school learners follow the same mainstream mathematics curriculum while from Grade 10-12 learners have a choice to take Mathematics or Mathematical Literacy (ML). In the first phase of schooling (Grades 1 to 3), children are introduced to collecting, sorting and representing data. In the Intermediate Phase (Grade 4-6) this is extended to analysing and reporting on data represented in different forms. In this phase some simple probability experiments are also introduced with the aim of counting and comparing frequency of outcomes. In the Senior phase, students are involved in summarising data using measure of central tendency as well as organising and recording data using stem and leaf displays as well as a variety of graphs using technology. In terms of probability students in this phase, consider situations with equally probable outcomes and consider relative frequencies and probabilities for outcomes of events.

In the next phase (Grades 10–12), for ML, data handling and probability form two out of five application topics, which involve working with real life problems that can be solved using concepts studied up to Grade 9. ML students work through the six interconnected stages of a statistical process using data sets that increase in complexity in higher grades. For probability, students are meant to focus more on interpreting situations involving probability (such as games, lotteries and risk assessments), than on the mathematical calculation of probability.

For students who pursue core Mathematics in the FET phase (Grades 10–12) the probability topics include Venn diagrams, contingency tables and tree diagrams as a means of solving probability problems. In Grade 12 they work with the fundamental counting principle. For data handling, the core mathematics students learn about representing measures of central tendency by using ogives, and calculations of variance and standard deviations, which may require technology for large data sets. They also work with various types of graphs and are introduced to the correlation coefficient as well as calculating the linear regression line that best fits bivariate numerical data.

Overall, it is evident that the curriculum emphasizes the development of statistical literacy skills that will enable learners to interpret and critically examine statistics that they encounter. However the support that could help teachers implement these ideas is limited. Resources such as specialised software, games, manipulatives etc., are important components of a data driven approach, but most teachers in South Africa do not have access to sufficient resources (Reddy et al., 2020). Schools based in more affluent areas generally obtain higher outcomes in mathematics (Bansilal & Lephoto, 2022); this may be because it is easier for teachers at such schools to access additional resources because of the additional funds that the schools can raise from fees and other activities (Reddy et al., 2020). Schools based in poorer areas are not as fortunate and are dependent on the resources provided by the government, which are mainly in the form of textbooks and online teacher and learner materials.

A few years ago an initiative called the Math4stats campaign was launched by the national statistics office and run independently in each province. The KZN Maths4stats Lecture series was launched as a collaboration between a KZN university, Statssa and the KZN department which aimed to familiarise teachers with the statistics and probability content in the curriculum and to help improve their pedagogic content knowledge. The materials were designed using a data driven approach, and teachers were provided with classroom resources and materials that could help them implement that approach in their classrooms. The workshops proved to be successful in that the participant teachers were committed to attending all the sessions over the five-week period. Teachers confidence in teaching statistics topics also improved over the time. Most teachers also perceived that their learners would be able to cope with the statistics and probability topics (North et al., 2014).

Since then there have been no largescale professional development interventions to support teachers in the teaching of statistics. This is despite the fact that research consistently points out that teachers need support in developing their personal knowledge of statistics concepts, broadening their teaching approach to statistics and accessing classroom materials and resources that they can use to enable their learners to be statistically literate (Umugiraneza et al., 2017, 2022; Kalobo, 2016; Lampen, 2015; Wessels & Nieuwoudt, 2011).

Wessels and Nieuwoudt (2011) found that most teachers in their sample still use traditional methods for teaching data handling topics instead of employing data-driven methods focusing on the development of statistical reasoning. Similarly Umugiraneza et al. (2017) found that although teachers did try to engage in progressive methods in their classrooms, teacher-led instruction methods were still their first choice. Naidoo and Mkhabela (2017) found that foundation phase teachers were keen to use demonstrations, or concrete examples to illustrate concepts and expressed that data handling concepts provided opportunities for teachers to bring in real life data. It is clear that teachers need some more support in helping their learners develop key ideas in statistical literacy and reasoning. It is necessary for teachers to first understand and to be able to describe the statistical learning outcomes themselves before they can develop their learners' statistical literacy skills and statistical reasoning (Kalobo, 2016). Many teachers are uncertain about how to assess learners' proficiency in statistical reasoning and statistical thinking (Kalobo, 2016). Teachers who understand the concepts and tools of statistics from a mathematical perspective engage their students in discourses that lead to limited understanding of the statistical concepts (Lampen, 2015). Furthermore teachers are not as confident about teaching concepts in statistics and those which require critical thinking skills as they are about teaching mathematics concepts which have traditionally appeared in the curriculum for many decades (Umugiraneza et al., 2022).

It is clear that mathematics teachers need support in developing innovative teaching strategies beyond that of teacher-centered discourses. Professional development workshops for teachers, focusing on developing the key ideas of statistical literacy and statistical reasoning, together with classroom support and relevant teaching resource, need be prioritised, if we want to make sure our learners develop the statistical literacy skills they need.

References

Bansilal, S., & Lephoto, T. (2022). Exploring particular learner factors associated with South African mathematics learners' achievement: Gender gap or not. *African Journal of Research in Mathematics, Science and Technology education*. https://doi.org/10.1080/18117295.2022.2057730

Department of Basic Education (DBE), Republic of South Africa. (2011). *Curriculum and assessment policy statement grades 4–6: Life skills*. DBE, Republic of South Africa.

Department of Education. (1997). *Curriculum 2005 lifelong learning for the 21st century*. National Department of Education.

Department of Education. (2002). *Revised national curriculum statement grades R-9 (schools): Mathematics*. Department of Education.

Kalobo, L. (2016). Teachers' perceptions of learners' proficiency in statistical literacy, reasoning and thinking. *African journal of research in mathematics, Science and Technology Education, 20*(3), 225–233. https://www.tandfonline.com/doi/full/10.1080/18117295.2016.1215965

Lampen, E. (2015). Teacher narratives in making sense of the statistical mean algorithm. *Pythagoras, 36*(1). https://doi.org/10.4102/pythagoras.v36i1.281

Naidoo, J., & Mkhabela, N. (2017). Teaching data handling in foundation phase: Teachers' experiences. *Research in Education, 97*(1), 95–111. https://doi.org/10.1177/0034523717697513

North, D., Gal, I., & Zewotir, T. (2014). Building capacity for developing statistical literacy in a developing country: Lessons learned from an intervention. *Statistics Education Research Journal, 13*(2), 15–27.

Reddy, V., Winnaar, L., Juan, A., Arends, F., Harvey, J., Hannan, S., et al. (2020). TIMSS 2019: Highlights of south African grade 9 results in mathematics and science. In *Achievement and achievement gaps. Pretoria: Department of Basic Education*.

Umugiraneza, O., Bansilal, S., & North, D. (2017). Exploring teachers' practices in teaching mathematics and statistics in KwaZulu-Natal schools. *South African Journal of Education, 37*(2). https://doi.org/10.15700/SAJE.V37N2A1306

Umugiraneza, O., Bansilal, S., & North, D. (2022). An analysis of teachers' confidence in teaching mathematics and statistics. *Statistics Education Research Journal, 21*(3). https://doi.org/10.52041/serj.v21i3.422

Wessels, H., & Nieuwoudt, H. (2011). Teachers' professional development needs in data handling and probability. *Pythagoras, 32*(1). https://doi.org/10.4102/pythagoras.v32i1.10

Western Cape Education Department (WCED). (1998). Guideline document: National examination 2001 mathematics paper 1 & 2. *Higher Grade and Standard grade. Unpublished*. WCED.

Statistics in the School Level in Turkey

Sibel Kazak

Abstract In Turkey, statistics is taught as part of the mathematics curriculum beginning in elementary grades, and this report describes the intended content focus for each grade level and the resources, curricular materials and approaches to teaching statistics. The discussion includes a comparison of the curriculum to what is needed for statistical literacy in today's world and a reflection on the realities and challenges in teaching statistics in Turkey's schools.

Keywords Data handling · On-line learning environment · Probability · Statistical literacy

In view of the advancements in science and technology as well as the changing needs of individuals and society, the national curriculum for mathematics (Grades 1–12) in Turkey was revised in 2018 (Milli Eğitim Bakanlığı [MEB], 2018a, b). Like in many other countries, statistics is taught as part of mathematics. The mathematics curriculum is comprised of the following learning domains: (1) In primary school (Grades 1–4, ages 6 to 9): Numbers & Operations, Geometry, Measurement, Data Handling; (2) In middle school (Grades 5–8, ages 10 to 13): Numbers & Operations, Algebra, Geometry & Measurement, Data Handling, Probability; and (2) In high school (Grades 9–12, ages 14 to 17): Numbers & Algebra, Geometry, Data, Counting & Probability. The curricular framework provides an overview of the structure and content for each learning domain. These learning domains are divided into sub-learning domains in which the learning objectives are described by grade level, but there are no guidelines for how teachers should approach specific content. Student learning content for the statistics and probability is briefly described by grade levels in next sections.

S. Kazak (✉)
Faculty of Education, Department of Mathematics and Science Education, Middle East Technical University, Ankara, Turkey
e-mail: skazak@metu.edu.tr

G. F. Burrill et al. (eds.), *Research on Reasoning with Data and Statistical Thinking: International Perspectives*, Advances in Mathematics Education,
https://doi.org/10.1007/978-3-031-29459-4_8

1 Data Handling in Primary School (Grades 1–4)

The data handling domain at primary grades aims to help students develop knowledge and skills in collecting, organizing, displaying, and interpreting data from tables and various graphs. Key content addressed in the learning objectives by each grade level is provided below (MEB, 2018a):

Grade 1: Reading simple data tables for one or two groups.
Grade 2: Collecting data based on the formulated statistical question, organizing data (tree diagrams, tally charts, and frequency tables), and making picture graphs.
Grade 3: Reading the information from a picture graph, conversions from graph to tally chart and frequency table; solving problems involving comparison, addition, and subtraction using the information given in graphs or making graphs; reading and interpreting simple data tables and organizing data from table.
Grade 4: Making interpretations and predictions from bar graphs; using different representations (bar chart, picture graph, table, tree diagram) to display data collected; solving problems using the information from data displays.

2 Data Handling and Probability in Middle School (Grades 5–8)

The data handling domain at middle grades aims to engage students in all data investigation phases (formulating question, collecting data, analyzing data, and interpreting results), comparing data groups using graphs and measures of central tendency and spread. The objective of the probability domain is that students are able to identify all possible outcomes of an event, investigate events with equally likely outcomes, and calculate the probability of simple events. Key focal points addressed by each grade level are as follows (MEB, 2018a):

Grade 5: Formulating statistical questions; collecting data (limited to one variable and discrete data) and displaying data with frequency table and bar graph; interpreting the results from these data displays.
Grade 6: Formulating statistical questions to compare two groups and collecting (discrete) data; displaying data using frequency tables and bar graphs; analyzing and interpreting the data to compare the groups using the arithmetic mean and range.
Grade 7: Constructing and interpreting appropriate data displays (bar graph, pie chart, and line graph) and making conversions between them; calculating and interpreting the mean, median, and mode of a data set).
Grade 8: Interpreting line graphs and bar graphs for up to three data groups; displaying data using bar graphs, pie charts, and line graphs, and making conversions between them; calculating the probability of simple events.

3 Data and Probability in High School (Grades 9–12)

The data domain in Grade 9 aims to engage students in analyzing data and interpreting results involving comparison of data groups. The objectives of the counting and probability domain in Grades 10 and 11 are to introduce combinatorics and to expand students' knowledge of probability. Below is the key content addressed in the learning objectives (MEB, 2018b):

Grade 9: Calculating and interpreting measures of central tendency and spread (range and standard deviation); representing and interpreting data using appropriate graphs (bar graph, pie chart, line graph, and histogram); comparing data groups.

Grade 10: Solving combinatorics problems involving arrangement and selection; Pascal's triangle; computing the probability of simple events.

Grade 11: Solving conditional probability problems; computing the probability of compound events; linking experimental probability and theoretical probability.

4 Resources, Curricular Materials, and Approaches to Teach Statistics

The state-approved mathematics textbooks are the main resources in teaching. Moreover, the Digital Education Platform of Turkey, known as EBA (www.eba.gov.tr) founded by the Ministry of National Education, provides an online learning environment for K–12 students as well as curricular materials and professional development content for teachers. With their personal EBA accounts, students and teachers can access to the digital learning content including interactive books, applications, lesson videos, tests etc. Problem solving, question and answer and discussion are the primary methods of teaching statistics reported in teachers' annual unit plans (Final Report Turkey, 2018). The assessment questions related to data and probability in the mathematics textbooks tends to focus on procedural knowledge (Tarim & Tarku, 2022). Although the curriculum recommends the use of technology tools for organizing, summarizing, and representing data, teachers' use of technology appears to be limited to using EBA to show relevant video or test questions on the classroom interactive board in teaching statistics (Gökce, 2019).

5 The Role of Statistical Literacy and the Nature of Probability in the Curriculum

The scope of statistics content in the curriculum is arguably narrow for the development of statistical literacy needed for today's data-driven society. When evaluated by the *Pre-K–12 Guidelines for Assessment and Instruction in Statistics Education*

(GAISE) framework including the statistical process components and the three developmental levels (A, B, C) for statistical literacy (Franklin et al., 2005), the content of the learning outcomes related to statistics in grades 1–12 was found to be at level A (60%) and level B (39%) (Batur et al., 2021). Essential concepts for statistical literacy in Burrill (2020), e.g., random sampling, two-way tables, non-standard graphs, correlation, etc., are also not addressed in the curriculum. The findings about middle and high school Turkish students' low statistical literacy level in relation to particular content, such as variation, sample selection, and evaluating inappropriate statistical claims (Batur & Baki, 2022; Çatman-Aksoy & Işıksal-Bostan, 2021; Yolcu, 2012) seem to reflect inadequate attention to statistical literacy in teaching and curriculum.

Probability is first introduced in Grade 8 at the basic level involving computation of theoretical probability of simple events. Then, in Grade 11 students are expected to solve problems involving compound events and conditional probability, and to build the connection between theoretical probability and experimental probability. Probability is treated in isolation from statistics and thus it is not conceptualized as an essential tool in statistics as suggested in the GAISE report (Franklin et al., 2005).

6 Reflection on the Realities and Challenges for Teaching Statistics

In the twenty-first century, citizens need skills and creativity to make sense of big, complex, and messy data they encounter in daily life and make informed decisions by using them. In preparation for that, students should be able to understand key statistical concepts and ideas and to develop statistical/data literacy skills during their school education. More recently, GAISE II report (Bargagliotti et al., 2020) highlights the importance of considering new data types involving text, picture, sound, and video, using multivariate thinking, and incorporating technology in statistical problem-solving process. With a focus on data science education, the new framework is enhanced with the concepts and experiences expected in each developmental level (A, B, and C) laying the foundation for understanding more complex statistical problems, data collection and analysis methods in the modern world. The current Turkish mathematics curriculum appears to be procedural and formula based and does not as much focus on making sense of data in today's world that requires statistical/data literacy skills mentioned in the GAISE II report.

The current undergraduate mathematics teacher education programs in Turkey have compulsory courses on probability, statistics, and teaching of probability and statistics. Even after the completion of these courses, pre-service teachers' content knowledge about probability can be limited to procedural knowledge (Birel, 2017) and their pedagogical content knowledge in data tends to be less adequate compared to the other content domains (Ertaş & Aslan-Tutak, 2021).

In addition to the need for incorporating aspects of data science in the school mathematics curriculum, reforms in the classroom practices are also needed. That means further efforts to be implemented in teacher preparation programs and professional development of teachers in terms of statistical content knowledge and technological pedagogical content knowledge.

References

Bargagliotti, A., Franklin, C., Arnold, P., Gould, R., Johnson, S., Perez, L., & Spangler, D. (2020). *Pre-K-12 guidelines for assessment and instruction in statistics education (GAISE) report II*. American Statistical Association and National Council of Teachers of Mathematics.

Batur, A., & Baki, A. (2022). Examination of the relationship between statistical literacy levels and statistical literacy self-efficacy of high school students. *Eğitim ve Bilim, 47*(209), 171–205.

Batur, A., Özmen, Z. M., Topan, B., Akoğlu, K., & Güven, B. (2021). A cross-national comparison of statistics curricula. *Turkish Journal of Computer and Mathematics Education, 12*(1), 290–319.

Birel, G. K. (2017). The investigation of pre-service elementary mathematics teachers' subject matter knowledge about probability. *Mersin Üniversitesi Eğitim Fakültesi Dergisi, 13*(1), 348–362.

Burrill, G. (2020). Statistical literacy and quantitative reasoning: Rethinking the curriculum. In P. Arnold (Ed.), *Proceedings of the roundtable conference of the International Association for Statistical Education (IASE)*.

Çatman-Aksoy, E., & Işıksal-Bostan, M. (2021). Seventh graders' statistical literacy: An investigation on bar and line graphs. *International Journal of Science and Mathematics Education, 19*, 397–418.

Ertaş, G., & Aslan-Tutak, F. (2021). Mathematics teacher education in Turkey through the lens of international TEDS-M study. *REDIMAT–Journal of Research in Mathematics Education, 10*(2), 152–174.

Franklin, C., Kader, G., Mewborn, D. S., Moreno, J., Peck, R., Perry, M., & Scheaffer, R. (2005). *Guidelines for assessment and instruction in statistics education (GAISE) report: A pre-K–12 curriculum framework*. American Statistical Association.

Gökce, R. (2019). *Ortaokul matematik öğretmenlerinin istatistiksel akıl yürütmeye ilişkin alan ve pedagojik alan bilgilerinin incelenmesi* [*Examining middle school mathematics teachers' content and pedagogical content knowledge of statistical reasoning*] (Unpublished doctoral dissertation). Pamukkale University.

Local Report Turkey. (2018). *Strategic partnership for innovative in data analytics in schools project*. https://spidasproject.org.uk/about/research

MEB (2018a). *Matematik Öğretim Programı (İlkokul ve Ortaokul 1, 2, 3, 4, 5, 6, 7 ve 8. Sınıflar) [Mathematics Curriculum (Primary and Middle Schools Grades 1–8)]*. MEB Yayınları.

MEB (2018b). *Matematik Öğretim Programı (9, 10, 11 ve 12. Sınıflar) [Mathematics curriculum (Grades 9–12)]*. MEB Yayınları.

Tarim, K., & Tarku, H. (2022). Investigation of the questions in 8th grade mathematics textbook in terms of mathematical literacy. *International Electronic Journal of Mathematics Education, 17*(2). https://doi.org/10.29333/iejme/11819

Yolcu, A. (2012). *An investigation of eighth grade students' statistical literacy, attitudes towards statistics and their relationship* (Unpublished master's thesis). Middle East Technical University, Ankara.

United States Statistics Curriculum

Christine Franklin

Abstract This paper describes the role of national organizations in bringing coherence to the statistical curriculum in the United States, which has no national curriculum, and the evolution of documents that shaped the status of statistics in US schools today. The latest documents recognize the changing nature of the type of data we encounter and the technology available to analyze data, and some states are currently revising their curriculum guidelines to reflect these realities. Some resources are mentioned as well as the need for educators at the school level and teacher educators to have some training in statistics.

Keywords Data · Data science · GAISE II · Teacher educators

Living and working in the twenty-first century requires the ability to evaluate and synthesize information from many global issues, which elevates the importance at the school level for statistics and data science. In the United States, school level mathematics education is organized and controlled by each of the 50 states. Since there is no defined national authority, much of the leadership for needed curriculum reform is provided by national mathematical and statistical organizations; in particular, the American Statistical Association (ASA) and the National Council of the Teachers of Mathematics (NCTM). The collaboration of ASA and NCTM began in 1968 with the formation of the ASA/NCTM Joint Committee on K-12 Education in Statistics and Probability. Through the work of this committee, the ASA and NCTM have promoted and established statistical reasoning as a key component in the school curriculum. Scheaffer and Jacobbe (2014) describe some of the curriculum and policy documents that have originated from this joint committee. This includes the groundbreaking Quantitative Literacy (QL) project (Scheaffer, 1990) in the

C. Franklin (✉)
American Statistical Association, University of Georgia, Watkinsville, GA, USA
e-mail: chris_franklin@icloud.com

G. F. Burrill et al. (eds.), *Research on Reasoning with Data and Statistical Thinking: International Perspectives*, Advances in Mathematics Education,
https://doi.org/10.1007/978-3-031-29459-4_9

1980's and the increased emphasis on data analysis in the NCTM curriculum standards for school mathematics (NCTM, 1989, 2000). The long-range impact of these efforts can be seen in the inclusion of data analysis and statistics in typical text book series (e.g., Illustrated Mathematics).

In the 1990's, College Board developed the Advanced Placement (AP) curriculum for Statistics with the first exam given in 1997. (https://apcentral.collegeboard.org/courses/ap-statistics). Secondary students who enroll in this course and perform successfully on the AP Statistics exam can receive post-secondary credit for an introductory statistics course. This course helped to jump start the consideration of more statistics at the K-12 school level in the US. According to a study of 2015 high school graduates, 25% of them had taken a statistics course and 8% of these took the AP exam (Change the Eq. 2016). A little over 183,000 US students took the AP statistics exam in 2021 (Packer, 2021).

After the publication of the 2000 *Principles and Standards for School Mathematics* that the ASA-NCTM Joint Committee followed up with the *Pre-K-12 Guidelines for Assessment and Instruction in Statistics Education (GAISE) Framework* (Franklin et al., 2007). The goal of the Pre-K-12 GAISE Framework was to provide enhanced guidance on the development of statistical reasoning concepts and skills from early grades to secondary graduation. The two-dimensional framework was structured around the statistical problem-solving process (formulate questions, collect data, analyze data, interpret results) and three developmental or evolutional levels of statistical reasoning (A, B, and C), where A = beginning, B = intermediate and C = advanced. These levels were not based on school grade but instead on student experience. Ideally, the future goal will be that level A would be equivalent to elementary grades, level B equivalent to middle grades, and level C equivalent to secondary grades. The Pre-K-12 GAISE has been an impactful document serving as the basis for curricula in statistics, the revision of state standards in the US and internationally, and the development of professional learning programs. In 2010, the Governor's Board commissioned a writing team to design a set of standards for mathematics that could be adopted by states in the United States. The vision was a united curriculum for mathematics across the US. This effort resulted in the Common Core State Standards for Mathematics (CCSSM)(NGACBP & CCSSO, 2010). The CCSSM or similar standards were adopted by most states. The GAISE framework was the foundation for the statistics standards in the CCSSM. GAISE has also been used extensively in statistics education research at the school level .

After the adoption of the CCSSM the Mathematical Education of Teachers (MET) document, (CBMS, 2012), written by a team of mathematics and statistics leaders in education including some writers of the CCSSM, made recommendations about the statistics that school level teachers needed to know to successfully implement the CCSSM. Given that most teacher preparation programs in mathematics included little to no statistics, the ASA/NCTM committee recognized the need for more guidance regarding the statistics education of teachers and in 2015 published the *Statistical Education of Teachers (SET)* (Franklin et al., 2015). SET made recommendations regarding programs of study, content based upon the framework learning outcomes from Pre-K-12 GAISE, assessment, and the mathematical practices through a statistical lens. The overarching goal was to encourage students and teachers to develop sound statistical habits of mind. The board of directors for both ASA and NCTM have the statistical education of teachers as a priority and endorsed a joint official position statement in response to this question, "What preparation and support do teachers need to successfully support students' learning of statistics and data science in the prekindergarten–grade 12 curriculum? (https://www.amstat.org/docs/default-source/amstat-documents/pol-jointasa-nctm-statement.pdf?v=1).

NCTM embarked on examining mathematics education with the publication of *Catalyzing Change in High School Mathematics* (2018) that was followed by versions for middle and elementary school. Key recommendations can be found at (https://www.nctm.org/uploadedFiles/Standards_and_Positions/Catalyzing_Change/CC-Recommendations-PK12-CatalyzingChange.pdf). The Catalyzing Change books recommend essential outcomes for developing statistical reasoning skills and also recommend that statistical knowledge in the curriculum be treated as equable with the other mathematical strands of algebra and geometry. The high school book included essential outcomes in quantitative literacy for the first time in a major curriculum document.

Since the publication of the Pre-K-12 GAISE in 2007, much has changed including the type of data we encounter and the technology available to analyze data. It is no longer acceptable for students to experience statistics by working with small data sets, collected first hand, and classifying data as either quantitative or category. Instead, they need to encounter data sets that are large and messy. Researchers often turn to secondary data sets collected for other purposes. Data might be photos, sounds, medical images, or twitter texts. Gould in his paper, Statistics and the Modern Student (2010) proposed the following goals for the statistics education community: teach about databases, improve access to data, develop new teaching tools, re-examine which fundamental concepts should be taught, integrate computation into the curriculum, and change the culture. Given the reliance on large secondary data sets in data science that is developing, Wise (2019) states, "…this expands the role of data scientist from one who analyzes a large data resource to produce information to one who must build a bridge between ill-defined questions and unstructured messy data that may (or may not) be fit to address them by assessing cleaning, organizing, integrating and visualizing data, … and communicating appropriate inferences in an ethical manner." (p.166).

Recognizing this new landscape and the evolving field of data science, the ASA/NCTM updated the original Pre-K-12 GAISE to account for these changes. The Pre-K-12 GAISE II was published in 2020 (Bargagliotti et al., 2020). The spirit and recommendations of the original GAISE remain. The statistical problem-solving process remains the core of statistical reasoning and making sense of data. The three developmental levels (A, B, and C) remain as experience based. Key enhancements in GAISE II are:

- An emphasis on recognizing and working with different data types including data from secondary sources that necessitate the management of larger messy data sets
- The importance of questioning throughout the four components of the statistical-problem solving process
- Multivariable thinking throughout the three developmental levels A, B, and C
- The role of probabilistic thinking throughout the three developmental levels A, B, and C
- The role of technology and computational thinking in statistics
- Assessment items that measure statistical reasoning not mathematical reasoning

Catalyzing Change in High School Mathematics (NCTM 2018) and *Pre-K-12 GAISE II* (2020) are being utilized by several states in revising their mathematics standards resulting in additional statistics and data science standards beginning in the early grades. For example, data science is included in the draft California mathematics framework, and Oregon, Virginia, and Ohio include data science as an optional third-year mathematics course. Groups such as TERC are offering after school or summer programs in data science (Rubin, 2021).

Because of the increased emphasis on statistics and data science in state standards, teacher preparation and professional development remain priorities for ASA, NCTM, and other organizations such as the Association of Mathematics Teacher Educators (AMTE). The joint ASA/NCTM committee has developed a wealth of open-source resources to help prepare teachers to implement statistics and data science in the classroom. All these resources can be found at: https://www.amstat.org/education/k-12-educators.This includes the book, *Statistics and Data Science for Teachers* (Bargagliotti & Franklin, 2021). (https://www.amstat.org/docs/default-source/amstat-documents/SDSTeacherBook-highres.pdf).

In summary, a key vision is creating an environment where teachers have the resources and educational preparation to develop a deep conceptual understanding of the statistics they teach and are able to plan lessons and activities to allow students to discover a structure or pattern in a set of data as they attempt to answer a statistical question. This is a lofty goal especially given that the majority of educators at the school level teaching statistics have little or no formal training in statistics. This leads to another challenge: to adequately prepare teacher educators, many who also have no formal training in statistics, to train school-level educators to teach statistics (both pre-service and in-service) (Tarran, 2020).

The vision presented in GAISE II is for a world where every individual is confident in reasoning statistically, grappling with fast-changing data and statistical information, and knowing how and when to bring a healthy skepticism to information gleaned from data. (Franklin, 2021) To achieve this vision, clarity must come to what are the learning outcomes for statistics and data science at the school level. There must also be a focus on student and teacher preparation in statistics and data science.

References

Bargagliotti, A., & Franklin, C. (2021). *Statistics and data science for teachers*. American Statistical Association. https://www.amstat.org/docs/default-source/amstat-documents/gaiseiiprek-12_full.pdf

Bargagliotti, A., Franklin, C., Arnold, P., Gould, R., Johnson, S., Perez, L., & Spangler, D. A. (2020). *Pre-K-12 guidelines for assessment and instruction in statistics education II (GAISE II)*. American Statistical Association and National Council of Teachers of Mathematics.

Change the Equation Analysis of 2015 National Assessment of Educational Progress. (2017). *Presentation at the National Council of Teachers of Mathematics Annual Meeting, Chicago IL*. https://schools.saisd.net/upload/page/0061/docs/NCTM_2017_Chicago_share.pdf

Conference Board of the Mathematical Sciences. (2012). *The mathematical education of teachers*. Providence.

Franklin, C. (2021). As Covid makes clear, statistics education is a must. *Significance, 18*(2), 35.

Franklin, C., Kader, G., Mewborn, D., Moreno, J., Peck, R., Perry, M., & Scheaffer, R. (2007). *Guidelines for assessment and instruction in statistics education (GAISE) report: A pre-K-12 curriculum framework*. American Statistical Association. https://www.amstat.org/docs/default-source/amstat-documents/gaiseprek-12_full.pdf

Franklin, C., Bargagliotti, A., Case, C., Kader, G., Scheaffer, R., & Spangler, D. A. (2015). *The statistical education of teachers*. American Statistical Association. https://www.amstat.org/docs/default-source/amstat-documents/edu-set.pdf

Gould, R. (2010). *Statistics and the modern student*. Department of Statistics, UCLA. Retrieved from https://escholarship.org/uc/item/9p97w3zf

National Council of Teachers of Mathematics (NCTM). (1989). *Curriculum and evaluation standards for school mathematics*. Reston.

National Council of Teachers of Mathematics (NCTM). (2000). *Principles and standards for school mathematics*. Reston.

National Council of Teachers of Mathematics (NCTM). (2018). *Catalyzing change in high school mathematics: Initiating critical conversations. National Council of teachers of mathematics*. Reston. https://www.nctm.org/change/

National Governors Association Center for Best Practices and Council of Chief State School Officers. (2010). *Common core state standards for mathematics*.

Packer, T. (2021). *AP statistics exam 2021 results*. https://allaccess.collegeboard.org/ap-statistics-exam-2021-results

Rubin, A. (2021). What to consider when we consider data. *Teaching Statistics, 43*(1), 23–33. https://doi.org/10.1111/test.12275

Scheaffer, R. (1990). The ASA-NCTM Quantitative Literacy Project: An overview. In the *Proceedings of the Third International Congress on Mathematics Education*. https://iase-web.org/documents/papers/icots3/BOOK1/A1-2.pdf?1402524941

Scheaffer, R. L., & Jacobbe, T. (2014). Statistics education in the K-12 schools of the United States: A brief history. *Journal of Statistics Education, 22*(2).

Tarran, B. (2020). Statistical literacy for all! *Significance, 17*(1), 42–43.

Wise, A. (2019). Educating data scientists and data literate citizens for a new generation of data. *Journal of the Learning Sciences, 29*(1), 165–181. https://doi.org/10.1080/10508406.2019.1705678

Part II
Data and Young Learners

Elementary Students' Responses to Quantitative Data

Karoline Smucker and Azita Manouchehri

Abstract In this work, relying on a teaching experiment methodology, we examined development of statistical thinking surrounding the creation of graphical displays amongst a cohort of elementary students. Students showed flexibility in adapting techniques from categorical data to create displays that took into account various wingspans of class members, though for some class members it was difficult to separate graphing from the "rules" which had been encountered in prior instruction. The context of using data collected by students led to increased engagement but also affected students' abilities to consider the overall characteristics of the distribution when analyzing their displays.

Keywords Elementary students' reasoning · Statistics education · Data analysis · Quantitative and categorical graphing techniques · Distributional features

1 Introduction

With the prevalence of data in the world today, understanding statistical information and interpreting or making conclusions based on this information are essential skills. Individuals are exposed to data through the media and other experiences in their daily lives (Wild et al., 2018). Because of this, there is agreement that students need to learn to evaluate, reason with, and communicate about quantitative information as early as the elementary grade levels (Ben-Zvi & Garfield, 2008; Shaughnessy, 2007).

K. Smucker (✉)
Eastern Oregon University, La Grande, OR, USA
e-mail: ksmucker@eou.edu

A. Manouchehri
The Ohio State University, Columbus, OH, USA
e-mail: manouchehri.1@osu.edu

G. F. Burrill et al. (eds.), *Research on Reasoning with Data and Statistical Thinking: International Perspectives*, Advances in Mathematics Education,
https://doi.org/10.1007/978-3-031-29459-4_10

One important area within the study of statistics is the creation and use of graphical representations (English, 2012; Watson, 2006). Recent standards in the United States, including the Common Core State Standards in Mathematics (National Governor's Association Center for Best Practices and Council of Chief State School Officers, 2010) and the Guidelines for Assessment and Instruction in Statistics Education: GAISE II (Bargagliotti et al., 2020) have emphasized the importance of data collection and display/interpretation in the elementary grades. According to these standards, young learners should be able to collect, display, and interpret categorical and quantitative data. However, research suggests that students may encounter data in very prescriptive ways based on prior instruction, viewing graph creation as the goal rather than graphs as a tool for understanding data (Friel et al., 2001; Watson, 2006). In addition, students may not be given opportunities to consider which displays make sense in different data contexts, leading to confusion when they are given data without specific instruction on what type of display may be appropriate (English, 2012; Lehrer & Schauble, 2004). It is essential that students engage with tasks and contexts where they are asked to create displays within the larger context of statistical inquiry, allowing them to consider what type of display may be appropriate to represent specific data sets.

To consider this issue, a teaching experiment with a group of third graders was conducted to investigate students' development towards representing and analyzing quantitative data. Of interest were student responses when encountering self-collected quantitative data after primarily being exposed to data based on categories. The goal was to explore how students approach the display of quantitative data prior to explicit instruction on quantitative techniques.

2 Literature Review

This research highlights the role of creating and interpreting graphical representations within statistical inquiry. The GAISE II framework for statistical inquiry suggests that young learners engage in investigations where they formulate questions, collect/consider data, analyze it, and interpret their results (Bargagliotti et al., 2020). This work focuses on the role of data collection and, in particular, investigates the representations students create when first exposed to quantitative data without explicit instruction on standard displays. While it is important that students are exposed to standard displays like dot plots and histograms in order to communicate their findings with others, research suggests that offering learners the opportunity and time to consider the data and make their own choices about how to represent it visually (potentially through non-traditional displays) will help them develop decision making skills when conducting statistical investigations in the future (Watson, 2006).

Konold and colleagues (2015) posit four different "lenses" that are used by students to view and interpret data. Since "perceiving, describing, and generalizing from aggregate features of data is what statistics is primarily about" (Konold et al., 2015, p. 306), researchers are interested in the ways that school age learners view data, and the extent to which their views match the aggregate approach that is essential for statistical inference. These include: data as pointer, data as case value, data as classifier, and data as aggregate. Those with a pointer perspective view data as a reference to the larger event from which the data came, while those who recognize data as case values focus on individual values and the attributes they represent (often those that represent themselves). A classifier perspective is indicated by those who sort the data values based on attributes and reference ideas related to frequency, while those with an aggregate perspective view data as a "unity with emergent properties such as center and shape" (Konold et al., 2015, p. 308).

Students' perceptions of center, shape, and spread (variation) are closely connected to research on distributions and distributional reasoning, which relates to the ability to reason about data as an aggregate versus focusing on individual cases (Bakker & Gravemeijer, 2004; Konold et al., 2015). Some of the important properties or measures that students may use as they navigate between a distributional approach and an approach focused on individual cases are center, spread, shape, density, and outliers (Bakker & Gravemeijer, 2004). Reasoning about distributions and their aggregate properties is essential in order to perform many statistical tasks, including comparing groups and making inferences.

Research has emphasized the importance of developing informal inferential reasoning (IIR) in classroom settings (Braham & Ben-Zvi, 2017; Makar et al., 2011; Paparistodemou & Meletiou-Mavrotheris, 2008; Pfannkuch, 2011). An informal statistical inference "uses available data as evidence to make an uncertain claim about the world" (Makar, 2014, p. 62). With younger learners, informal inferences may focus on "modal clumps" when considering what value is typical for a distribution or when comparing two distributions presented graphically (Frischemeier, 2019; Makar, 2014). Watson and English (2017) suggest that instruction should emphasize graphical representations in conjunction with values like the mean and median as students are asked to draw inferences regarding typical values. Activities that emphasize IIR have been promoted as essential to prepare students to encounter more formal methods of statistical inference at the secondary and tertiary levels (Garfield et al., 2015).

2.1 Graphical Representations

Watson (2006) proposes six stages that students undergo in their development of statistical literacy, including idiosyncratic, informal, inconsistent, consistent non-critical, critical, and critical mathematical. Within the realm of Representation, which includes the creation and analysis of statistical displays, students typically begin with "basic table and graph reading" before moving on to the creation and interpretation of displays (Watson, 2006, p. 253).

In many classrooms students are given limited opportunity to explore their own graphical techniques and are instead exposed to standard displays from textbooks and other curriculum (English, 2012). Students are offered "prescriptions" on how to construct various displays without focusing on "the reasons the graphs were created in the first place" (Friel et al., 2001, p. 132), along with textbook examples that are too preprocessed. Because of this, research with elementary students suggests the importance of grounding graph creation in the larger context of statistical inquiry, allowing students the opportunity to consider what might be useful to showcase for specific data based on its characteristics and their questions (Leavy, 2008; Watson & English, 2017, 2018). This approach may "promote a more flexible, fluid, and generalizable understanding of graphs and their uses" (Friel et al., 2001, p. 134). This is especially important as students are asked to analyze and/or create both written displays and displays using computer software.

Students of all ages tend to view tasks involving graphs very procedurally, considering graph creation, not analysis, as the goal (Friel et al., 2001). Cooper and Shore (2008), drawing on their research with introductory statistics students, argued that traditional instruction in K-12 may not provide students with the tools they need to coordinate measures and displays of center and spread effectively and that students need opportunities to not only create graphs but tasks where they must interpret a variety of displays with varying centers and spreads. They suggest that students need to develop "graph sense" by creating and interpreting graphs in a variety of contexts for problem solving.

According to English (2012), children may struggle with "imposing structure consistently", and often create displays which either "overlook important information" or "include redundant information" (p. 17). To address these difficulties, English suggests that young learners need experiences that require them to make their own decisions about how to structure data, so that they can analyze their displays and make revisions. In particular, elementary level students need more experience with both displaying and interpreting variation in a variety of data contexts (English, 2012; Lehrer & Schauble, 2004; Watson, 2006).

Students can use non-traditional displays and measures, as long as these displays are useful in making sense of data. Lehrer and Schauble (2004) found that, when fifth grade students were asked to create their own displays of plant height data, students first focused on case values (a column or row in the display for each measure). Next, students began to organize the data in groups of similar values. At the end of the activity, there was recognition of the importance of ordering the data and including gaps to display its true features, showing an informal appreciation of distribution. This was also found within a similar study with fourth grade students on student-created displays of the measure of class arm spans (English & Watson, 2015). Tasks where students create and then debate the usefulness of displays appear to be an important way for students to move from a case value to an aggregate approach to data (Konold et al., 2015; Lehrer & English, 2018).

In their work with secondary students investigating Old Faithful data, Shaughnessy and Pfannkuch (2002) asked middle school students to create displays that made sense to them for the geyser data. By allowing various groups to present their results, students were able to see that different characteristics of the data in terms of variation were evident based on the display chosen. For example, time plots showed the bimodal nature of the data more clearly, while those who created box plots missed out on this feature of Old Faithful entirely. Activities like this, which allow students to explore data to discover various properties, provide them the opportunity to think critically about what information can be extracted from different displays and discover patterns in the data (Leavy & Hourigan, 2018).

Technology also plays a key role in graphical representation and analysis. Computer applications can allow for the creation and manipulation of displays, along with giving learners the opportunity to quickly modify representations to investigate new ideas and questions (Paparistodemou & Meletiou-Mavrotheris, 2008). Frischemeier (2019) found that 10–11-year-old students could effectively use dot plot and hat plot comparisons created with technology to determine that heavy frogs jump further than light frogs. While similar comparisons can be made within the context of displays created by hand, data analysis software allows students to create and adjust displays of large data sets quickly so that they can focus on analysis and interpretation (Biehler et al., 2013; Chance et al., 2007). At the same time, Frischemeier (2018) suggests it is important that students also be given the opportunity to create displays by hand and that these experiences should be included alongside the use of technology in graphing.

2.2 *Context*

When working with data, the context surrounding it plays an essential role in developing students' understanding. Context is one feature that can separate the study of statistics from the study of mathematics. According to G. W. Cobb and Moore (1997), in mathematics "context obscures structure", while in statistics, "context provides meaning" (p. 803). Watson (2006) emphasizes the essential role that consideration of context plays in development of statistical literacy.

The use of information which comes from real settings is essential to engage students in inquiry around data. When "meaningless or nonexistent" context is provided for data, students will often apply incorrect on inappropriate procedures, while when the context is "interesting, and relevant to students' worlds" they are more likely to use statistical ideas (Doerr & English, 2003, p. 111). Makar (2014) found that eight year-old students were able to determine the reasonableness of both individual values and inferences surrounding hand span and height measurements by "self-checking" their results against what they observed of themselves and their peers (p. 68).

On the other hand, when students engage with the context too personally, it can hinder their ability to think about the overall distribution of data (Watson, 2006). Students may focus so closely on the story behind data that they make conclusions based on their informal knowledge of the data's context over patterns in the data itself. Pfannkuch (2011) found that, while the data's context helped secondary learners observe and identify meaningful patterns during a statistical investigation, at times they needed to disassociate from contextual features in order to make more general conclusions and connections to broader theoretical ideas. Makar et al. (2011)

suggest that students need experiences that challenge them to coordinate their statistical and contextual knowledge in order to be able to "integrate these knowledge bases" (p. 156).

In this work we considered the role self-collected measurement data played in a statistical investigation. Measurement data collected by students is frequently used with younger learners (i.e. English & Watson, 2015; Lehrer & Schauble, 2004; Watson & English, 2017, 2018), but more work is needed to understand how these contexts may impact students' representations, interpretations, and inferences. A secondary aim of this study was to understand how students approached representing quantitative data without explicit instruction on standard techniques. We studied elementary students' approaches to creating and analyzing graphical representations when investigating class wingspan data. Of interest were the following research questions:

1. How does a designed sequence of instruction involving collection, representation, and analysis of quantitative data without explicit instruction on quantitative techniques impact elementary students' approaches and analysis?
2. How do elementary students respond to the context of quantitative wingspan data representing themselves and their peers?

3 Methods

The setting for this study was a third-grade classroom in the Midwestern United States (8–9 year-old students). A focus group of five students from a class were recruited to participate in the research, which used a teaching experiment methodology (Steffe & Thompson, 2000) and principles of design research (P. Cobb et al., 2003). The students were selected from a regular education classroom whose teacher agreed to allow recruitment for the study and were chosen based on their availability to participate during the time when the study occurred. Prior to the teaching experiment, students had received instruction on the creation of categorical displays (specifically pictographs and bar graphs) but not on displays/analysis of quantitative data. While we had anticipated that such exposure could potentially impact the participants' mathematical practices the extent of the influence was not yet known.

3.1 Methodology

In this work we relied on a teaching experiment methodology (Steffe & Thompson, 2000). The methodology applies when the researcher aims to build a model of learners' thinking based on implementation of research-based activities and designed instruction. When teaching, the teacher-researcher is engaged in inquiry, tests hypotheses, and promotes development; while reflecting, the teacher-researcher engages in analysis, hypothesis generation, and model building. This leads to an analytical approach which includes both ongoing reflection and analysis of individual lessons or tasks, with more in-depth retrospective analysis of the progression as a whole at the end of the experiment. An initial proposed trajectory is modified as the research team changes their plans based on what is revealed during previous sessions. In their initial analysis the team considers what next steps may be necessary and adapts the sequence of instruction if needed. These conversations or interactions become part of the larger data set for the study. At the end of the study, a more in-depth retrospective analysis occurs (P. Cobb et al., 2003). These principles applied to the current study.

3.2 Study Design

The teaching experiment consisted of three 45-minute-long sessions over the course of one week. Data collected consisted of all small and large group discussions amongst a focus group of five students. All sessions were video recorded, along with student work throughout the task sessions. The videos were then transcribed to allow for more in-depth analysis.

3.3 Task Sequence

The designed task involved an investigation by students on documenting and representing the lengths of wingspans in their class. The following question guided the investigation: how long are the wingspans of third graders in Mrs. Jones' class? A description of each of the three sessions used is given below:

Session 1 Introduction/Data Collection: Students were introduced to the investigation question, discussed measurement techniques, practiced measuring wingspans within the focus group, and collected wingspan data from peers.

Session 2 Creation of Representations: Students, either in pairs or individually, represented the class wingspans graphically.

Session 3 Group discussion/Analysis: Students discussed the features of the displays they had created during the previous session and how they might answer the question about the length of wingspans in the class, along with follow up questions about what they might be able to say about another third-grade class, a fifth-grade class, or the wingspan of a new student who entered the class based on their displays.

3.4 Analysis

In accordance with the teaching experiment methodology, analysis was ongoing and followed two stages. First, the events of each teaching session were evaluated to make adjustments to plans for the next session. For example, after the first session, the team decided to offer the participants the opportunity to work individually or in pairs when creating their displays during the next session, rather than requiring them to work in pairs or groups of three. This instructional decision was made to give students more choice/freedom in how they engaged with the task to obtain more variety in displays presented during the class discussion.

At the end of the experiment, the transcripts and student work were coded to identify examples of students using their prior understandings and contextual features of the data during data collection, display creation, and interpretation. The codes which were developed and used, along with examples of each, are shown in Table 1.

This analysis led to the identification of several themes in terms of student approaches to graphing and interpreting self-collected measurement data after previous exposure to categorical techniques. These will be discussed in the following section.

Table 1 Code descriptions and examples

Code	Description	Example(s)
Graph types and characteristics	Moments where students referenced specific graphs and their features	"Like a pictograph?", "You're supposed to have a scale label"
Quantitative/ categorical	Evidence of quantitative and categorical considerations of data and display creation	"Maybe you could count by 10 s" (categorizing using a number line)
Context	Moments where students considered the context of the data (measurement data they had collected from themselves and their peers)	"Everybody's was in the 50 s but mine – But that kind of makes sense because I'm shorter than everybody", "It's not the same thing because Julie is missing a 40"
Interpretation or inference	Judgments/justifications of typical values, distributional characteristics for the class and/ or other groups of students	50–55 typical/average because "it's the highest", "Mrs. Gordon's class has some really tall kids" so their graph would be "higher"
Teacher-researcher	Interventions by the teacher-researcher relative to display creation, interpretation, and inference	"What if there was a person who was 59 and 3/4? How would we categorize that person?"

4 Findings

In the initial task session, the focus group collected wingspan measurement data from their class (including themselves). In the subsequent sessions, researchers found that when exposed to data that differed from those previously experienced (the quantitative wingspan data was different than the categorical data students were used to), students assessed and adapted strategies from several graphical methods. Students also engaged with the context of the data in several ways, which impacted their analysis and interpretation. These are explained in the following sections.

4.1 Focusing on the "Right" Graph and Features

When first confronted with quantitative data, all students focused on using what they perceived to be the "right" kind of graph. This was not unexpected, since "young children's typical exposure to data structure and displays has been through conventional instruction on standard forms of representation" (English, 2012, p. 17).

Frequently referenced displays included bar graphs and pictographs. When asked to create a display, two of the participants immediately asked questions about the type of graph they were expected to create. One student, Julie,[1] asked, "Like a bar graph?" while Jeff stated, "What kind of graph? A pictograph?" Students expected to be told explicitly what type of graph to make, and appeared unsure when the instructions asked them to create any display they felt was appropriate.

Students also engaged in describing the characteristics of both their graphs and those of their classmates but often showed little interest in the data itself, instead focusing on whether a display rigidly followed the prescriptions of graphing they had been previously taught. Thus, at times the goal of graphing was not to understand the data but simply an exercise in drawing and labeling. Students assessed their own or others' displays solely on whether they had all the "right" features. An instance of this occurred as students discussed one group's graph (see Fig. 1, Bob and Jeff's display). Katelyn said, "So you're supposed to have like a scale label – you're supposed to have three labels and they only have the title". Instead of discussing what she might be able to say about the class wingspans based on Bob and Jeff's representation, Katelyn's focus was on what was missing from the display. It was assumed that labeling the graph was necessary to establish correctness of the model. Indeed, students articulated that they considered certain display features to be incorrect because of what they recalled from prior classroom experiences. The following vignette (involving further discussion of Fig. 1) is illustrative of this instructional influence.

[1] All names are pseudonyms.

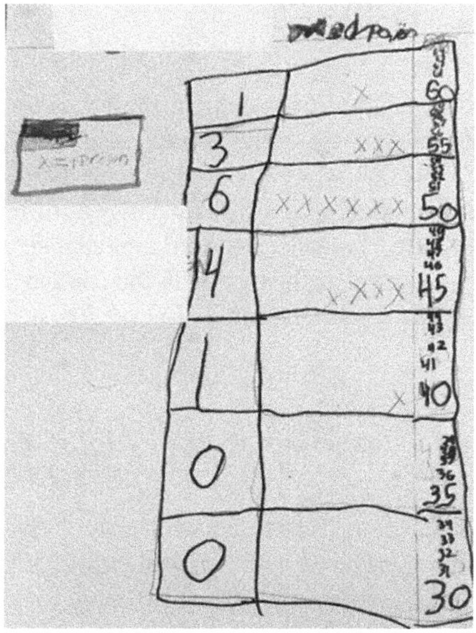

Fig. 1 Bob and Jeff's display of wingspan lengths

Katelyn: Mrs. Jones always says don't put the number on it cause that's basically giving away all the math

Researcher: Hmmm

Bob: I don't remember her saying that

Katelyn: Well she said that to me – she said never say the actual number – they need to do it by theirselves

Researcher: I have seen graphs where they put the numbers on top - for this graph it does give us an idea of the values in each category

Katelyn's statement that including the frequencies for each category (left-hand column of display) would "give away all the math" emphasizes the perception that graph creation is an exercise. She believed it was important to leave off the summary numbers in order to give people something to do "theirselves", not for any reason that would aid in analysis or interpretation.

In this instance, the teacher-researcher's intervention intended to stress that displays can look different as long as they help those reading the graph interpret the data. It is important that students learn to create standard displays clearly labeled for communication with a wider audience and it may be important to consider how much to include in a display to keep it from appearing cluttered. However, these instances highlight how a rigid focus on following the rules can keep students from engaging with data flexibly, a skill which will help them when they encounter data outside a classroom setting (Friel et al., 2001).

4.2 *Adaptation of Both Categorical and Quantitative Techniques*

In the first instructional session, students practiced measuring and recording each other's wingspans. As they recorded their own wingspans and talked about their measurement techniques, they immediately discussed how they might represent the data using a graph, as depicted in the vignette below. While the teacher-researcher had not planned to discuss graphing during this initial teaching session, it seemed important to pursue students' ideas and questions about graphing at this time.

Abbie:	I think the whole class would have different wingspans – How could we collect data and put it on like a graph?
Julie:	We could do it between inches, like everyone in between – Everyone between 60 inches
Researcher:	Oh, say a little more about that
Katelyn:	Oh yeah – I agree with Julie – I think what she's saying is if they are between say 60 inches let's say 60 and 65 – let's say Bob was 65 and Jeff was 62 they could be in a group together because they are so close together

It appeared the students recognized that the data they explored were not consistent with their prior experiences in graphing, which required categories to organize the data. Abbie was unsure how she would create a graph since she could not think of a way to categorize the data if each student had a different wingspan. This motivated Julie and Katelyn to suggest that one way to resolve this issue would be to cluster values that were "close together".

This adaptation of pictographs and bar graphs to include ranges of values instead of standard categories allowed students to use their prior knowledge in a new way. They proposed a strategy that led to production of modified versions of a histogram in the subsequent class session. While the research team had expected at least some students to initially create displays using case value bars (i.e. Lehrer & Schauble, 2004), perhaps because of the discussion in the initial session, all groups immediately decided to cluster their values into categories. Figures 2 and 3 showcase Katelyn and Julie's displays. Both utilized features of traditional categorical techniques (Katelyn adapted a pictograph; Julie adapted a bar graph), but the quantitative wingspan data required them to make some adjustments to their displays. They had to consider how to turn the numerical wingspan data into categories, and how to adjust their labels and axes accordingly.

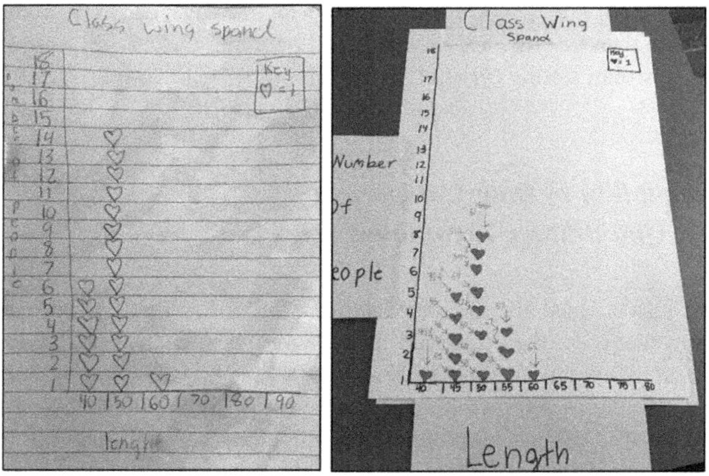

Fig. 2 Katelyn's initial display (*left*) and final display (*right*) of wingspan lengths

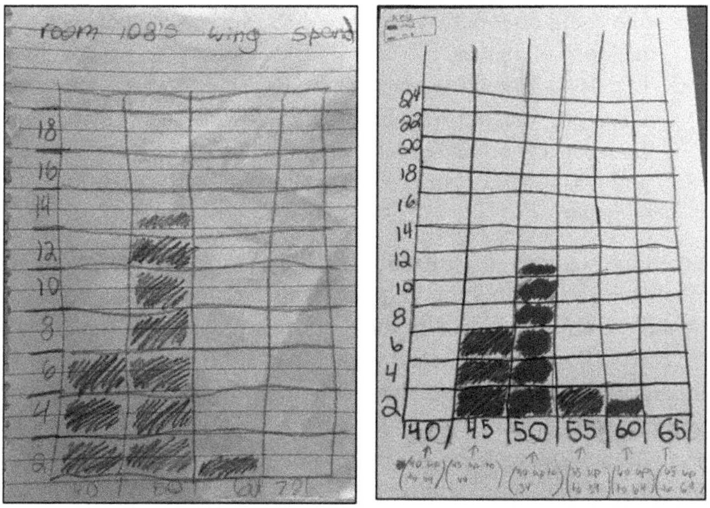

Fig. 3 Julie's initial display (*left*) and final display (*right*) of wingspan lengths

It seemed significant that students began creating their displays with labels like 40, 50, 60, and so on, with several participants (Julie and Katelyn) initially grouping all values in the 40 s above 40, all values in the 50 s above 50, etc. The other participants (Bob/Jeff and Abbie) initially grouped their data by 5 s, with values from 40–44 above the 40, and 45–49 above the 45. After creating their initial displays, Katelyn and Julie realized that most of their values would be in the 50 s and desired to create more categories. When asked by the teacher-researcher whether it was possible to create more categories both Katelyn and Julie decided to regroup their data by 5 s. It was unsurprising that this strategy was utilized since it broke each of their original groups in half. Interestingly, this grouping model was adopted by all participants.

Students were later questioned by the teacher-researcher about the data values represented in their categories, for example, if all the values above the 50 represented exactly 50 inches. This intervention appeared to impact students' approaches, leading them to adjust their displays to clarify what values were in their categories. Students made a variety of adaptations to their displays, as shown in Figs. 1, 2, and 3. Katelyn (Fig. 2, final display) decided to leave her categories as is (40, 45, 50, etc.) but individually labeled the lengths by drawing an arrow to each specific wingspan (heart). Julie (Fig. 3, final display), on the other hand, decided to change her categories to include a range of values. Jeff and Bob (Fig. 1) listed all values that were included in each category.

Although students adapted categorical techniques, they also used techniques and analysis typical of standard quantitative displays including line plots and histograms. During the initial discussion of data collection techniques that led to an unplanned conversation about graphing, Julie referenced line graphs and said, "Maybe you could count by 10s," to create categories, drawing a number line with values of 30, 40, 50, 60 and 70 to represent how she would divide her groups. Katelyn then suggested that they could show values in between each number on the number line using tick marks, though none of the students took up this suggestion when making their displays. Similarly, all students organized their groups from least to greatest when creating their graphs, another indication that their categories were viewed as markers rather than distinct from each other. In this way, students' adaptations of categorical techniques emulated many standard quantitative procedures.

However, when adapting their categorical techniques to the quantitative data, students occasionally included redundant elements in their displays, an issue noted by English (2012) when young leaners engage with data modeling. For example, Katelyn (Fig. 2) included a frequency axis to represent the number of students in each category while also including a key stating that each heart represents one person. Julie (Fig. 3) showed the same redundancy with her bar graph, and Bob/Jeff (Fig. 1) provided a key in addition to the category frequencies on the left of their display. When asked by the teacher-researcher whether the key was necessary to understand Katelyn's display, participants agreed that it might not be needed since a scale was included. This intervention appeared to help students recognize a structural redundancy.

In addition, participants did not appear to recognize that their categories included gaps in values when considering wingspan measures, for example the categories 55–59 and 60–64 inches would not explicitly determine how to categorize a value like 59 ¾ inches. This may explain why Julie's final graph (Fig. 3) does not have a value in the 40–44 category while the others do, since one student had a wingspan of 44 ¾ inches. However, another explanation for this difference is that students' data lists varied slightly due to measurement variability. The teacher-researcher attempted to tackle categorizing values like 59 ¾ inches with the following planned follow up question during the final discussion.

Researcher: What if there was a person who was 59 and 3/4? How would we categorize that person?
Katelyn: I would put it in 55 because it's not 60 yet, it's 1 quarter away from 60, but I could see some people putting it in 60 because it's closer to 60 but I wouldn't
Researcher: So it's closer but you wouldn't – what do you think Bob?
Bob: I feel like I would put it in the 60
Researcher: Why?
Bob Because it's closer to 60 than 59
Researcher: Ok so we're getting two opinions
Julie: I agree with Katelyn, because they are not in the 60 s yet they are still in the 50 s
Researcher: But when you wrote your categories you said 55–59
Julie: But you're over 59 though
Katelyn: It's over 59
Julie: So I would put it in 60

In this instance, the teacher-researcher attempted to push students to consider the values their current displays might not be able to account for. Students considered two potential solutions, rounding up the values to the next category (the 60 s) or keeping values in the lower category because, as Katelyn said, "It's not 60 yet," though for this strategy participants could not determine how they would communicate this in their current displays. This is likely why, when pressed by the researcher on which strategy they would utilize, all students chose the rounding option, though Katelyn stated that they could also have a vote in the class to determine how to categorize these tricky values. Thus, while students did utilize some quantitative techniques to create their displays, they at times appeared to view their categories as discrete/distinct sets of values as opposed to as intervals on the number line.

4.3 The Role of Context in Analysis and Inference

Context played a significant role in the way that students interacted with the wingspan data. While the use of self-collected data from the class led to a high degree of student engagement and excitement about the task, it also raised challenges as students were creating and discussing their displays. In addition, the data context, including knowledge of specific groups (other classrooms of third graders) and relationships between variables (wingspan and height) also impacted students' interpretations and inferences.

When students were prompted to consider common features of their graphs, they spent much of their time labeling and "fixing" differences between their displays. Participants had used their own measurements which differed slightly for several of their classmates, so students were intent on tracking these differences and noting them in each other's graphs. During the initial task discussion where students practiced collecting wingspan data, they appeared to recognize and accept that these small differences could occur (measurement variability), but it also seemed important to participants to know exactly which students may have been recorded with slightly different measures. While precision in graphical representation is important, the personal nature of the data made it difficult for students to move beyond minor differences in their displays.

In considering all graphs produced by group members, two students reported that they noticed "they're all different". In an attempt to move students towards a discussion of the more general/aggregate characteristics of their displays, the teacher-researcher asked them to compare the shapes of their graphs. This intervention inspired students to look more deeply at the content missing or present in displays as evidenced in the following exchange.

Researcher:	Look at these side by side and look at the shape – What do you notice?
Abbie:	Oh, it's the same
Jeff:	Oh, they're the same
Bob:	It's the same thing
Jeff:	It's the same thing except these are pictures and those are like bars (pointing)
Abbie:	Wait, wait, wait!
Researcher:	What do you mean it's the same thing?
Abbie:	It's not the same thing – it's not the same thing
Katelyn:	It's not the same thing because Julie is missing a 40

Although students' graphs had similar structures, Julie's graph (Fig. 3) does not include a value in the 40–44 inch category (as was noted previously). Thus, an exchange that initially seemed to be moving students towards a discussion of the more general characteristics of their displays ("It's the same") transitioned back to a discussion of how their graphs differed for individual cases. At this point, Abbie and Katelyn appeared to be utilizing a case-oriented view of the data which was focused on how each value related to an individual student (Konold et al., 2015). This led the teacher-researcher to highlight the similarities between the graphs more explicitly, as shown below:

Researcher:	But if I said just glance at them and don't count everything
Abbie:	Then yeah -
Researcher:	And look at them side by side, what would you say about them?
Julie:	They look pretty-
Katelyn:	They look like – they all go up and down
Julie:	They look pretty similar
Researcher:	(highlighting Bob and Jeff's graph, Fig. 1, and turning horizontally) This one looks kind of different, but what if I turned it this way?
Bob:	It's the same, looks the same
Jeff:	Looks the same

This intervention appeared to help students move towards a more aggregate view of the data. Follow up questions about what they could say about wingspans in their classroom or other classrooms (another third-grade classroom or a fifth-grade classroom) created a context for students to consider the data more holistically. A consensus was reached that the "average" for the class was "50–55 inches" because it was "the highest" value amongst the data set. Thus, students' conception of average in this case appeared to align with modal clumps in the data, a phenomenon previously reported by researchers who had considered students of the same age group (Frischemeier, 2019; Makar, 2014; Mokros & Russell, 1995).

Students determined that potential values could be anywhere between 40 and 65 inches, and believed this to apply to third graders in general. While they envisioned that the graph for another class might look different in terms of "average", they also expressed that the wingspans of another class would fall within this range, demonstrating awareness of reasonable variation between samples.

At the same time, the data context caused students to reflect on populations in other known classrooms when articulating how the graphs would be similar or different. Abbie noted that "Mrs. Gordon's class has some really tall kids", and that she believed the graph for Mrs. Gordon's class would be "higher" than the one for their class. While this might be the case (if Mrs. Gordon's class had taller students they likely would have larger wingspans), it may have impeded Abbie from considering what she might be able to say about any third-grade class based on their class wingspan data.

In another instance, participants considered the connection between wingspan and height to frame how they might predict the wingspan of a new classmate. This contextual connection had been highlighted in the first session as students practiced collecting wingspan data amongst themselves. Julie recognized that her wingspan was the least, and she was also the shortest individual. Participants appeared to accept this relationship throughout the task sessions, and would often note the role of being "taller" or "shorter" in having a larger or smaller wingspan (as Abbie referenced with Mrs. Gordon's class). This impacted the kinds of inferences participants would make about wingspans more generally, as highlighted below:

Researcher:	So if a new student came to class-
Bob:	Ok
Researcher:	For the last week of school – and you were going to guess what their wingspan was-
Abbie:	50–54
Researcher:	You would say that range – why would you say that Abbie?
Abbie:	Because if most people are in that range, then that would probably still mean –
Julie:	But if he's really short or really tall then it's probably going to be in the 60 s or the 40 s
Researcher:	Oh ok, so if you didn't know the student – they weren't in the classroom yet – what would you guess?
Bob:	Like 50

Katelyn: Because it's the average
Researcher: But then once they got to the classroom what might you do to change it?
Katelyn: Higher or lower
Bob: Off the height

This exchange emphasizes the role of contextual considerations (the relationship that participants had established both based on their data and prior experience between wingspan and height) influencing students' analysis. Initially, Abbie utilized the proposed average for her prediction for a new student, but Julie noted that if they could actually see the individual represented they could adjust their prediction based on their height. She would adjust her prediction to the 60 s for a tall student and the 40 s for a short student based on the range of their class data. Students considered their collected data along with other contextual features in order to make an informal inference about an unknown student.

5 Discussion

The creation and analysis of graphical displays is an important component of stochastic reasoning. However, in the classroom, students are frequently told the specific display expected for given data, leading to a rigid understanding of what constitutes a "correct" graph. Students may then view graphing as an exercise in labeling and drawing, as opposed to a strategy for determining the features of data (Friel et al., 2001). In addition, this type of instruction does not prepare students for encounters with data outside the classroom.

Findings of the current research suggest that students can be flexible in adapting known techniques to new types of data. Students were able to create displays which were meaningful to them when given the opportunity, using features of both categorical and quantitative displays, without explicit instruction in quantitative methods. This strategy also allowed students to engage with important questions surrounding the differences between categorical and quantitative data. This is in agreement with other research suggesting tasks that allow students to make their own decisions about representation can be an effective way to frame instruction surrounding data modelling and analysis (Lehrer & Schauble, 2004; Shaughnessy & Pfannkuch, 2002). However, at times students still appeared to view graph creation as an exercise as opposed to as a tool for analysis and interpretation (English, 2012).

Students were able to use their non-traditional displays to interpret the wingspan data and make informal inferences about other classrooms and individuals, but their interpretations were impacted by the context of the data. In particular, the fact that the data involved measurements they had collected from themselves and their friends may have led to a focus on individual measurements and made it more difficult for students to consider the data holistically (Pfannkuch, 2011). At the same time, the familiarity of the context may have made students more willing to engage in informal inferential reasoning (Makar, 2014), in particular as they considered the relationship between wingspan and height.

Their views of data ranged from an initial case perspective which focused on the individual wingspan values, to a more aggregate view which considered typical wingspans based on modal clumps (Frischemeier, 2019; Konold et al., 2015). As students debated what a graph for another third-grade class or a fifth-grade class might look like, they referenced distributional characteristics like center (i.e. a fifth-grade class would be "higher"), shape (the data goes "up and down"), and reasonable variation (i.e. another third-grade class's graph would have a similar range of wingspans), emphasizing many of the characteristics of distributional reasoning highlighted by Bakker and Gravemeijer (2004).

We further note the impact of specific interventions as students engaged with quantitative data for the first time. Questions about what values would be represented by the groups students envisioned and how they might categorize values like 59 ¾ inches appeared to impact student approaches and lead to further adaptation of categorical techniques to fit the quantitative data. In addition, a specific moment when students were asked to compare the shapes of their displays appeared to lead them to contemplate a more aggregate view of the data.

6 Implications

The study reported on approaches to a designed activity amongst a small group of third grade students. Due to the small sample of students involved in the investigation, questions can be raised regarding the generalizability of the results. As such, additional work is needed to determine whether the findings reported in this paper may apply more generally. In addition, students in this report had recently received instruction from their classroom teacher on categorical graphing techniques (pictographs and bar graphs), which likely impacted their approaches. Future work could investigate how elementary learners may engage with the task sequence prior to instruction on standard categorical displays. Students' motivation to interpret features of the wingspan data may also have been impacted by the lack of practical rationale for their investigation. Future iterations of this study could frame the investigation in a way that would make it more meaningful; for example the need for wingspan lengths could be grounded in the context of designing a class sweatshirt.

This study also emphasizes the important role that context plays in statistical investigations, something that has been shown in other research studies (Pfannkuch, 2011; Watson, 2006). Since many investigations with elementary students begin with data collected at the classroom level (i.e. English & Watson, 2015; Makar, 2014, 2018; Watson & English, 2017, 2018), researchers must consider the impact of these kinds of activities. Our research suggests that using personal data can lead to high engagement and new questions (i.e. the relationship between wingspan and height), but that students may have trouble viewing the data holistically/impartially when it represents themselves and their peers. Future research might investigate how elementary students analyze and interpret data from a variety of sources (some self-collected, some from outside the classroom) in order to gain insight about the impact of data from varying sources on their statistical thinking and inquiry.

Finally, students involved in this study created their displays by hand, with some participants creating multiple drafts as they considered how to best represent the wingspan data. Research, along with recent standards, suggest the important role that technology can play in exploratory data analysis, leading to dynamic investigations where learners can easily modify their displays as new questions emerge (Bargagliotti et al., 2020; Biehler et al., 2013; Chance et al., 2007; Paparistodemou & Meletiou-Mavrotheris, 2008). These environments also allow students to collect and organize large amounts of data quickly and efficiently. Investigating the wingspan context by hand led to important questions about grouping and bin size which may not have emerged if students had created their displays using a technological environment. At the same time, the static nature of their displays and the time and effort that were necessary to create a new graph may have impacted students' investigations and analysis. Considering the impact of the wingspan investigation with varying access to technology could be another line of investigation for future research.

References

Bakker, A., & Gravemeijer, K. (2004). Learning to reason about distribution. In D. Ben-Zvi & J. Garfield (Eds.), *The challenge of developing statistical literacy, reasoning and thinking* (pp. 147–168). Kluwer Academic Publishers. https://doi.org/10.1007/1-4020-2278-6_7

Bargagliotti, A., Franklin, C., Arnold, P., Johnson, S., Perez, L., & Spangler, D. A. (2020). *Pre-K-12 guidelines for assessment and instruction in statistics education II: (GAISE II): A framework for statistics and data science education*. American Statistical Association.

Ben-Zvi, D., & Garfield, J. (2008). Introducing the emerging discipline of statistics education. *School Science and Mathematics, 108*(8), 355–361. https://doi.org/10.1111/j.1949-8594.2008.tb17850.x

Biehler, R., Ben-Zvi, D., Bakker, A., & Makar, K. (2013). Technology for enhancing statistical reasoning at the school level. In M. A. Clements, A. J. Bishop, C. Keitel, J. Kilpatrick, & F. K. S. Leung (Eds.), *Third international handbook of mathematics education* (pp. 643–689). Springer Science and Business Media. https://doi.org/10.1007/978-1-4614-4684-2_21

Braham, H. M., & Ben-Zvi, D. (2017). Students' emergent articulations of statistical models and modeling in making informal statistical inferences. *Statistics Education Research Journal, 16*(2), 116–143. https://doi.org/10.52041/serj.v16i2.187

Chance, B., Ben-Zvi, D., Garfield, J., & Medina, E. (2007). The role of technology in improving student learning of statistics. *Technology Innovations in Statistics Education, 1*(1), 1–24. https://doi.org/10.5070/T511000026

Cobb, G. W., & Moore, D. S. (1997). Mathematics, statistics, and teaching. *The American Mathematical Monthly, 104*(9), 801–823. https://doi.org/10.1080/00029890.1997.11990723

Cobb, P., Confrey, J., diSessa, A., Lehrer, R., & Schauble, L. (2003). Design experiments in educational research. *Educational Researcher, 32*(1), 9–13. https://doi.org/10.3102/0013189x032001009

Cooper, L. L., & Shore, F. S. (2008). Students' misconceptions in interpreting center and variability of data represented via histograms and stem-and-leaf plots. *Journal of Statistics Education, 16*(2), 1. https://doi.org/10.1080/10691898.2008.11889559

Doerr, H. M., & English, L. D. (2003). A modeling perspective on students' mathematical reasoning about data. *Journal for Research in Mathematics Education, 34*(2), 110–136. https://doi.org/10.2307/30034902

English, L. D. (2012). Data modelling with first-grade students. *Educational Studies in Mathematics, 81*(1), 15–30. https://doi.org/10.1007/s10649-011-9377-3

English, L. D., & Watson, J. M. (2015). Exploring variation in measurement as a foundation for statistical thinking in the elementary school. *International Journal of STEM Education, 2*(1), 1–20. https://doi.org/10.1186/s40594-015-0016-x

Friel, S. N., Curcio, F. R., & Bright, G. W. (2001). Making sense of graphs: Critical factors influencing comprehension and instructional implications. *Journal for Research in Mathematics Education, 32*(2), 124–158. https://doi.org/10.2307/749671

Frischemeier, D. (2018). Design, implementation, and evaluation of an instructional sequence to lead primary school students to comparing groups in statistical projects. In A. Leavy, M. Meletiou-Mavrotheris, & E. Paparistodemou (Eds.), *Statistics in early childhood and primary education* (pp. 217–238). Springer Nature. https://doi.org/10.1007/978-981-13-1044-7_13

Frischemeier, D. (2019). Primary school students' reasoning when comparing groups using modal clumps, medians, and hatplots. *Mathematics Education Research Journal, 31*(4), 485–505. https://doi.org/10.1007/s13394-019-00261-6

Garfield, J., Le, L., Zieffler, A., & Ben-Zvi, D. (2015). Developing students' reasoning about samples and sampling variability as a path to expert statistical thinking. *Educational Studies in Mathematics, 88*(3), 327–342. https://doi.org/10.1007/s10649-014-9541-7

Konold, C., Higgins, T., Russell, S. J., & Khalil, K. (2015). Data seen through different lenses. *Educational Studies in Mathematics, 88*(3), 305–325. https://doi.org/10.1007/s10649-013-9529-8

Leavy, A. (2008). An examination of the role of statistical investigation in supporting the development of young children's statistical reasoning. In O. Saracho & B. Spodek (Eds.), *Contemporary perspectives on mathematics education in early childhood* (pp. 215–232). Information Age Publishing.

Leavy, A., & Hourigan, M. (2018). Inscriptional capacities and representations of young children engaged in data collection during a statistical investigation. In A. Leavy, M. Meletiou-Mavrotheris, & E. Paparistodemou (Eds.), *Statistics in early childhood and primary education* (pp. 89–108). Springer Nature. https://doi.org/10.1007/978-981-13-1044-7_6

Lehrer, R., & English, L. (2018). Introducing children to modeling variability. In D. Ben-Zvi, K. Makar, & J. Garfield (Eds.), *International handbook of research in statistics education* (pp. 229–260). Springer International Publishing. https://doi.org/10.1007/978-3-319-66195-7_7

Lehrer, R., & Schauble, L. (2004). Modeling natural variation through distribution. *American Educational Research Journal, 41*(3), 635–679. https://doi.org/10.3102/00028312041003635

Makar, K. (2014). Young children's explorations of average through informal inferential reasoning. *Educational Studies in Mathematics, 86*(1), 61–78. https://doi.org/10.1007/s10649-013-9526-y

Makar, K. (2018). Theorising links between context and structure to introduce powerful statistical ideas in early years. In A. Leavy, M. Meletiou-Mavrotheris, & E. Paparistodemou (Eds.), *Statistics in early childhood and primary education* (pp. 3–20). Springer Nature. https://doi.org/10.1007/978-981-13-1044-7_1

Makar, K., Bakker, A., & Ben-Zvi, D. (2011). The reasoning behind informal statistical inference. *Mathematical Thinking and Learning, 13*((1–2)), 152–173. https://doi.org/10.1080/10986065.2011.538301

Mokros, J., & Russell, S. J. (1995). Children's concepts of average and representativeness. *Journal for Research in Mathematics Education, 26*(1), 20–39. https://doi.org/10.2307/749226

National Governor's Association Center for Best Practices & Council of Chief State School Officers. (2010). *Common Core state standards in mathematics*. Authors.

Paparistodemou, E., & Meletiou-Mavrotheris, M. (2008). Developing young students' informal inference skills in data analysis. *Statistics Education Research Journal, 7*(2), 83–106. https://doi.org/10.52041/serj.v7i2.471

Pfannkuch, M. (2011). The role of context in developing informal statistical inferential reasoning: A classroom study. *Mathematical Thinking and Learning, 13*(1–2), 27–46. https://doi.org/10.1080/10986065.2011.538302

Shaughnessy, J. M. (2007). Research on statistics learning and reasoning. In F. K. Lester Jr. (Ed.), *Second handbook of research on mathematics teaching and learning* (pp. 957–1010). Information Age Publishing.

Shaughnessy, J. M., & Pfannkuch, M. (2002). How faithful is old faithful? Statistical thinking: A story of variation and prediction. *The Mathematics Teacher, 95*(4), 252–259. https://doi.org/10.5951/MT.95.4.0252

Steffe, L. P., & Thompson, P. W. (2000). Teaching experiment methodology: Underlying principles and essential elements. In R. Lesh & A. E. Kelly (Eds.), *Research design in mathematics and science education* (pp. 267–307). Erlbaum.

Watson, J. M. (2006). *Statistical literacy at school: Growth and goals*. Routledge. https://doi.org/10.4324/9780203053898

Watson, J. M., & English, L. (2017). Reaction time in grade 5: Data collection within the practice of statistics. *Statistics Education Research Journal, 16*(1), 262–293. https://doi.org/10.52041/serj.v16i1.231

Watson, J. M., & English, L. (2018). Eye color and the practice of statistics in grade 6: Comparing two groups. *The Journal of Mathematical Behavior, 49*, 35–60. https://doi.org/10.1016/j.jmathb.2017.06.006

Wild, C. J., Utts, J. M., & Horton, N. J. (2018). What is statistics? In D. Ben-Zvi, K. Makar, & J. Garfield (Eds.), *International handbook of research in statistics education* (pp. 5–36). Springer International Publishing. https://doi.org/10.1007/978-3-319-66195-7_1

Reading and Interpreting Distributions of Numerical Data in Primary School

Daniel Frischemeier

Abstract The ability to read and interpret distributions is a cornerstone in analyzing data. In primary school, the opportunities to read and interpret distributions are often limited to reading and interpreting categorical data displayed in tallies, pie charts, and bar graphs. This chapter reports on a teaching-learning unit that has the aim to support primary school students (age 10–11) when extracting information from graphs displaying numerical data. Within the realization of this teaching-learning unit, throughout the study we collected several types of data from participating students, including a pre-and post-test, written work sheets, and an evaluation survey at the end of the teaching-learning unit. The results show that the competencies of the young learners concerning reasoning about distributions of numerical data improved over time throughout the teaching-learning experiment.

Keywords Early statistical thinking · Distribution · Numerical data · Primary school · Tinkerplots

1 Introduction

Data play a dominant role in society, and especially considering the COVID-19 pandemic, it is more important than ever to develop an adequate data understanding. In addition, being data literate is important to enable students to become engaged citizens as part of a vibrant democracy (Engel, 2017). Data literacy is intended not only to be developed in selected grades in school, but rather the aim should be to develop data literacy by means of a spiral curriculum (Biehler et al., 2018). Concerning statistical thinking at the elementary and primary levels, some researchers recommend that statistical reasoning and thinking should be developed as early as possible (e.g.,

D. Frischemeier (✉)
WWU Münster, Münster, Germany
e-mail: dfrische@uni-muenster.de

G. F. Burrill et al. (eds.), *Research on Reasoning with Data and Statistical Thinking: International Perspectives*, Advances in Mathematics Education,
https://doi.org/10.1007/978-3-031-29459-4_11

Ben-Zvi, 2018). Ideas on how to enhance statistical reasoning and thinking with young students can be found in Leavy et al. (2018). Relevant components of data analysis in the primary school curriculum can also be identified within the phases of the well-known PPDAC cycle (Wild & Pfannkuch, 1999): *posing statistical questions*, *collecting data*, *displaying data*, and *reading and interpreting data*.

While in many countries data analysis activities have found their way into the curriculum and are also realized across the grades in school, in Germany some selective elements of data analysis can be found in textbooks or the curriculum. From a German perspective, primary school students (in the most federal states of Germany primary school is from grade one to four; the age range of primary school students in Germany is generally from 6 to 11 years) are supposed to know…

- how to record data about objects or events.
- how to document data, especially if they are fleeting (transitory).
- how to clearly present the data collected in tables and diagrams.
- that it may be helpful or even necessary to further process the data to increase its information value
- how to extract information from data representations and then use it (adapted and translated from Hasemann & Mirwald, 2012 p. 145).

According to Biehler et al. (2018), these recommendations, understanding and reasoning about data, are the elements of the DAC phases (Data, Analysis, Conclusion) of the PPDAC-cycle of Wild and Pfannkuch (1999). However, the component *reading and interpreting data* especially for distributions of numerical variables like "Minutes_to_school" may be very challenging for young students. In Germany, categorical data dominates in the school books when looking for data exploration activities. Very few activities tackle numerical data in the data exploration activities. This chapter concentrates on how primary students can be guided to understand numerical data distributions. More specifically, the goal of the project reported in this paper is to design a teaching-learning unit to support young students in learning to read and interpret graphs of distributions of numerical variables – note that the focus of this manuscript is on univariate data only. During this teaching-learning unit, amongst other things, pre-and post-tests and intermediate tests of the participants were collected to document their process, progress and success in reading and interpreting numerical data. The first results from an analysis of the data gathered during the teaching-learning unit are reported in Sect. 4.

2 Theoretical Background

2.1 Primary School Students' Understanding of Distributions of Numerical Data

The ideas of representation and distribution are fundamental in statistics and statistics education (Burrill & Biehler, 2011). One fundamental aim is that learners develop a so-called global perspective on distributions (Biehler, 2007). Several representations can be used to display the distribution of numerical variables (Konold & Higgins, 2003); on the one hand case-value bar graphs, for example, can be an adequate first starting point for the first visualization of a distribution of a numerical variable like "minutes_to_school". On the other hand case value bar graphs have the disadvantage that they are only applicable for small data sets. More convenient are stacked dot plots, which allow displaying distributions of numerical variables in large datasets and also promote using precursor-conceptions to develop a global view on distributions.

In some cases young learners may be guided by outliers or individual data points and see them as characteristics for the distribution. This behavior can be seen as a local view of data (Bakker & Gravemeijer, 2004); in contrast taking into account characteristics like shape, density, center, or spread can be seen as a step towards a more global perspective on data and distributions. Bakker and Gravemeijer (2004) and Konold et al. (2015), among others, provide further differentiations between a local and a global view of data. While Bakker and Gravemeijer (2004) distinguish between the local view ("data as individual points") and the global view ("data as an entity") with respect to data and distributions, Makar and Confrey (2002), for example, distinguish between a local view, a global view ("aggregate view") and the intermediate level of "mini-aggregates". The so-called "mini-aggregates" (e.g. modal clumps) can be addressed in mathematics lessons as early as the primary level. Finally, Konold et al. (2015) distinguish between four perspectives on data and distributions. Exemplified by the distribution of six balls (three red balls, two green balls, one blue ball), Konold et al. (2015) consider a statement like "We said our favorite colors" as "data as pointer" because it shows no relation to the distribution of the feature color of the balls in this distribution. According to Konold et al. (2015), a statement like "Juan loves red" would emphasize the single case (namely Juan) and therefore be categorized as "data as case value". Finally according to Konold et al. (2015), a statement like "three children love the color red" would fall into the category "classifier view on data" and a statement like "half of the children love red" would fall into the category "aggregate view on data".

But how can especially young learners, develop a global perspective on distributions? Konold and Pollatsek (2002) and Konold et al. (2002) suggest that so-called modal clumps can serve as precursor representations and give young students a first sense of center and spread when reading and interpreting distributions of numerical data. These concepts can also lay the foundations for further elaborated statistical activities like group comparisons. For instance, Frischemeier and Schnell (2021), Frischemeier (2019), Fielding-Wells (2018) and Makar and Allmond (2018) found in their work with young learners (ages 8–11) that modal clumps served as valuable concepts and data models to help students grasp information from numerical data or to compare distributions of numerical data concerning characteristics of center and spread. In addition to that Frischemeier (2019) has shown that the primary school students in one study did not focus on outliers when comparing two groups of numerical data (jump lengths of paper frogs) but rather took into account modal clumps and hatplots (which are similar to boxplots but divide the distribution in a lower quarter, a middle half and an upper quarter) as the basis for their comparison, showing a global perspective on the distributions.

Concerning the quality of the (comparison) statements or statements when describing statistical displays in general, the framework of Friel et al. (2001) is helpful to distinguish between different levels of understanding distributions of numerical data: *reading the data*, *reading between the data*, and *reading beyond the data*. When *reading the data*, learners for example tend to lift "information from the graph to answer explicit questions for which the obvious answer is in the graph (e.g., How many boxes of raisins have 30 raisins in them?)" (Friel et al., 2001, p. 130). In the intermediate stage of *reading between the data*, the learner integrates information that is presented in a graph and "completes at least one step of logical or pragmatic inferring to get from the question to the answer (e.g., How many boxes of raisins have more than 34 raisins in them?)" (Friel et al., 2001, p. 130). In the level of *reading beyond the data* learners tend to extend, predict or infer from the graph to answer questions – "the reader gives an answer that requires prior knowledge about a question that is related to the graph (e.g., If students opened one more box of raisins, how many raisins might they expect to find?)" (Friel et al., 2001, p. 130). These levels can also be found in the *graphical competence* defined by González et al. (2011), who see the emergence of the following three facets as groundwork for *graphical competence*:

> The ability to extract data from different sorts of graphs and to interpret meanings from them by reading between, beyond, and behind the data displayed to form hypotheses about the phenomena represented in the graph;
>
> The capacity to select and create appropriate graphs for specific situations, with or without the support of technology; and.
>
> The ability to critically evaluate graphs and to distinguish the relative strengths and limitations of particular graphical representations, recognizing that creating a graph involves an interpretation of the original data. (González et al., 2011, p. 190)

In the following and taking into account the definition of *graphical competence* of González et al. (2011), several perspectives to create a framework for primary school students to read and interpret numerical data were integrated.

2.2 Integrating the Perspectives: Framework for Primary School Students Reading and Interpreting Numerical Data

Since, in reasoning about numerical data, developing a global perspective on distributions is crucial, both perspectives were merged and integrated – the levels of graphical understanding of Friel et al. (2001) and the local/global perspectives on distributions of Bakker and Gravemeijer (2004) and Konold et al. (2015), and specifically, I differentiated the level *reading between the data* between an *intermediate* and a *global* perspective. From my perspective identifying the minimum, maximum or the mode of distribution of numerical data means that a local view is not enough – learners have to take into account the whole distribution to identify the minimum, maximum, or mode of the distribution of numerical data. However, the values themselves can be regarded as rather local, and it is not possible to make deep and elaborated statements about the center and spread of the distributions using these values. Therefore I decided the identification of center or spread is as important an issue as *reading between the data* in the sense of a global perspective. In addition to this, I see the use of modal clumps, the identification of subgroups, or summarizing different classes into one class in numerical data distributions as *reading between the data* with a rather global perspective because the learners focus on intervals in the distributions rather than on single data points or values. The integrated framework with descriptions and definitions can be found in Table 1.

Table 1 Adapted framework from Friel et al. (2001), enriched with the perspective on distributions local/global (Bakker & Gravemeijer, 2004)

Level	Definition
Reading the data Local view (cf. Bakker & Gravemeijer, 2004)	Learners read from given information; they refer to one or more of the answers to the following questions: What is represented? Which values occur? How many students are in category X?
Reading between the data intermediate	Learners establish relationships between numerical data (focusing on intermediate issues). This level requires mathematical skills/abilities, comparing quantities, performing elementary arithmetic operations to discover and interpret relationships in data, for example: What is the most frequent value? What is the largest/smallest value?
Reading between the data global	Learners establish relationships between numerical data (focusing on global issues). Requires mathematical skills/abilities, comparing quantities, performing elementary arithmetic operations to discover and interpret relationships in the data, for example: Where is the data located? (Statements about the center of the distribution) How are the data scattered? (Statements about the spread of the distribution) Are there indications of subgroups? (Statements about modal clumps) Summarizing classes into one class
Reading beyond the data global	Learners predict/conclude with the inclusion of background knowledge. Which questions can be answered with the help of the graph? Which ones can't? What might the distribution of the characteristics look like in a different situation?

2.3 *TinkerPlots: A Digital Tool to Enhance Early Statistical Thinking*

Digital tools are inevitably in data exploration processes, especially in the case of exploring large and multivariate data. For an overview of different digital tools in statistics education across the school levels see Biehler et al. (2013). According to them, TinkerPlots

> is designed for creating many simulation models without the necessity of using symbolic input. In addition, TinkerPlots meets the third requirement of Biehler's (1997) framework by making students participate in the construction and evaluation of methods by providing a graph construction tool for young students who can invent their elementary graphs, whereas most other tools provide only a readymade selection of standard graphs (Biehler et al., 2013, p. 658)

In particular, TinkerPlots uses the data operations *stack*, *separate* and *order* to allow young learners to create stacked bar graphs or stacked dot plots by separating and stacking and to connect these operations in the software with the operations which can be realized with data cards. One fundamental point is that TinkerPlots allows learners to manage large and multivariate datasets and therefore can reduce the extraneous load for learners when exploring data. In this sense, TinkerPlots allows young students to transfer their data operations (separating, stacking, …) used with small data sets to larger data sets and also enables them to create conventional diagrams using larger data sets (Konold, 2006). In addition, TinkerPlots makes it possible to explore multivariate data analysis by the so-called color gradient feature that can help learners to read and interpret distributions of numerical data (Konold, 2002). Furthermore, TinkerPlots realizes a step-by-step transformation from the case value bar plot to the stacked dot plot (Bakker, 2004; Cobb, 1999). Building upon this, TinkerPlots also allows young primary school learners to use precursor concepts like modal clumps (Konold et al., 2002) or hatplots (Watson et al., 2008) to read and interpret distributions of numerical data and to compare distributions. From a general perspective, TinkerPlots realizes many "potentials" for the use of digital tools and resources in primary school classrooms discussed by Walter (2018) such as interactive visualizations, connecting representations, reducing extraneous cognitive load, and displaying mental operations visually.

3 The Study

3.1 *Research Questions*

The aim of the study described in this chapter is to develop, realize, and evaluate a teaching-learning unit to support young students' statistical thinking concerning numerical data (e.g. in the form of stacked dot plots). The research questions for this study are:

1. In which way does statistical thinking of the students concerning understanding numerical data improve within and after the teaching-learning unit?

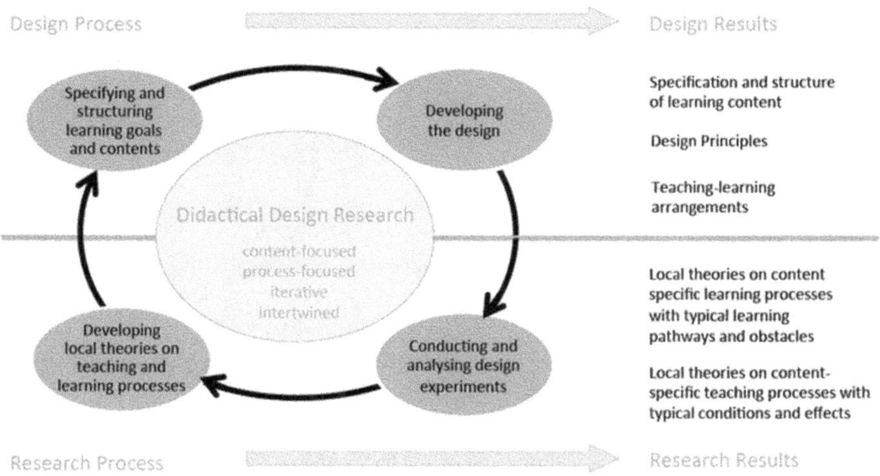

Fig. 1 Process of didactical design research. (Prediger & Zwetzschler, 2013)

2. In which way does the teaching-learning unit have an impact on the attitudes of the participants?

3.2 *Methodology*

The teaching-learning unit has been developed using a didactical design research setting (Prediger & Zwetzschler, 2013) by the author and by two preservice teachers (Bock, 2017; Stein, 2019) as they framed their bachelor theses (see Fig. 1).

As can be seen in Fig. 1 didactical design research consists of four fundamental phases, which are intertwined and iterative and is content- and process-focused. The following section provides details of selected phases of didactical design research in developing, realizing, and evaluating our teaching-learning unit to lead young learners to understand distributions of numerical data.

4 Developing, Realizing, and Evaluating a Teaching-Learning Unit to Support Young Students' Statistical Thinking Concerning Numerical Data

4.1 *Specifying and Structuring the Learning Content*

The goal was that students were to develop a conceptual understanding of distributions of numerical data so they have a global perspective on distributions and can read and interpret distributions of numerical data concerning the levels *of reading the data* (local) and *reading between the data* (intermediate and global) (see Table 1). But reading and interpreting data is not supposed to be seen in an isolated

Fig. 2 The investigative cycle within the PPDAC cycle of Wild and Pfannkuch (1999)

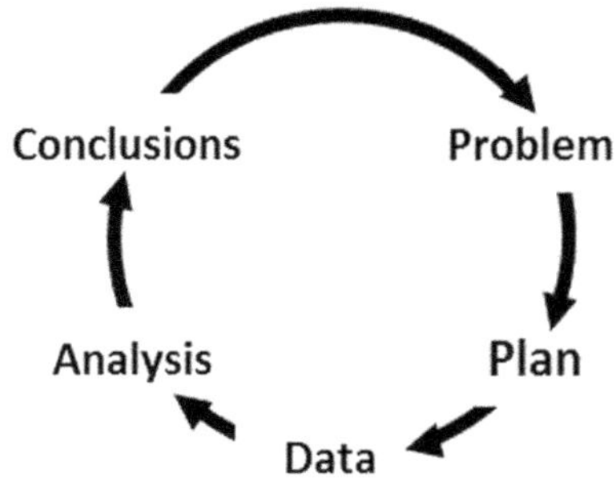

way. It has to be included in the data analysis process and the five phases of the PPDAC cycle as in Fig. 2 (Wild & Pfannkuch, 1999).

With a focus on the "Analysis" part in primary school, case-value bar graphs and stacked dot plots are common representations to visualize the distribution of numerical data. The transformation process from case-value bar graphs to stacked dots on the one hand and the reading and interpreting of stacked dot plots and focusing on a global perspective on distributions on the other hand were the main learning goals of this teaching-learning unit.

4.2 (Re-)Developing the Design

The teaching-learning unit was embedded in the frame of the PPDAC cycle (Wild & Pfannkuch, 1999) so that the young learners could experience several data exploration phases on their own. The design elements of our teaching-learning unit were oriented towards the design ideas of Statistical Reasoning Learning Environments suggested by Garfield and Ben-Zvi (2008) such as the use of real and meaningful data, the use of digital tools (TinkerPlots), and the implementation of collaborative working elements. Using meaningful and real data enables young students to use their contextual knowledge to draw inferences from the data.

As a first step in the learning process data operations like separate and stack were experienced when exploring small datasets with data cards (Harradine & Konold, 2006) and then transformed and applied to larger datasets using TinkerPlots. The framework in Table 1 adapted from Friel et al. (2001) and Bakker and Gravemeijer (2004) was used to discuss and evaluate the young students' different levels of reading and interpreting numerical data.

4.3 Realizing and Evaluating Design Experiments

This paper reports on the second cycle (Stein, 2019) of the design experiments in which 21 primary school students (age 10–11) participated. In this chapter this second and more current cycle will be in the focus. The third cycle is in preparation for the future.

Table 2 Overview of the content of lessons and learning goals of the teaching learning-unit

No	Lesson content	Learning goals: Participants will be able to
1	Pretest, different representations of data, reading pie charts	Use fundamental data operations like separate and stack; Read pie graphs on the levels reading the data and reading between the data
2	Creating and reading a bar graph, case-value bar graph, important features of diagrams	Create a bar graph and a case-value bar graph; Identify important features of diagrams like a headline, axis, and scale.
3	Collecting data from schoolmates	Collect numerical data (variable "minutes_to_school") from their class- and schoolmates
4	Creating stacked-dot plots with data from schoolmates (minutes_to_school)	Create case-value bar graphs of the variable "minutes_to_school" in a small data set (class data, $n = 19$); Transform a case-value bar graph into a stacked dot plot in a small data set (class data, $n = 19$)
5–6	Learning to read and interpret stacked-dot plots	Create stacked dot plots in TinkerPlots (school data); Read stacked dot plots on the levels reading the data and reading between the data (school data); Use modal clumps in TinkerPlots to describe informal ideas of center and spread of the distributions (school data)
7	Posttest	

4.3.1 Realizing the Design Experiments

Participants

According to the teacher the 21 primary school students had no specific pre-knowledge. They had not been taught statistics or data analysis before attending our teaching-learning unit – note that on some days of the teaching unit (especially the days on which data were collected) not all students were able to attend the lessons and the data collection (e.g., due to illness). The selection of the participants was related to the circumstances that each participant had to have consent from their parents that they were allowed to take part in the teaching unit and in the study described here.

Teaching-Learning Unit

The teaching-learning unit consisted of seven lessons (each lesson lasting 45 mins). The overview of the contents of each lesson and the corresponding learning goals can be seen in Table 2.

First the students were introduced (lessons 1 and 2) to different data operations and data representations of categorical data (pie charts and bar graphs). For this purpose the students wrote their birth month on a data card and created a diagram of the distribution of their months of birth for their class (Fig. 3).

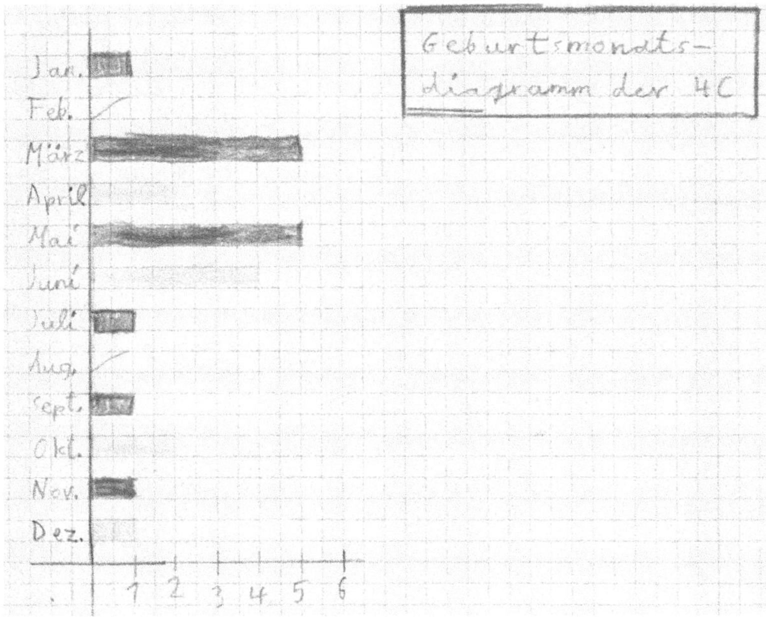

Fig. 3 Diagram displaying the distribution of month of birth in class

Together with the teacher several statements on the levels *reading the data* ("Five of us are born in March") and *reading between the data* ("more of us are born in June compared to July") were discussed. Next students used data cards to collect data from their classmates on the way they get to school and how long it takes. In this case students separated the data cards and stacked the them to create data card bar graphs to have a distribution of categorical data. To enable and facilitate the young learners to describe (read and interpret) distributions of numerical data, scaffolding materials like examples of sentences were provided to give them support in documenting and communicating their findings.

In lesson 3, the students collected data from their schoolmates on how they got to school and how many minutes they need to get from their home to their school. Lessons 4–6 were then devoted to introducing the young learners to read and interpret distributions of numerical variables like "minutes_to_school" in the classroom and the school data sets. The teacher first (lesson 4) used data collected in class ($n = 19$) to display the distribution of the variable "minutes_to_school" in the class and on the board (see Figs. 4 and 5). The names are fantasy names chosen by the students.

As proposed in Cobb (1999), Konold and Higgins (2003) and Bakker (2004), the teacher first displayed the data in form of a case-value bar graph (Fig. 4, upper-left) and then ordered the value bars (Fig. 4, upper-right) and replaced the end of the value bars with dots (Fig. 4, lower-left) to finally come to stacked dot plots (Fig. 4

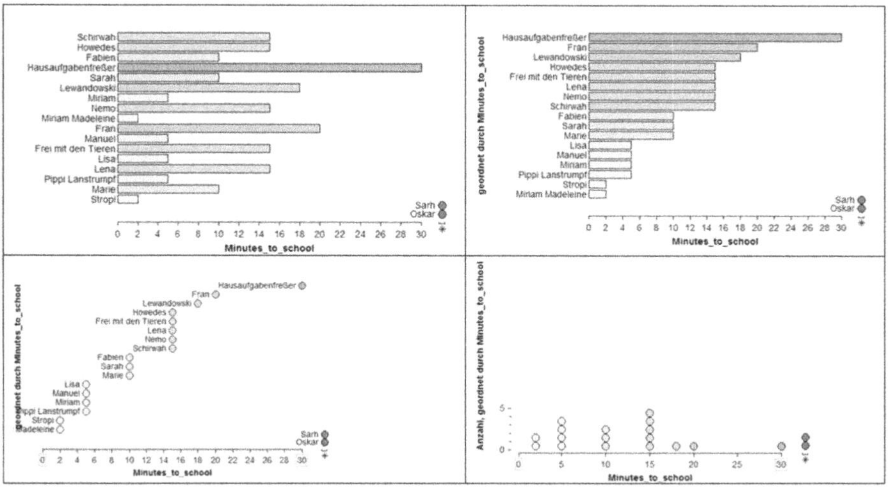

Fig. 4 Transformation from value bars to stacked dot plots

lower-right). This was done by the teacher on the board in an explain-the-screen setting (Drijvers, 2012) where the teacher discussed what the students can extract from the unordered value bars and how the representation can be improved. The students concluded that it would be helpful to order the value bars by the number of minutes the students need to come to school (see Fig. 4 upper-right). In this situation, it is still difficult to develop a global perspective on the distribution, so in the next step (see Fig. 4 lower-left) the teacher used TinkerPlots to represent the end of the bars with dots, each dot symbolizing a case, a student. Finally, the stack function was used in TinkerPlots to create a stacked dot plot (see Fig. 4 lower-right).

For this example, in Fig. 4 and also for lessons 5–6 with a larger dataset containing all schoolmates, the teacher discussed with the students how to read the visualizations using the levels of *reading the data* and *reading between the data* of Friel et al. (2001) – e.g., "Fran needs 20 minutes to get to school" (Fig. 4 upper-left) or "Eight of us need 15 or more minutes to get to school" (Fig. 4 lower-right).

Together with the teacher, the students discussed important features of a graph (headline, adequate labeling of scales and categories) and how to extract information from stacked-dot graphs in the sense of *reading the data* ("two of us need 17 minutes to get to school") and *reading between the data* ("3 of us need more than 30 minutes to get to school") like the distribution of the variable minutes to school from the neighbor class – see Fig. 5.

To support students in developing a global perspective on stacked dot plots, the teacher finally introduced modal clumps (see Fig. 6) as a pre-cursor representation of center and spread (Konold et al., 2002).

The teaching unit concluded with the post-test.

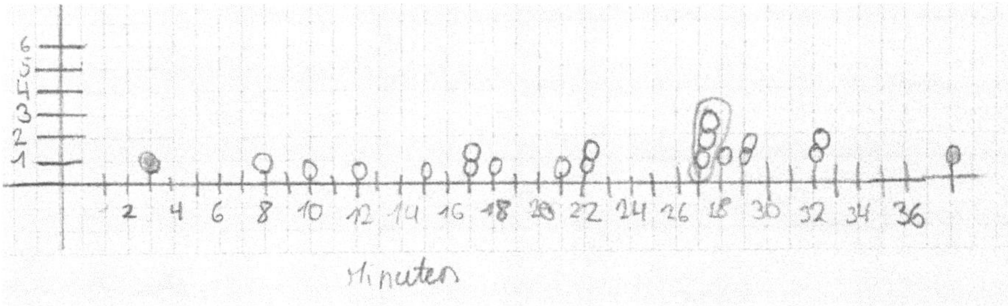

Fig. 5 Diagram displaying the distribution of minutes to school in class (data set from neighbor class)

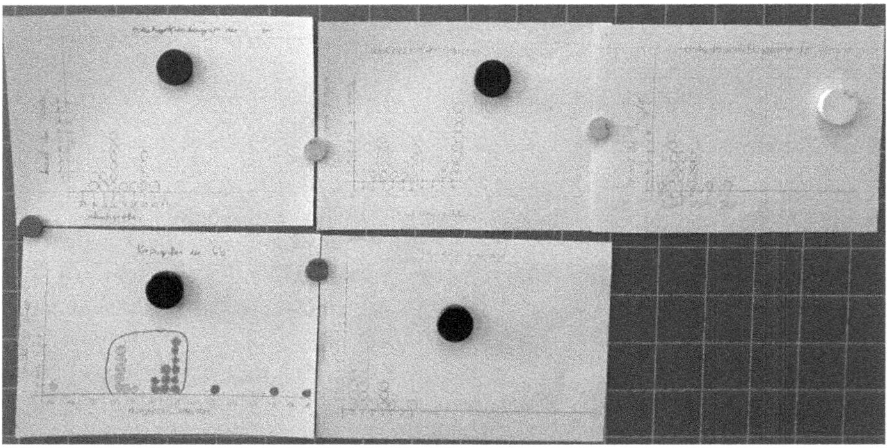

Fig. 6 Diagram displaying the distribution of minutes to school in class using modal clumps

4.3.2 Evaluating the Design Experiments

Data Collection

This section focuses on the (1) analysis of the pre-and post-test, (2) results from analyzing student work sheets from selected classroom activities, and (3) an evaluation of the survey on our students' affective attitudes.

Analysis of the Pre-test and Post-test

All 19 participants participating in the data collection were given a stacked dot plot of the distribution of the variable "number of books read per month" (see Fig. 7) before participating in the teaching-learning unit and the same visualization after having participated in the teaching-learning unit.

Their precise task consisted of four subtasks: (a) What is the highest/what is the lowest number of books read per month by anyone? (*Reading between the data - intermediate*); (b) How many children read 10 books per month? (*Reading the data*); (c) How many children read two, three, or four books per month? Is this more than half of the distribution? (*Reading between the data – global*); (d) Find a

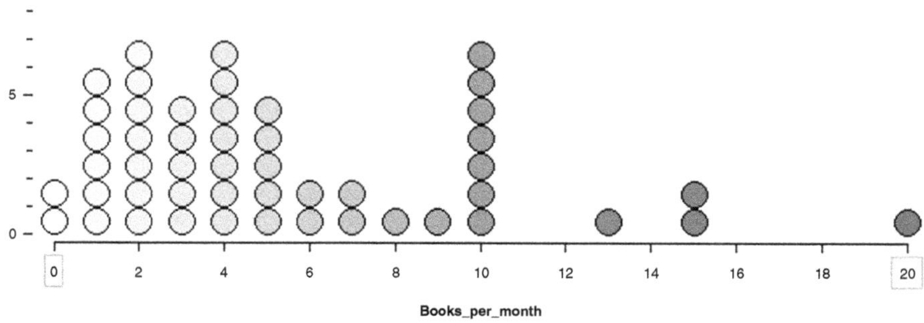

Fig. 7 Distribution for task from pre/post-test for participants (the color intensity of the dots is proportional to the number of books read by month)

headline that fits the stacked dot plot. (*Reading between the data*). All written pre- and post-tests ($n = 19$) were analyzed using qualitative content analysis according to Mayring (2015). The subtasks (a), (b) and (d) were coded 0 points (incorrect or missing solution) or 1 point (correct solution). Subtask (c) was coded 0 points (incorrect or missing solution), 1 point (number of children read two, three, or four books), 2 points (number of children read two, three, or four books, and comparing it to half of the number of cases).

For example, Caro's answer to sub-task (a) "2 children read 0 books per month and 1 child reads 20 books per month" was coded with 1 point because Caro identified correctly that two children read no books per month and one child reads 20 books per month. The answer of Caro in sub-task (b) was "7 children read 10 books per month" which was coded with 1 of 1 point because Caro has mentioned correctly that seven children read 10 books per month. Caro's answer to sub-task (c) "7 children read 2 books per month, 5 children read three books per month, 7 children read 4 books per month" was coded with one of two points because Caro has identified correctly that 7 children read two books per month, 5 children read three books per month, 7 children read four books per month but she has not reflected this to the amount of the number of cases in the distribution. Caro's headline in sub-task (d) "Book diagram of the children per month" was coded with 1 point because Caro identified a headline that fits the stacked dot plot and which we regard as adequate. Table 3 displays the distribution of the coding.

One can see there is a considerable overall improvement between the pretest and posttest. In the pretest, the participants answered nearly half of the subtasks (46%) correctly. After the posttest, more than two-thirds (70%) answered the subtasks correctly. We can say that our participants did well even before the teaching-learning unit – given the fact that they were not taught statistical issues before, 46% of the points at the beginning of the teaching-learning unit seem to be a very good start.

Looking at the analysis of the several subtasks, the largest improvement was observed in subtask (a) in which the students were asked to identify the maximum/minimum. In the pretest, only two participants were able to identify the minimum and the maximum of the distribution, in the posttest eleven participants were able to do this. All in all, the results in Table 3 suggest that the teaching-learning unit

Table 3 Frequency analysis of coding pre-and posttest

	Pretest (of 19)	Posttest (of 19)	Change
Level (*Reading between data* – intermediate): identify max and min	02 (11%)	11 (58%)	+09 (+47)
Level (*Reading the data*): identify single absolute frequency	14 (74%)	18 (95%)	+04 (+21)
Level (*Reading between the data*– global). Identify the frequency of different values, reflecting whether this amount is the half of the number of data cases	10 (53%)	11 (59%)	+01 (+06)
Level (*Reading between the data*: headline)	09 (47%)	13 (68%)	+04 (+21)
Overall	**35 (46%)**	**53 (70%)**	**+18 (+24)**

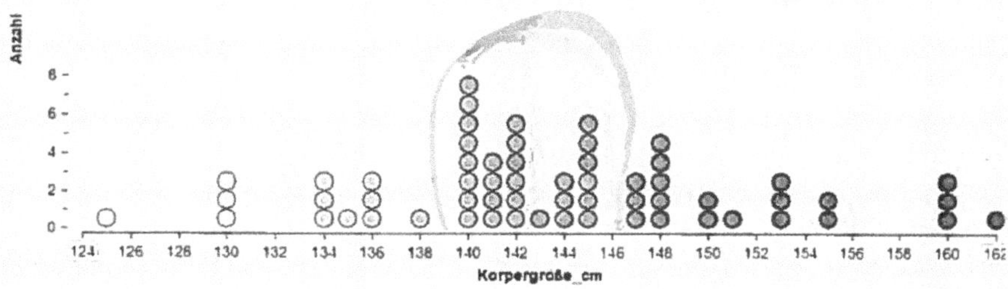

Fig. 8 Stacked dot plot of variable height in cm with a marked modal clump of student Jonas

developed the statistical reasoning and thinking of young students and supported them in extracting information from distributions of numerical variables.

Analysis of the Students' Documents During the Teaching-Learning Unit
In the selected task in the middle of the teaching-learning unit, the students were given a stacked dot plot of the variable body height (see Fig. 8) and were asked to (task 1) describe the stacked dot plot in Fig. 8 with a maximum of five sentences and (task 2) to mark the main interval of the diagram "In which interval are the majority of the children?"

Analysis of the Students' Work During the Teaching-Learning Unit–Task 1
The coding scheme in Table 4 was adapted from the framework developed in Table 1 with the perspective of different reading levels (Friel et al., 2001) and different views on distributions (Bakker & Gravemeijer, 2004). The complete written statements of the students concerning task 1 were coded as a whole. Thus, if several levels occurred in the written statements (e.g. *reading the data* and *reading between the data*), the highest level for the written statement was coded, in this case: *reading between the data*.

More than half of the students (11 of 18 students) were able to extract information from the distribution of numerical data on *reading between the data* level. Eleven of 18 participants (note that three of the 21 students missed the lesson and

Table 4 Coding scheme for the analysis of task 1 (note the examples are translated from German)

Example Reading the data—Local view
The written description of Gregor: "The stacked dot-plot is about height. The dots represent the children. Below the dots is the x-axis. Below the x-axis are the heights. On the side from bottom to top the number of children is mentioned" (only description of graph elements, therefore this statement has been coded *reading the data*)
Definition
Learners read from given information – they refer to one or more of the answers to the following questions: What is represented? Which values occur? How many students are in category X?
Example Reading between the data—intermediate
The written description of Jamie: "140 cm are the most children. 125 cm, 135 cm, 138 cm, 143 cm, 151 cm and 162 cm are all equal to one point. There are fewer children which are 124 cm tall than children which are 130 cm. 130 cm and 134 cm are the same amounts of the children. 134 cm are fewer children than 140 cm." (Jamie identifies the mode, therefore this statement is coded *reading between the data* – intermediate)
Definition
Learners establish relationships between numerical data (focusing on intermediate issues like mode or the minimum or maximum of a distribution). This level requires mathematical skills/abilities, comparing quantities, performing elementary arithmetic operations to discover and interpret relationships in data, for example: What is the most frequent value? What is the largest/smallest value?
Example Reading between the data—global
The written description of Leonie: "No child is 158 cm tall. The main interval is between 140 cm and 148 cm. There are more children which are 160 cm tall compared to children which are 162 cm tall. Most children are 140 cm tall. 4 children are 141 tall." (Leonie used modal clumps and identified the main part/interval of the distribution, therefore this statement is coded *reading between the data* – global)
Definition
Learners establish relationships between numerical data (focusing on global issues of the distributions like intervals, modal clumps, center, spread). Requires mathematical skills/abilities, comparing quantities, performing elementary arithmetic operations to discover and interpret relationships in the data, for example: Where is the data located? (Statements about the center of the distribution) How does the data scatter? (Statements about the spread of the distribution) Are there indications of subgroups? (Statements about modal clumps)

data collection due to absence) used *reading between the data* elements, compared frequencies between different classes, or even used intervals to describe the stacked dot plot in five sentences. Five of 18 students used *reading between the data* from a global perspective by identifying the main interval in the distribution of numerical data. This can be seen as a tendency to a global perspective on distributions; see the statement of Leonie in Table 4. Finally seven of 18 participants (39%) showed a rather local view on distributions of numerical data as we see exemplarily in the written notes of Gregor (Table 4).

Analysis of the Students' Documents During the Teaching-Learning Unit – Task 2

The written statements of 18 participants of the teaching-learning unit were analyzed using qualitative content analysis according to Mayring (2015). Table 1 was adapted for the coding scheme to evaluate the answers of the participants (see Table 5) and provided examples from the analysis of our data. The students' written statements concerning task 2 were coded as a whole unit. If several levels occur (e.g. *reading the data* and *reading between the data*), the highest level was coded for the written statement.

Fifteen of 18 participants showed *reasoning between the data* when being asked about the main interval in the distribution of numerical data. Students who showed rather intermediate *reasoning between the data* identified the mode (140 cm) in the distribution of the variable body height (for example the written statements by Merle or Madline), students who tended to show *reasoning between the data* from a rather global perspective used modal clumps to correctly identify the main interval and therefore a precursor idea of center of the distribution as in the written statement by Lukas (see Table 5). All in all, it is a positive finding that 15 of 18 students showed *reasoning between the data*, although no student showed *reasoning the data*, and three of 18 students did not answer this task.

Analysis of the Evaluation Form of Students' Attitudes

All 21 students took part in this survey (see Table 6) which took place at the end of the teaching-learning unit.

Table 6 indicates that the students show a positive attitude towards the contents of the teaching-learning unit and also reflect their learning in a positive way. So only groupwork, creating/drawing diagrams and distinguishing between diagrams (items 3–5) are compared to the other items a bit lower.

5 Summary and Outlook

This chapter describes a teaching-learning unit to support primary school students in reading and interpreting displays of numerical data. Concerning the research question "In which way does statistical thinking of the students concerning understanding numerical data improve within and after the teaching-learning unit?", one can say that it is possible to develop statistical thinking of primary school students in the sense of understanding numerical data in a teaching-learning unit as described in this chapter.

The first analysis of a pre/posttest and the intermediate tasks indicates that the teaching-learning unit improves the skills of the participants of the teaching-learning unit and allows them to extract information from distributions of numerical data. More specifically it can be said that from the teaching-learning unit the transformation from case-value bar graphs to stacked dot plots (as it can be found in Cobb, 1999 and Bakker, 2004) was a valuable step for our participants to connect both

Table 5 Overview of the codes on the analysis of students' documents in task 2 (note, no example is given for reading the data)

Example: Reading between the data—intermediate

The written answer of Madline "The most children are 140 cm tall".

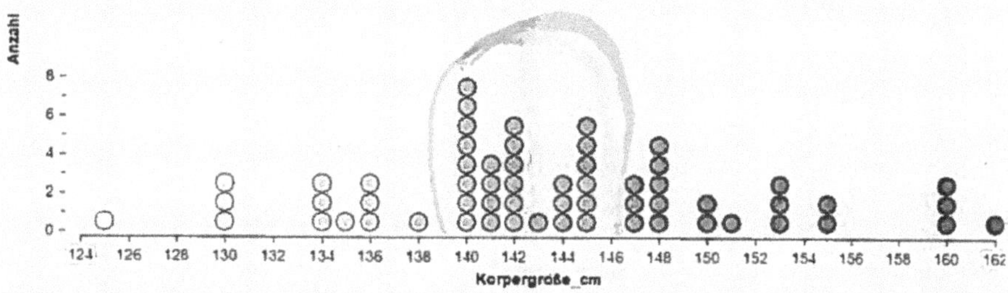

The written answer of Merle

Definition

Learners establish relationships between numerical data (focusing on intermediate issues like mode or the minimum or maximum of a distribution).
This level requires mathematical skills/abilities, comparing quantities, performing elementary arithmetic operations to discover and interpret relationships in data, for example:

- What is the most frequent value?
- What is the largest/smallest value?

Example: Reading between the data—global

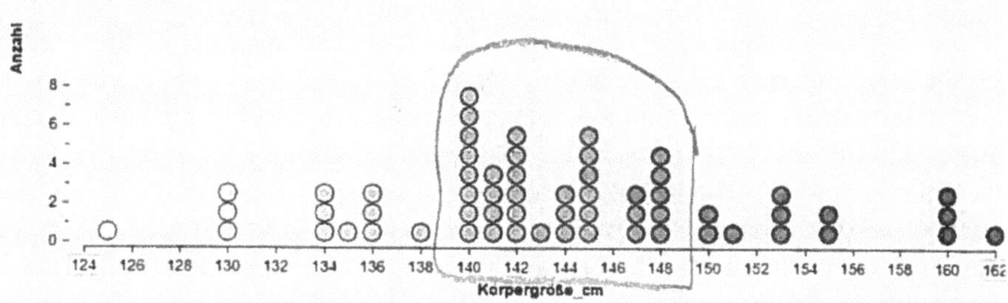

The written description of Lukas: "The main part of height is from 140 cm to 148 cm."

Definition

Learners establish relationships between numerical data (focusing on global issues of the distributions like intervals, modal clumps, center, spread).
Requires mathematical skills/abilities, comparing quantities, performing elementary arithmetic operations to discover and interpret relationships in the data, for example:

- Where is the data located? (Statements about the center of the distribution)
- How does the data scatter? (Statements about the spread of the distribution)
- Are there indications of subgroups? (Statements about modal clumps)

Table 6 Overview of responses from students' final survey

	Good	Okay	Bad
1. I found the project overall…	19	2	
2. I felt comfortable during the work phases.	18	3	
3. I found the group work…	15	6	
4. Creating/drawing diagrams is easy for me…	14	7	
5. I can distinguish between types of diagrams.	13	7	1
6. I can read stacked dot plots.	19	2	
7. I found writing down key statements concerning the diagrams	15	5	1
8. I found collecting my data…	19	2	
9. I found presenting our collected and analyzed data…	17	4	
10. I could always ask questions and they were answered for me.	17	4	

representations and develop a rather global perspective on the data and the distribution of numerical data. So TinkerPlots may have served as a digital tool that helped the teacher and the students to visualize this transformation process. In addition, TinkerPlots seemed to enable learners to realize this step and analyze numerical data in larger data sets (lessons 5–6) on their own. The framework of Friel et al. (2001) enriched with the local/global perspective on distributions allows us to monitor the learning processes of young learners and to evaluate how far they can extract information from distributions of a numerical variable. The framework also helped learners and teachers set a norm for describing and interpreting statistical diagrams from the perspectives *reading the data* and *reading between the data.*

The analysis of the pre/post-test additionally shows that the young learners in the study bring some statistical pre-knowledge, although they had not been taught statistical content so far. Three fourths of the participants were already able to identify single absolute frequencies in distributions of numerical data in the pretest. Furthermore, for *reading between the data*, nearly half of the young learners were able to provide the correct solution in the pre-test (e.g. to identify frequencies summarized in different classes, task 3 or to provide an adequate headline to describe the diagram, task 4). The teaching-learning unit seemed to support our young participants and the improvement could be observed from pre- to post-test. The largest improvement could be identified in task 1 of the pre-test, e.g. the identification of the minimum or maximum.

In addition, the results of the two intermediate tasks were also positive. Concerning the analysis of intermediate task 2, one may state that modal clumps (see also Konold et al., 2002; Makar & Allmond, 2018; Frischemeier, 2019; Frischemeier & Schnell, 2021) proved to be helpful to identify precursor stages of center when looking at distributions of numerical variables. Hat plots (Watson et al., 2008) can be the next step to take also into account spread issues of distributions.

From an affective point of view (and concerning the second research question "In which way does the teaching-learning unit have an impact on the attitudes of the participants?") the participants showed positive attitudes towards the contents of the teaching-learning unit. But concerning task 1 (description of the stacked dot plot) at least six of 21 students stated on the final survey that they had problems writing down key findings from stacked dot plots, which suggests that young learners have to be supported with expressions and statements in the sense of scaffolding to help them to describe statistical displays, especially in the form of distributions of numerical data. Although the students were offered some scaffolds to describe and compare distributions, this has to be emphasized more in a re-designed teaching-learning unit. A further step and another aspect of the re-design may be to confront young learners with statements on different levels and to discuss and reflect on the quality of these statements to establish a normative view on reading and interpreting statistical displays of numerical data.

The modified framework in Table 1 may be helpful in other settings of teaching and research to monitor the reading and interpreting of statistical information in distributions of numerical data and may serve as a framework to set up adequate tasks for reading and interpreting statistical displays showing the distribution of numerical data. However, this study is just a first step and does not allow drawing generalizations beyond the participants of the teaching-learning unit. To get a deeper insight into the cognitive processes of young learners, interviews were conducted at the end of the teaching-learning unit. The next step is to analyze the interview data and take into account all insights from the data to re-design the teaching-learning unit and realize an implementation of the teaching-learning unit in a third cycle.

Acknowledgements I am very grateful to Gail Burrill for very helpful and constructive comments and feedback on this manuscript and also for her support with formatting and language issues. I am also very grateful to the anonymous reviewers for their constructive comments and advices to improve this manuscript. Furthermore, I am very grateful to Lena Bock and Anna-Lena Stein who have been collaborators on the design and the realization of the instructional unit on reading and interpreting numerical data.

References

Bakker, A. (2004). *Design research in statistics education – On symbolizing and computer tools* [Dissertation, University of Utrecht].

Bakker, A., & Gravemeijer, K. (2004). Learning to reason about distributions. In D. Ben-Zvi & J. Garfield (Eds.), *The challenge of developing statistical literacy, reasoning and thinking* (pp. 147–168). Kluwer Academic Publishers. https://doi.org/10.1007/1-4020-2278-6_7

Ben-Zvi, D. (2018). Foreword. In A. Leavy, M. Meletiou-Mavrotheris, & E. Paparistodemou (Eds.), *Statistics in early childhood and primary education* (pp. vii–viii). Springer.

Biehler, R. (1997). Software for learning and for doing statistics. *International Statistical Review, 65*(2), 167–189. https://doi.org/10.1111/j.1751-5823.1997.tb00399.x

Biehler, R. (2007). Denken in Verteilungen – Vergleichen von Verteilungen (Thinking in distributions – comparing distributions). *Der Mathematikunterricht, 53*(3), 3–11.

Biehler, R., Ben-Zvi, D., Bakker, A., & Makar, K. (2013). Technology for enhancing statistical reasoning at the school level. In M. A. Clements, A. J. Bishop, C. Keitel-Kreidt, J. Kilpatrick, &

F. K.-S. Leung (Eds.), *Third international handbook of mathematics education* (pp. 643–689). Springer Science + Business Media). https://doi.org/10.1007/978-1-4614-4684-2_21

Biehler, R., Frischemeier, D., Reading, C., & Shaughnessy, M. (2018). Reasoning about data. In D. Ben-Zvi, K. Makar, & J. Garfield (Eds.), *International handbook of research in statistics education* (pp. 139–192). Springer International. https://doi.org/10.1007/978-3-319-66195-7_5

Bock, L. (2017). *Design, Durchführung und Evaluation einer Unterrichtseinheit zur Förderung des Lesens und Interpretierens von eindimensionalen Streudiagrammen für lernschwache Kinder einer vierten Klasse unter Verwendung kooperativer Lernformen* (Design, implementation and evaluation of a teaching unit to promote the reading and interpretation of one-dimensional scatter plots for children with learning difficulties in a fourth-grade class using cooperative learning methods). (Bachelor of Education), University of Paderborn.

Burrill, G., & Biehler, R. (2011). Fundamental statistical ideas in the school curriculum and in training teachers. In C. Batanero, G. Burrill, & C. Reading (Eds.), *Teaching statistics in school mathematics-challenges for teaching and teacher education* (pp. 57–69). Springer. https://doi.org/10.1007/978-94-007-1131-0_10

Cobb, P. (1999). Individual and collective mathematical development: The case of statistical data analysis. *Mathematical Thinking and Learning, 1*(1), 5–43. https://doi.org/10.1207/s15327833mtl0101_1

Drijvers, P. (2012). Teachers transforming resources into orchestrations. In G. Gueudet, B. Pepin, & L. Trouche (Eds.), *From text to ´lived´ resources: Mathematics curriculum materials and teacher development* (pp. 265–281). Springer.

Engel, J. (2017). Statistical literacy for active citizenship: A call for data science education. *Statistics Education Research Journal, 16*(1), 44–49. https://doi.org/10.52041/serj.v16i1.213

Fielding-Wells, J. (2018). Dot plots and hat plots: Supporting young students emerging understandings of distribution, center and variability through modeling. *ZDM, 50*(7), 1125–1138. https://doi.org/10.1007/s11858-018-0961-1

Friel, S. N., Curcio, F. R., & Bright, G. W. (2001). Making sense of graphs: Critical factors influencing comprehension and instructional implications. *Journal for Research in Mathematics Education, 32*(2), 124–158.

Frischemeier, D. (2019). Primary school students' reasoning when comparing groups using modal clumps, medians, and hatplots. *Mathematics Education Research Journal, 31*(4), 485–505. https://doi.org/10.1007/s13394-019-00261-6

Frischemeier, D., & Schnell, S. (2021). Statistical investigations in primary school–the role of contextual expectations for data analysis. *Mathematics Education Research Journal*, 1–26. https://doi.org/10.1007/s13394-021-00396-5

Garfield, J., & Ben-Zvi, D. (2008). *Developing students' statistical reasoning*. Springer.

González, M. T., Espinel, M. C., & Ainley, J. (2011). Teachers' graphical competence. In C. Batanero, G. Burrill, & C. Reading (Eds.), *Teaching statistics in school mathematics-challenges for teaching and teacher education* (pp. 187–197). Springer. https://doi.org/10.1007/978-94-007-1131-0_20

Harradine, A., & Konold, C. (2006). How representational medium affects the data displays students make. *Seventh international conference on teaching statistics, Salvador.*

Hasemann, K., & Mirwald, E. (2012). Daten, Häufigkeit und Wahrscheinlichkeit (Data, frequency and chance). In G. Walther, M. van den Heuvel-Panhuizen, D. Granzer, & O. Köller (Eds.), *Bildungsstandards für die grundschule: Mathematik konkret* (Educational standards for elementary school: Mathematics in concrete terms) (pp. 141–161). Cornelsen Scriptor.

Konold, C. (2002). Alternatives to scatterplots. *Paper presented at the sixth international conference on teaching statistics, Cape Town.*

Konold, C. (2006). Designing a data analysis tool for learners. In M. Lovett & P. Shah (Eds.), *Thinking with data: The 33rd annual Carnegie symposium on cognition*. Lawrence Erlbaum Associates.

Konold, C., & Higgins, T. L. (2003). Reasoning about data. In J. Kilpatrick, W. G. Martin, & D. Schifter (Eds.), *A research companion to principles and standards for school mathematics* (pp. 193–215). National Council of Teachers of Mathematics.

Konold, C., & Pollatsek, A. (2002). Data analysis as the search for signals in noisy processes. *Journal for Research in Mathematics Education, 33*(4), 259–289. https://doi.org/10.2307/749741

Konold, C., Robinson, A., Khalil, K., Pollatsek, A., Well, A., Wing, R., & Mayr, S. (2002). Students' use of modal clumps to summarize data. *Paper presented at the sixth international conference on teaching statistics, Cape Town.*

Konold, C., Higgins, T., Russell, S. J., & Khalil, K. (2015). Data seen through different lenses. *Educational Studies in Mathematics, 88*(3), 305–325. https://doi.org/10.1007/s10649-013-9529-8

Leavy, A., Meletiou-Mavrotheris, M., & Paparistodemou, E. (2018). *Statistics in early childhood and primary education: Supporting early statistical and probabilistic thinking*. Springer. https://doi.org/10.1007/978-981-13-1044-7

Makar, K., & Allmond, S. (2018). Statistical modelling and repeatable structures: Purpose, process and prediction. *ZDM, 50*, 1–12. https://doi.org/10.1007/s11858-018-0956-y

Makar, K., & Confrey, J. (2002). Comparing two distributions: Investigating secondary teachers' statistical thinking. *Sixth international conference on teaching statistics, Cape Town.*

Mayring, P. (2015). Qualitative content analysis: Theoretical background and procedures. In A. Bikner-Ahsbahs, C. Knipping, & N. Presmeg (Eds.), *Approaches to qualitative research in mathematics education* (pp. 365–380). Springer.

Prediger, S., & Zwetzschler, L. (2013). Topic-specific design research with a focus on learning processes: The case of understanding algebraic equivalence in grade 8. In T. Plomp & N. Nieveen (Eds.), *Educational design research: Illustrative cases* (pp. 407–424). SLO.

Stein, A.-L. (2019). *Planung, Durchführung und Evaluation einer Unterrichtsreihe zur Förderung der Datenkompetenz in Klasse 4 unter besonderer Berücksichtigung des Lesens und Interpretierens von ein-dimensionalen Streudiagrammen* (Planning, implementation and evaluation of a series of lessons to promote data literacy in grade 4 with special consideration of reading and interpreting one-dimensional scatter plots). (Bachelor of Education), University of Paderborn.

Walter, D. (2018). *Nutzungsweisen bei der Verwendung von Tablet-Apps* (Use patterns when using tablet apps). Springer. https://doi.org/10.1007/978-3-658-19067-5

Watson, J., Fitzallen, N., Wilson, K., & Creed, J. (2008). The representational value of HATS. *Mathematics Teaching in Middle School, 14*(1), 4–10. https://doi.org/10.5951/MTMS.14.1.0004

Wild, C. J., & Pfannkuch, M. (1999). Statistical thinking in empirical enquiry. *International Statistical Review, 67*(3), 223–248. https://doi.org/10.1111/j.1751-5823.1999.tb00442.x

Young Learners Experiencing the World Through Data Modelling

Stine Gerster Johansen

Abstract This chapter addresses how young learners can experience aspects of their own life-world through data modelling and hence gain new insights through informal statistical reasoning. A teaching experiment was conducted in a Danish third grade (aged 9–10) classroom where students performed all parts of the data modelling process. Video recordings were made and transcribed. This paper presents empirical data showing examples of statistical reasoning. It includes posing questions relevant to data modelling, reasoning about how to structure data and reasoning about how the data modelling process sheds new light on the chosen topic. Finally, it discusses how students' experiences can be potential for their *Allgemeinbildung*.

Keywords Data modelling · Statistical reasoning · Young learners · Allgemeinbildung

1 Introduction

The term *Allgemeinbildung* has roots in German educational philosophy and has been on the agenda for centuries, mostly in Germany and Scandinavian countries. In its broadest sense, it addresses the formation of human beings into societies (Niss, 2000). Since the origin of the concept, self-determination and autonomy have been at the core of how this process should take place (e.g., Biehler, 2019; Jahnke, 2019). In the literature, the term is primarily discussed theoretically and is often vaguely defined (Biehler, 2019). Nevertheless, the German concept is deeply

S. G. Johansen (✉)
Department of Educational Theory and Curriculum Studies, Danish School of Education, Aarhus University, Copenhagen, Denmark
e-mail: sgj@edu.au.dk

G. F. Burrill et al. (eds.), *Research on Reasoning with Data and Statistical Thinking: International Perspectives*, Advances in Mathematics Education,
https://doi.org/10.1007/978-3-031-29459-4_12

interwoven in the curricular aims of the Danish School system (Børne- og Undervisningsministeriet, 2022), which is the context of this study. However, the concept's complexity and ambiguity make it difficult for teachers to operationalise its tenets into concrete teaching practice.

In the Danish literature, both mathematical modelling and modelling with data are emphasised as important for students' development of *Allgemeinbildung* (e.g., Alrø et al., 2000; Andersen & Weng, 2019; Blomhøj, 2001). In Alrø et al.'s (2000) study, statistical modelling in secondary school established the anchorage point of a discussion of its contribution to students' *Allgemeinbildung.*

In recent years, the statistics education community has produced profound research on statistical and probabilistic learning focusing on young learners (e.g., Leavy et al., 2018). Thus, the aim of this study is to empirically discuss how data modelling with young learners can serve the aim of *Allgemeinbildung* and hence can be incorporated into a concrete mathematics teaching practice in Danish primary mathematics education.

In this chapter, a short section suggesting how (1) mathematics education and (2) modelling relate to *Allgemeinbildung* is presented. This is followed by a section suggesting how this can be interpreted in the context of data modelling with young learners. The presentation of an empirical example follows, where two student groups' ways through the data modelling process are dissected. Finally, how this experience can be interpreted in the light of *Allgemeinbildung* is discussed.

2 Mathematics Education and Allgemeinbildung

The aim of *Allgemeinbildung* requires students' self-development and the development of autonomy (Niss, 2000; Winter 1995). Historically, mathematics education has been considered an important contribution to this aim, and the connection has been discussed in both Denmark, and more intensively, Germany. Winter is a prominent voice in German literature. As translated by Biehler (2019, p. 153), he specified three so-called basic experiences in mathematics that support the development of *Allgemeinbildung*. Students should have experiences that enable them to do the following:

1. To perceive and understand the phenomena of the world around us in nature, society and culture in a specific way,
2. To get to know and to apprehend mathematical objects and facts represented using language, symbols, images or formulae as intellectual creations and as a deductively organised world of its own
3. To acquire by working on tasks capabilities of problem-solving which go beyond mathematics (heuristic competencies) (Winter 1995 p. 37).

Mathematical modelling and the application of mathematics comprise potential that can contribute to *Allgemeinbildung*, but students have to experience this as a creative and interactive process, and it is crucial for students to experience how the process works (Biehler, 2019). If the application of mathematics is trivialised, the potential for enlightenment can be missed, and with it, the potential for *Allgemeinbildung*. It should provide insights of a more general nature. Examples from real life must be experienced, and how the modelling works and which kind of enlightenment can be evoked must also be experienced (Biehler, 2019; Neubrand, 2015). Describing the challenges of mathematical modelling, Niss and Blum (2020) stated that 'the person at issue only becomes a modeller once he or she has identified a context, situation and questions worth dealing with' (Niss & Blum, 2020, p. 22), and engaging in mathematical modelling, among other, presupposes 'mathematical' self-confidence and perseverance (Niss, 2010).

3 Data Modelling with Young Learners

The potential connection between mathematical modelling and *Allgemeinbildung* is described in the previous section. However, mathematical modelling and modelling with data differ. According to English and Watson (2018 p. 104), data modelling is 'a process of inquiry involving comprehensive statistical reasoning that draws upon mathematical and statistical concepts, contexts and questions' (p. 104). What characterises modelling with data (as opposed to mathematical modelling) is handling the random variation in the real world to connect data, chance and context (e.g., Patel & Pfannkuch, 2018; Wild, 2006). Based on the work by Lehrer and Schauble (2000), English (2010) defined data modelling with young learners as 'a developmental process that begins with young children's inquiries and investigations of meaningful phenomena, progressing to deciding what is worthy of attention … and then moving towards organizing, structuring, visualizing and representing data' (p. 27). According to Lehrer and English (2018), modelling with young learners comprises the following (Fig. 1):

The first step in the data modelling process is to pose a statistical question. Wild and Pfannkuch (1999) explain the thinking process transitioning between contextual knowledge to statistical knowledge when posing a statistical question. They described a path from having an initial idea, posing a broad question and posing a precise question to arriving at a plan for data collection. Arnold and Franklin (2021) studied what makes a good statistical question. They distinguished between interrogative and investigative questions and the act of posing and asking questions. Posing investigative questions is the first step in investigating a statistical problem of interest. They presented a list of criteria for a good statistical question, such as a clear variable, clear population or group of interest, clear intent, the answer should be possible using data, the question should be worth investigating and interesting and the question should allow for the analysis of the whole group.

In this study, there is an emphasis on letting students experience the statistical modelling process in its entirety, from posing questions in a meaningful context to

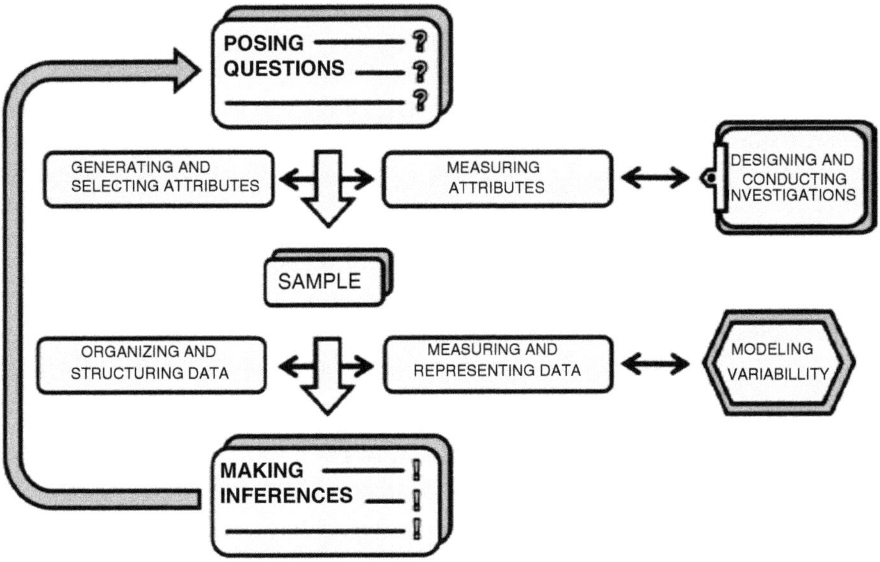

Fig. 1 Data modelling (Lehrer & English, 2018 p. 232). (a) Posing statistical questions within meaningful contexts that highlight variability; (b) generating, selecting and measuring attributes that vary in light of the questions posed; (c) collecting first hand data so that children encounter decisions about the design of investigations; (d) representing, structuring, and interpreting sample and sampling variability; and (e) making informal inferences in light of all these processes. (Lehrer & English, 2018, p. 235)

drawing conclusions. To explore the potential for *Allgemeinbildung*, this process must be experienced as non-trivial, and the idea is that students should investigate a context of their own interest. This will potentially create an opportunity that allows students to experience statistics as a way to gain new insight, to experience how modelling with data works and to experience how it can be of relevance in their own lives. This paper thus examines the research question: *How can a specific emphasis on moving from a context of students' interest to data in the data modelling process create potential for experiences of* Allgemeinbildung *for young learners?*

4 Task Design and Method

The teaching experiment was conducted in a Danish third grade class (ages 9–10) from a mid-level public school. The experiment was set in a frame of critical research (Skovsmose & Borba, 2004). The lesson plan was organised as four sessions (90 min per session) that framed the students' participation throughout the entire data modelling process. The class's mathematics teacher conducted the teaching, and the researcher was a participating observer. Four principles were used when designing the teaching experiment, which emphasised that students should experience the modelling process as a creative and interactive process and that they should have opportunities to perceive the world around them through their own statistical reasoning. The aim was to enable students to cope with the demanding process of moving from the context of interest to data and to experience data modelling as a process through which they can obtain new insights about interesting phenomena in their own lifeworld. The four principles are as follows:

Principle 1: *The students should go through the entire data modelling process.* This is to allow students to experience a coherent process and to create an opportunity to experience statistics as a lens through which to understand phenomena in their own lives.

Principle 2: *Students should model from a context within their own lifeworld.* For students to experience data modelling as a way of understanding phenomena in the world around them, modelling must be based on their curiosity and questions about these phenomena. To accentuate this possibility, students must reason freely based on data and not with the requirement to include certain statistical descriptors.

Principle 3: *Students must answer their own questions based on their own explorations of the data.* Under this principle, it is essential that teachers not guide students towards specific conclusions or require the use of particular descriptors or techniques. Instead, teachers must let students explore data as governed by their own informal concepts. The teacher must guide the students based on their initial curiosities and support them with the concepts needed to explore the data.

Principle 4: *The design should create a supportive learning environment so that students can convert their curiosities into statistical questions.* As movements between the context and the statistical basis are demanding, the design should create room for time, space and support from the teacher when the students make this conversion.

Some weeks before the first session, the teacher introduced the students to a way of using statistics to gain useful insights. Every student was given a piece of cardboard with a drawing representing their favourite sweets. The students then collected data about their peers' favourites to decide what to serve for their next birthday celebration. They then conducted a second round where they answered honestly instead of what was on the cardboard. In the first session of their own data modelling some weeks later, the students generated ideas for questions that could be investigated with the use of statistics. This phase was meticulous. First, the teacher created an idea bank of students' ideas on the blackboard and then collected the ideas and sorted them into themes before presenting the two themes of greatest interest to the students. The students continued with the two themes *family* and *students' well-being*. In their groups, the students discussed which theme they wanted to proceed with and continued working to pose a statistical question. The teacher presented the process they were about to go through on the blackboard while referring to the birthday investigation.

The second session involved planning the investigation in small groups. The teacher initiated a classroom discussion along the way for the students to share their initial ideas and struggles. In the third session, students collected, structured, visualised and represented their data. In the fourth session, the students analysed and interpreted the data. They also planned a presentation of their investigations for the rest of the class.

Group discussions, classroom discussions and the students' presentations were video recorded and analysed using the seven-step model outlined by Powell et al. (2003) for analysing students' mathematical ideas and reasoning through video data. This methodology was considered appropriate for the study because it enabled variety in the articulation of students' informal reasoning. Two groups' transcripts were chosen to present and discuss the students' statistical reasoning and to investigate whether they articulated something that could be identified as a way of experiencing their world in a new way, hence, an experience of *Allgemeinbildung*. The groups were chosen to represent one theme and different processes, respectively. One group experienced a smooth process, whereas the other experienced huge difficulties along the way. The dialogues were translated into English from Danish for this paper.

5 Students' Way Through the Data Modelling Process

In this section, dialogue from the two student groups illustrates the students' way through the data modelling process. The two groups had different approaches to the process and encountered different kinds of difficulties. The discussion will include how informal reasoning can form the basis for later conceptual understandings and be identified as the experiences of *Allgemeinbildung*.

5.1 How Old Are Your Parents?

One group chose to investigate the question: *How old are your parents?* The following excerpt comes from the classroom presentations in Session Two, during which the teacher initiated a classroom discussion. The students presented their initial ideas, comments and advice to each other. Student 4 was not a part of the group but a peer from another group offering suggestions. The excerpt illustrates the movement from an initial interest to posing a question that not only reflects their curiosity but also asks a question requiring a statistical answer. The excerpt shows their way from asking '*How many have a father older than 39*'*?* to a more suitable question.

01.	T:	Okay, tell me again why you changed it. It was you who asked *how many in our class have a dad older than 39?* [She writes it on the blackboard.] Why did you change it?
02.	S2:	You get more…. Well, it was like this: are you over or under 39? Here, you get more like 21, 22, 23, 24, 25…
03.	T:	Okay, so you found out that when you ask how many have a father over 39…. How many options are there?
04.	S2:	2.
05.	T:	2.
06.	S2:	Or 3. You can also be 39.
07.	T:	Yes, my dad can be over 39, under 39 or my dad is 39.
08.		What did you think then? Why was the question not good with that few options?
09.	S3:	It was difficult to find out how old they were.
10.	T:	So, it was difficult to find out how old they were…. Then you said something [points at S2]. What did you do, then?
11.	S2:	We did it by putting…. We elaborated on this.
12.	T:	How did you elaborate on it?
13.	S2:	Well, we wrote how old they were.
14.	T:	Okay, so now it says [the teacher writes on the blackboard] *how old are you?* Was it only 'dad'?
15.	S2:	No, also mum.
16.	T:	Okay, so it is actually…could you say two investigations? Yes? Okay, so how should you answer that?
17.	S2:	We would write 21, 22, 23…. [The teacher repeats this and writes on the blackboard what was said.]
18.	T:	How far up?
19.	S3:	Up to 64.
20.	T:	Okay, all the way up to 64. Why?
21.	S2:	Because E's dad is 60 something.
22.	S1:	He is 62.

This dialogue highlights the fact that it is a somewhat complicated process to decide which parts of the context are worth attention. The students went from asking, '*Is your father over 39*'? to asking, '*How old are your parents*'? as they did not think their first attempt would tell them much. They decided that they would instead like to learn something more general about the age of the parents. This points to the subprocesses of determining what is worthy of attention and identifying the properties of the phenomenon that provide insight. This also shows the non-linear path to determining the group of interest (all parents or only the fathers) and identifying what the variable is.

Another consideration was raised about grouping or not grouping data.

23.	T:	So, you thought we should at least include E's father. May I ask you, have you thought about it? Can there be a challenge with this, too? Or maybe you got there yourself?
24.	S3:	It will be a little long.
25.	T:	It will be a little long. What will be long?

26. S3: If you should put a markin in all ages.
27. S1: I have an idea….
28. T: It is definitely a large diagram you have to make because there are quite a few numbers from 21 to 64. So I have to know, did you manage to find out for yourself if you could do it smartly in some way? .
29. S4: Well, instead of writing all the ages, you could maybe write from 20 to 30 and then from 31 to 40, from 41 to 50....
[The teacher repeats this and writes it on the blackboard.]
30. T: How many options are there now?
31. S4: There are five.
32. T: What do you think about that?
33. S2: We want it to be more precise.
34. T: You want it to be more precise?
35 S3: Yes, and we are not going to change it now. It is written in pen.

The dialogue reflects that the students wanted to know the exact ages of parents, but the teacher was worried about whether the students would get data that does not tell them very much. The students were reluctant, but in the group discussion, they still took note of their classmate's suggestion to group data in the intervals of 20–30, 31–40, 41–50, etc., which can be seen in their graphical representation.

The students dealt with a valid consideration about whether to group data. On the one hand, the students considered reducing the data (five options); on the other hand, they were concerned that they could not see the parents' exact ages. This is an important discussion in terms of developing their statistical reasoning abilities. As seen in their representations of the data, they chose not to reduce the amount of information, as they found an alternative method. These stemmed from debates about organising, visualising, structuring and representing data, and these ways of structuring and organising data are precisely what allows the complexity of data to be reduced in a way so that one becomes able to make unifying statements about the data. Figures 2 and 3 show that the students experimented with the two ways of representing data.

The last excerpt from this group is from their final presentation of their investigation.

36. S3: We chose to investigate how old our parents in the third grade class are. [S2 points on the board and explains how they first grouped the data (Fig. 2) and afterwards chose not to group data (Fig. 3).]
37. S3: From the information on the mothers, you can see that a whole lot of them are between 40 and 47, and there is no one older than 55. And then from the information on the fathers (…)
38. S4: Yes, as you can see, there were two who were 32, and there was no one at all who had fathers who were 33, 34, 35, 36 or 37. [They continue to the mothers.] Okay, it is kind of the same. There were also two who were 32, but there were more who had parents in their 30s. And then there were…the most we had was mothers who were 43.
39. S3: Yes, those were the ones we had the most of. (…)
40. S5: We thought that the most common would be from 34 to 42, both for mothers and fathers.

41. T: Both mothers and fathers? Then, you found out it was not quite like that? (…)
42. S4: As you can see, there were many, many, many, many, many more mothers, actually many more, who were in their forties. There were many who were, yeah. Many were in their forties.
43. R: Was it different with the fathers? (…)
44. S2: Yes, there were a whole lot between 50 and 59.
45. R: But there were not that many mothers in their fifties or what?
46. S2, S3, S4: No!

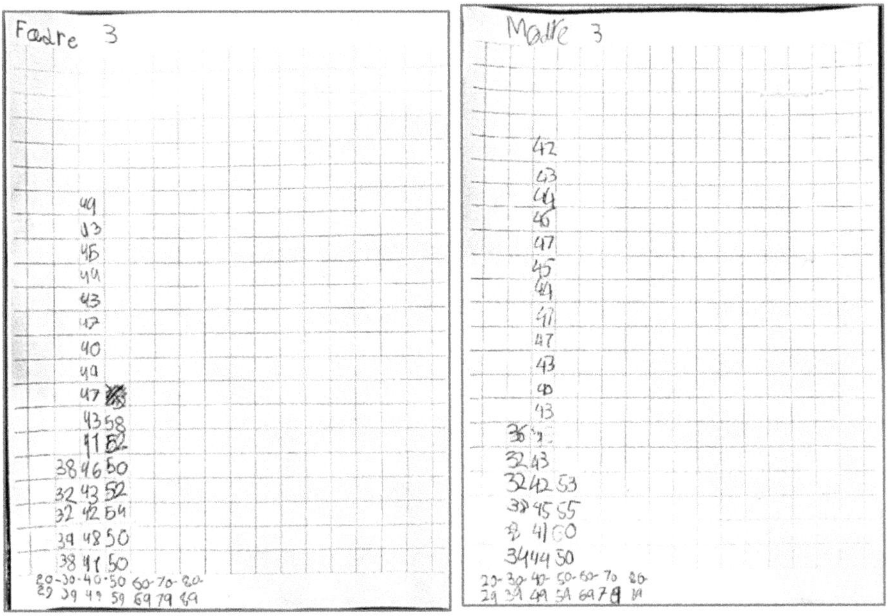

Fig. 2 The students' representations of the ages of their fathers (*left*) and their mothers (*right*). They found a way to group data without losing knowledge of the exact ages

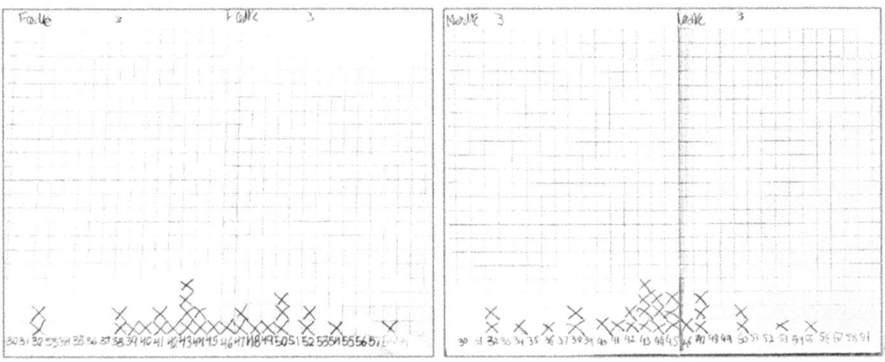

Fig. 3 The group's second representation of the fathers' ages (*left*) and the mother's ages (*right*). In this version, they left out the exact ages of the parents. Here they had a better picture of the difference of ranges between the genders

This excerpt is connected to the process of analysing and interpreting data. The students summarised what their investigations could tell them about the context. The students had expected a smaller range in the parents' ages than they found in the data. They also expected the ages of the mothers and fathers to have the same distribution. They found that the fathers' ages had a larger range than the mothers' ages. The students also addressed central tendencies in the data. They found the mode and could tell that many parents were in their forties.

In summary, the students investigated a context that was meaningful to them, despite the messiness of the context. They discovered that the initial question should be revised to make a meaningful statistical investigation. They identified which attributes were worthy of attention. They structured and organised their data, for example, whether to group the data. They tried out different representations and reasoned about which ones could tell them something meaningful, and they reasoned about the loss of information. The students analysed the data in a way that enabled them to draw conclusions about the phenomenon.

5.2 How Often Do Your Parents Argue?

The following discussion was from the first session. The dialogue was between the researcher and a group, in which the researcher tried to support the students in posing a statistical question. This particular group of students experienced huge difficulties during the process compared to the other groups. The students were concerned about the well-being of the class and wanted to revolve their investigations around that.

47.	S6:	I think it is important that we are all well.
48.	S7:	We already wrote that.
49.	R:	How can you investigate that with statistics?
50.	S6:	By asking others if they are doing well.
51.	R:	Yes? Then you have two categories: how many are doing well and how many are not.
52.	S6:	Are we supposed to make two spaces?

In the dialogue, the researcher tried to guide the students from their aim of inquiring about students' well-being to asking a quantifiable question. After some time, the students came up with the subject: '*How often do parents argue'?* (See Fig. 4). The next excerpt shows that it was challenging for the students to design an investigation and structure the data.

53.	R:	If you choose the one with parents who argue, then you could…. There must be some who never argue, and then there must be some who argue almost every day?
54.	S6:	Then there are some who argue sometimes.
55.	S8:	I think it is sometimes for me, like 3–4 times a week.

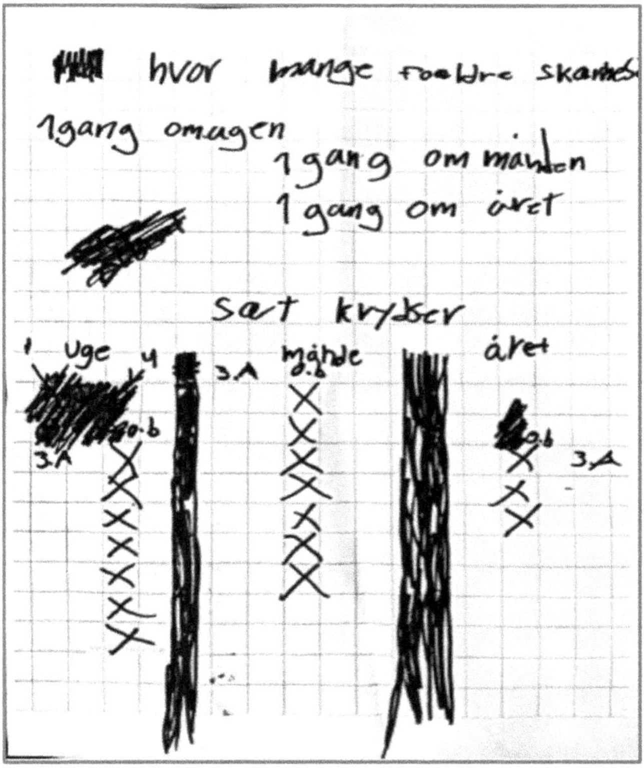

Fig. 4 How often do your parents argue? The data are structured based on weekly (the crosses in the left side), monthly (in the middle) and yearly (to the right)

The above excerpt is about posing questions. In Line 47, Student 6 put forward the purpose of the investigation instead of posing a statistical question. This might be due to a lack of statistical knowledge, as they had no prior experience with this. Despite this, they had an idea of how to collect data: 'By asking others…'. Their problem was that they did not pose a question to which answers could be quantified. This shows how difficult it can be to move from a meaningful context to data for young and inexperienced learners. In Lines 53–55, the students came up with a question that could be quantified. They then moved to the next part of the modelling process: generating, selecting and measuring attributes. The teacher started this reasoning process. In Line 54, Student 6 continued reasoning about how to quantify parents' arguments. In Line 55, Student 8 shared his own experience with the topic. Knowledge of the context affected the design of the investigation (see Fig. 4). This shows how the modelling process is an interplay between knowledge about the context and ideas of quantification, even with a minimal amount of statistical knowledge.

The next excerpt is from the students' final presentation. The teacher asked the students if they had learned something new about the phenomenon.

56. T: What did you discover in your investigation?
57. S6: It surprised us that some answered yearly. We had not counted on that.
58. T: You had not counted on that?
59. S7: I did not even count on monthly.
60. T: Okay, so you thought more parents argued weekly. Or what?
61. S6: No, most....
62. S7: Most weekly...monthly.

The teacher consistently asked all the groups if there was something that had surprised them and who else could use their investigation. This allowed the students to articulate the insights they had gained through their investigation. This excerpt shows that the frequency of parents arguing was somewhat different from what they thought it would be.

After the presentation, the students were asked if they would adjust their investigation if they had to repeat it:

63. S8: Yes, this is not very precise. You don't know if this cross [points to a cross in the category 'weekly'] is one day or more days.

In Line 63, Student 8 reasoned about the loss of specific knowledge regarding the individual data points when grouping data. This was a reflection on structuring and representing the data. After this, the students argued that their investigation might be interesting for other students experiencing parents arguing or even parents to understand how much children think parents argue.

6 Discussion

6.1 Posing Questions in a Meaningful and Interesting Context

The two groups engaged in the process in different ways. Still, both groups went through the entire data modelling process with support from the teacher and their peers. Both groups engaged in relevant statistical reasoning throughout the different phases of the process. Both groups found the process of moving from their initial curiosity to posing a statistical question difficult. The first group went from asking how many fathers were older than 39 to broadening their investigation to a topic they found more interesting. However, the second group had trouble when moving from the overall aim of assessing the overall well-being of the students in their class to posing a statistical question. To pose a quantifiable question, support from the teacher and researcher was necessary, but the students succeeded at last. Of course, the grouping of data blurred some of the information that they found relevant to some extent. Knowledge about both the context and the properties of statistical concepts is a prerequisite for engaging in statistical modelling. However, in this group, guidance was provided not for complex statistical concepts but to refine the initial curiosity to something quantifiable. This shows that this method of modelling might be suitable for students who are not yet far along in their development of sophisticated statistical knowledge. However, teachers have a difficult role, as they must balance guidance and support. On the one hand, the investigation must stay interesting and authentic to the students; on the other hand, it is necessary to guide the students into posing statistical questions.

The students' experiences, and hence, context knowledge of their own parents' arguing supported them in structuring the investigation. Because they found a topic meaningful to them, they were able to use their knowledge about the context to design and revise the investigation. The closing discussion showed that they chose a purpose for their statistical investigation that became more insightful than their original stated purpose.

Niss (2010) pointed out that engaging in the modelling process requires 'mathematical' self-confidence and perseverance. Principle 4 is important in this regard. The design must create room for time, support and recognition so that students can translate their curiosity into questions that can be answered with statistics. Recognition of the process as difficult becomes crucial in supporting self-confidence and perseverance. In Session 2, the teacher highlighted their struggles as an exemplary process and invited other students to give supportive suggestions.

Due to the design principles, the students were not asked to use specific statistical measures in their investigations. However, the later development of the language to describe the median, mode and range is necessary to refine their statistical reasoning. This is a prerequisite for developing students' opportunities to understand the world around them in a more competent way.

6.2 Data Modelling with Young Learners and Allgemeinbildung

On the one hand, it would be a false claim to have observed or measured *Allgemeinbildung*, as it is a life-long individual developmental process. On the other hand, it is an important issue to discuss in a specific way. This project showed small indications that the students gained a new understanding of the world around them through data modelling. The students handled the messiness of the modelling process with support from the teacher and their peers, who acknowledged and welcomed openness. However, the openness of the task was demanding for the students, and the task could have been more manageable if the teacher had pre-phrased the initial question. Nevertheless, because of the specific emphasis on involving students' own lifeworlds and experiences in the data modelling process, the task comprised both relevant statistical reasoning and gave rise to a new understanding of a phenomenon in the world around them.

Shedding new light on already known phenomena seems to have some potential. Still, it is important for students to investigate unfamiliar phenomena to expand their insights into the world around them. In addition, the potential of data modelling to make predictions and reason about a larger population was, for most students absent in this study. These aspects could provide new ways to experience the world with the use of statistics. However, starting in the context of students' interest might be a good starting point to view data modelling as a way to gain new insights.

7 Conclusion

The initial research question was, '*How can a specific emphasis on moving from a context of students' interest to data in the data modelling process create potential for experiences of* Allgemeinbildung *for young learners?*

The study reported in this chapter shows that modelling with data is a suitable informal modelling experience for young learners and can be a way to gain a new understanding of the world around them. It showed how the students gained new insights into an already familiar phenomenon and hence experienced how the modelling process worked. Still, it is difficult for students to move from context to pose a statistical question. Overcoming these challenges requires a learning environment that ensures that the teacher and peers provide support and invite students to grapple with ideas during the process.

References

Alrø, H., Blomhøj, M., Skovsmose, O., & Skånstrøm, M. (2000). Farlige små tal : almendannelse i et risikosamfund [Dangerous small numbers : Allgemeinbildung in a risk society]. *Kvan – et tidsskrift for læreruddannelsen og folkeskolen, 20*(56), 17–27.

Andersen, M. W., & Weng, P. (2019). Dannelse gennem meningsfulde oplevelser med matematik [Bildung through meaningful experiences with mathematics]. In I. J. Hansen, M. Rønø, S. E. Soneff, & A. H. Yates (Eds.), *Dannelse i alle fag* (pp. 85–100). Dafolo.

Arnold, P., & Franklin, C. (2021). What makes a good statistical question? *Journal of Statistics and Data Science Education, 29*(1), 122–130. https://doi.org/10.1080/26939169.2021.1877582

Biehler, R. (2019). Allgemeinbildung, mathematical literacy, and competence orientation. In H. N. Jahnke & L. Hefendehl-Hebeker (Eds.), *Traditions in German-speaking mathematics education research* (pp. 141–170). Springer International Publishing. https://doi.org/10.1007/978-3-030-11069-7_6

Blomhøj, M. (2001). Hvorfor matematikundervisning? : Matematik og almendannelse i et højteknologisk samfund (why mathematics education? Mathematics and Allgemeinbildung in a high-tech society). *Centre for Research in Learning Mathematics, 24*, 218–246.

Børne- og Undervisningsministeriet. (2022, 2/2/2019). *Folkeskolens historie: Et kort rids over folkeskolens lange historie* [The history of the Danish school: A short outline of a long history]. https://emu.dk/grundskole/uddannelsens-formaal-og-historie/folkeskolens-historie

English, L. D. (2010). Young children's early modelling with data. *Mathematics Education Research Journal, 22*(2), 24–47. https://doi.org/10.1007/BF03217564

English, L. D., & Watson, J. (2018). Modelling with authentic data in sixth grade. *ZDM, 50*(1), 103–115. https://doi.org/10.1007/s11858-017-0896-y

Jahnke, H. N. (2019). Mathematics and Bildung 1810 to 1850. In H. N. Jahnke & L. Hefendehl-Hebeker (Eds.), *Traditions in German-speaking mathematics education research* (pp. 115–140). Springer International Publishing. https://doi.org/10.1007/978-3-030-11069-7_5

Leavy, A., Meletiou-Mavrotheris, M., & Paparistodemou, E. (Eds.). (2018). *Statistics in early childhood and primary education: Supporting early statistical and probabilistic thinking*. Springer Nature. https://doi.org/10.1007/978-981-13-1044-7

Lehrer, R., & English, L. (2018). Introducing children to modeling variability. In D. Ben-Zvi, K. Makar, & J. Garfield (Eds.), *International handbook of research in statistics education* (pp. 229–260). Springer International Publishing. https://doi.org/10.1007/978-3-319-66195-7_7

Lehrer, R., & Schauble, L. (2000). Inventing data structures for representational purposes: Elementary grade students' classification models. *Mathematical Thinking and Learning, 2*(1–2), 51–74. https://doi.org/10.1207/S15327833MTL0202_3

Neubrand, M. (2015). Bildungstheoretische Grundlagen des Mathematikunterrichts [Education theory foundations of mathematics teaching]. In R. Bruder, L. Hefendehl-Hebeker, B. Schmidt-Thieme, & H.-G. Weigand (Eds.), *Handbuch der Mathematikdidaktik [Handbook of mathematics education]* (pp. 51–73). Springer. https://doi.org/10.1007/978-3-642-35119-8_3

Niss, M. (2000). Gymnasiets opgave, almen dannelse og kompetencer [The task of upper secondary school, Allgemeinbildung and competencies]. *Uddannelse: Undervisningsministeriets tidsskrift, 33*(2), 23-33.

Niss, M. (2010). Modeling a crucial aspect of students' mathematical modeling. In R. Lesh, P. L. Galbraith, C. R. Haines, & A. Hurford (Eds.), *Modeling students' mathematical modeling competencies: ICTMA 13* (pp. 43–59). Springer US. https://doi.org/10.1007/978-1-4419-0561-1_4

Niss, M., & Blum, W. (2020). *The learning and teaching of mathematical modelling*. Routledge. https://doi.org/10.4324/9781315189314

Patel, A., & Pfannkuch, M. (2018). Developing a statistical modeling framework to characterize year 7 students' reasoning. *ZDM, 50*(7), 1197–1212. https://doi.org/10.1007/s11858-018-0960-2

Powell, A. B., Francisco, J. M., & Maher, C. A. (2003). An analytical model for studying the development of learners' mathematical ideas and reasoning using videotape data. *The Journal of Mathematical Behavior, 22*(4), 405-435. https://doi.org/10.1016/j.jmathb.2003.09.002

Skovsmose, O., & Borba, M. (2004). Research methodology and critical mathematics education. In P. Valero & R. Zevenbergen (Eds.), *Researching the socio-political dimensions of mathematics education: Issues of power in theory and methodology* (pp. 207–226). Springer. https://doi.org/10.1007/1-4020-7914-1_17

Wild, C. (2006). The concept of distribution. *Statistics Education Research Journal, 5*(2), 10–26. https://doi.org/10.52041/serj.v5i2.497

Wild, C. J., & Pfannkuch, M. (1999). Statistical thinking in empirical enquiry. *International Statistical Review, 67*(3), 223–248. https://doi.org/10.1111/j.1751-5823.1999.tb00442.x

Winter, H. (1995). Mathematikunterricht und Allgemeinbildung [Mathemathics teaching and Allgemeinbildung]. *Mitteilungen der Gesellschaft für Didaktik der Mathematik, 61*, 37–46.

Part III
Data and Simulation to Support Understanding

Investigating Mathematics Teacher Educators' Conceptions and Criteria for an Informal Line of Best Fit

Jale Günbak Hatıl and Gülseren Karagöz Akar

Abstract In this study, we investigated eleven mathematics teacher educators' (TEs) subject matter knowledge for teaching an informal line of best fit. We scrutinized what conceptions and placement criteria for an informal line of best fit TEs used and discussed the relationship between TEs' conceptions and criteria to draw implications for teaching and research. We used basic qualitative research design and conducted approximately one-hour-long task-based interviews. Results showed that TEs' dominant conceptions were representer, predictor and signal. While TEs having the representer conception described an informal line of best fit as visually focusing on the best representation of the data, the ones having the predictor considered making close predictions for unknown independent values. Further, the TEs having the signal conception focused on the data at hand, the predicted values on the best fit line, and the residuals simultaneously and holistically. These results implied that signal conception reflects and includes the ideas of both representer and predictor conceptions. Results also indicated that the TEs' most dominant criteria for placement of an informal line of best fit were closest, sum deviation and ellipse and area.

Keywords Association · Informal line of best fit · Least squares regression · Statistical knowledge for teaching

J. G. Hatıl (✉)
Koç University, Office of Learning and Teaching (KOLT), Istanbul, Türkiye
e-mail: jhatil@ku.edu.tr

G. K. Akar
Boğaziçi University, Department of Mathematics and Science Education, Istanbul, Türkiye
e-mail: gulseren.akar@boun.edu.tr

G. F. Burrill et al. (eds.), *Research on Reasoning with Data and Statistical Thinking: International Perspectives*, Advances in Mathematics Education,
https://doi.org/10.1007/978-3-031-29459-4_13

1 Introduction

Statistical association is one of the fundamental ideas that many researchers and statistics educators agree should be included in all levels of education (Garfield & Ben Zvi, 2004; Burrill & Biehler, 2011). Researchers define statistical association as "the correspondence of variation of two statistical variables" (Moritz, 2004, p. 228). Researchers also suggest that statistical association is at the heart of learning subsequent statistical methods such as regression and correlation. However, research has shown that students may not reach understanding of these methods due to heavy emphasis on procedural knowledge in educational settings (Garfield & Ben Zvi, 2004; Batanero & Díaz, 2010). Hence, there is a trend in research to introduce association informally as a starting point for the development of conceptual understanding behind formal procedures for these methods (i.e. correlation and regression) (Zieffler et al., 2008; Kazak et al., 2021).

Zieffler et al. (2008) defined informal knowledge as "either a type of everyday real-world knowledge that students bring to their classes based on out-of-school experiences, or a less formalized knowledge of topics resulting from prior formal instruction" (p. 42). It is important to study and consider the role of informal knowledge as a starting point for the development of formal understanding of a particular topic (Zieffler et al., 2008). As an example of informal approach in statistics, Casey (2015) refers to an informal line of best fit as "fitting a line to data displayed in a scatter plot by eye, without using calculations, or technology to place the line" (p. 2). Working on an informal line of best fit could also provide teachers with ideas about their students' prior knowledge for a line of best fit. Thus, some researchers recommended an informal approach to be taught as an initial step for learning linear regression and statistical association (Casey, 2015; Casey & Wasserman, 2015; Sorto et al., 2011). An informal line of best fit is also included in some national curricula such as the Australian Curriculum, Assessment and Reporting Authority (ACMEM, 2012); England's Qualifications and Curriculum Authority (2007); and the United States Common Core State Standards (National Governors Association Center for Best Practices & Council of Chief School Officers, 2010). For instance, in the United States, students in the eighth grade are expected to "informally fit a straight line, and informally assess the model fit by judging the closeness of the data points to the line" (CCSSI, 2010, p. 56). Similarly, in the Australian curriculum, senior secondary level students are expected to describe association between two quantitative variables and also determine an informal line of best fit according to the objective, "find the line of best fit by eye" (ACMEM, 2012).

Albeit its importance in different curricula, few studies have examined students' and teachers' conceptualizations of an informal line of best fit (Casey, 2015; Casey & Wasserman, 2015; Sorto et al., 2011). Results from Casey's (2015) study showed that students' previous knowledge of lines in mathematics courses often interfered with their thinking processes in accurately interpreting a line of best fit. Similarly, Casey and Wasserman's (2015) study showed that teachers have varying conceptions and criteria for line placement, indicating significant conceptual gaps in their content knowledge. Their results, aligned with Batanero and Díaz (2010), suggested that understanding how teachers think about and improving their knowledge of the line of best fit are necessary conditions to enhance students' conceptualizations. They concluded that conducting an expert-level study in addition to the teacher-level and student-level studies on this concept could be fruitful to improve the teaching and learning processes related to fitting lines to data. Accordingly, we investigated mathematics teacher educators' conceptualization of an informal line of best fit.

Mathematics teacher educators (TEs) are the educators responsible for the education and development of mathematics teachers (Beswick & Goos, 2018) and are in charge of teaching statistical content. In particular, while teachers are expected to have the subject matter knowledge they teach, TEs are expected to have more than the subject matter knowledge mathematics teachers are required to have (Beswick & Goos, 2018). Several studies indicated that mathematics teachers focus more on procedural aspects, rather than developing students' statistical thinking and conceptual understanding in most statistics classroom instruction (Garfield & Ben Zvi, 2004; Sorto et al., 2011; Henrequies & Oliveira, 2013). This points to the importance of fostering teachers' and preservice teachers' statistical knowledge for teaching. The following sections first define the constructs associated with the informal line of best fit, then elaborate on the need for studying it. We then discuss previous research and the corresponding conceptual framework of the study.

2 Constructs Related with Informal Line of Best Fit and Need for Studying It

The concept of association might possibly be considered as an extension of the deterministic concept of mathematical function with an appreciation of the role of variation (Engel & Sedlmeier, 2011; Batanero et al., 1996). An association between two variables uses information about one variable to understand, explain and predict values of the other variable (Garfield & Ben Zvi, 2004). A line of best fit is a model that represents the association between two quantitative variables if the trend of the points in a scatterplot of paired numerical data is similar to a linear shape (Franklin et al., 2005). So, the purpose of regression is to find a line of best fit for a given set of data to explain and predict relationships between numerical variables (Garfield & Ben Zvi, 2004) while minimizing inherent noise in the dataset. Although many different straight lines can be drawn that seem to be passing through the "center" of the scattered data, it is crucial to find the line that gives the best fit to the data points according to some criteria (Gravetter & Wallnau, 2013). The typical approach in school mathematics is to identify residuals, defined as the deviations in the "y" direction between the points in the scatterplot and a given line (Franklin et al., 2005), and to make the sum of the squares of the residuals a minimum. This is known as the least squares method to find a line of best fit or regression line. Although students are able to use the formulae or statistical software to get the regression line, the meaning of the line and the relationship between variables are not easily understood (Engel & Sedlmeier, 2011). Particularly, in most statistics classroom, students have difficulties in correctly reasoning about fundamental big ideas regarding association (Garfield & Ben Zvi, 2004; Sorto et al., 2011). Common misconceptions in reasoning about statistical association are thinking of the relationship as deterministic, unidirectional, local, and causational (Batanero et al., 1998; Estepa & Sánchez Cobo, 2001; Engel & Sedlmeier, 2011).

Students having a *deterministic conception of association* misconception believe that a correspondence between two variables means assigning only a single value to the response variable for each value of the independent variable. This kind of correspondence is in fact a property of the function from a mathematical view. However, in a statistical setting, it is important to observe several values for the response variable corresponding to the same value for the independent variable because of the inherent variation (Engel & Sedlmeier, 2011). That is, understanding regression depends on an appreciation of the role of variation in data. In addition, students having *the unidirectional conception of association* perceive the association only when the correlation coefficient sign is positive. Otherwise, they interpret inverse association as independence (i.e., no association) (Batanero et al., 1998; Engel & Sedlmeier, 2011). Moreover, students *with a local conception of association* use only part of the data to make a judgment about a given relationship and generalize it to the complete dataset (Casey, 2014). Finally, students having a *causation conception of association* believe that if there is association between the variables, then there is a causal relationship between them. This belief may be a result of instruction students receive in mathematics, science and other areas in which all phenomena are given with a causal explanation (Estepa & Sánchez Cobo, 2001). Researchers emphasize that understanding statistical association requires both students and teachers to identify the distinction between association and causation such that association describes a general trend, but associated variables are not necessarily dependent (Casey, 2008).

In order to recognize and overcome the aforementioned misconceptions on association, Sorto et al. (2011) suggested using informal lines of best fit for class discussions before studying formal regression. They stated that without a discussion comparing students' tendencies about fitting a line with standard mathematical methods, students might have difficulty in understanding and interpreting the meaning of regression. This is important because informal linear regression underpins students' study of formal linear regression and regression of other function forms such as quadratic and exponential (CCSSI, 2010). Similarly, understanding students' conception of an informal line of best fit engages teachers in anticipating students' ideas prior to the introduction of the topic and allows them to plan their instruction accordingly (Casey, 2014). This is important as what teachers know shape and determine what their students might possibly know (Ball et al., 2008). However, since statistics is a relatively new subject in different curricula, many teachers have not had an opportunity to learn underlying concepts and principles for teaching practices of data analysis (Franklin et al., 2005). It is in this regard that we studied this topic with TEs to learn more about and characterize how they conceptualize an informal line of best fit as well as how they place a line of best fit informally on a given data set.

3 Previous Research and Conceptual Framework

Our work was mainly informed by the findings from three studies (Casey, 2015; Casey & Wasserman, 2015; Sorto et al., 2011). The purpose of the Sorto et al. (2011) study, conducted with eighteen university students, was to reveal these students' conceptions and misconceptions related to linear regression before being introduced to the concept in their statistics classes. Building on Sorto et al. (2011), Casey (2015) studied eighth grade (ages 11–14) students' understanding of the informal line of best fit. Casey and Wasserman (2015) also investigated nineteen teachers' subject matter knowledge for teaching an informal line of best fit. Results showed that both students and teachers had similar conceptions (See Table 1).

Casey and Wasserman (2015) stated that teachers having a model conception described the line of best fit as a line that "closely approximates the actual true relationship as possible" (p. 17) and used this to make an estimation for the general relationship between variables in the population. Teachers having a *representer* conception referred to the line of best fit as the line that represents all the data at hand. In particular, the best fitting line could be used to make estimations for where the data points are situated. Teachers with the *predictor* conception considered that the line of best fit should get close predictions for independent values not in the dataset allowing estimations for unknown or future values. Finally, a *typical* conception is related to the data's central position in a distribution of values (Konold & Pollatsek, 2004). For instance, teachers with this conception considered that the line of best fit is "a line that averages all the data points" (Casey & Wasserman, 2015, p. 17). Also different from the students' dominant conceptions, two teachers had a *signal* conception. The *signal* in noisy processes can be thought of as the stability in a variable system or the certainty in situations involving uncertainty (Konold & Pollatsek, 2002). In particular, according to Engel and Sedlmeier (2011), the trend in the data expressed as linear regression represents the explained part of the variation whereas the deviation between model and data represents the unexplained variation. Thus, data can be thought of as "a structural component plus residuals, that is, Data = Signal + Noise" (Engel & Sedlmeier, 2011, p. 25). Their results also showed that, although teachers and students had similar conceptions, teachers were able to give better explanations for their responses because of their background knowledge and

Table 1 Conception categories for lines of best fit (Casey & Wasserman, 2015 p. 17)

Conception	Description of line of best fit
Model	The model to show the general relationship between two variables in the population.
Signal	The meaningful signal for data that eliminates the inherent noise.
Typical	The bivariate equivalent for determining middle or typical values; sometimes named as average, other times as median.
Representer	The best representation of all the sample data, where the line accounts for the data at hand rather than a more general relationship.
Predictor	The line that enables predictions for independent values not in the data set.

experience. In fact, on a conceptual level, teachers are expected to construct a bridge between "a deterministic view of a function and a statistical perspective that appreciates variation" (Casey & Wasserman, 2015, p. 25) in the analysis of bivariate numerical data (Engel & Sedlmeier, 2011). However, in the study with teachers, only two out of nineteen teachers had the signal-noise conception supporting functional explanation with appreciation of the role of variation in statistical data. Casey and Wasserman (2015) stated that "teachers' varying conceptions point to some significant gaps in their knowledge" to make sense of statistical data (p. 27).

Results regarding both students' and teachers' criteria and methods for placement of the informal line of best fit also showed similarities (See Table 2), but their most preferred criterion differed as teachers used sum deviation or closest and students utilized equal number.

Based on the aforementioned research results and considering TEs as experts, in this study, we further investigated TEs' preferred criteria and methods for placement of the informal line of best fit and explanations of underlying reasons for their choices. In this regard, in accordance with Casey and Wasserman (2015), this study mainly drew on Statistical Knowledge for Teaching Framework (SKT) (Groth, 2007, 2013). Based on Mathematical Knowledge for Teaching (MKT) (Hill et al., 2008), the SKT framework includes both subject matter knowledge and pedagogical content knowledge with regards to statistics. In what follows we describe how we conceptualized the framework in this study.

Subject matter knowledge includes three categories: common content knowledge, specialized content knowledge and horizon knowledge (Hill et al., 2008). With regards to statistics, common content knowledge is related to competencies developed in conventional statistics courses and used in the work of many occupations along with the use of statistics in teaching (Groth, 2013). In this study, accurately reading scatterplots and determining a line of best fit were considered common content knowledge for statistics. Regarding specialized content knowledge, which relates to teachers' knowledge of how to explain and justify the reasoning behind mathematical ideas (Ball et al., 2008), this study considered TEs' explanations and justifications related to conceptualization of a line of best fit and

Table 2 Criteria categories for placement of line of best fit (Casey & Wasserman, 2015 p. 17)

Criterion	Description of criteria to place line of best fit
Equal Number	Equal number of points above and below the line
Pairs	Each point above the line has a symmetric pair below the line so that the distance to the line is the same for each member of the pair; pairing off individual data points to determine line of best fit
Closest	The line that is closest to all points (implies total distance from points to line is minimized)
Sum Deviation	Deviations for points above the line sum to the same as the sum of deviations for points below the line (they did not necessarily make measurements, but referenced this as they placed their line).

association between variables on eight tasks as indicators of their specialized content knowledge. Additionally, specialized content knowledge also relates to statistical issues in teaching contexts as "appraising students' unconventional methods for solving problems, and constructing and evaluating multiple representations for concepts" (Groth, 2007, p. 428). Finally, Groth (2013) stated that horizon knowledge for teaching statistics requires teachers to "know statistics beyond the prescribed curriculum" (p. 9) to provide a learning basis for later grade levels. For example, Casey and Wasserman (2015) stated that dealing with non-associated data could be classified as horizon content knowledge for teachers since the line in this condition is not for describing the present data but for making statistical inferences about association or non-association based on slope. However, making statistical inferences using linear regression is included in the content of university level courses. So, we considered this knowledge as part of specialized content knowledge for TEs. Acknowledging this, therefore, in this study, we considered subject matter knowledge for the informal line of best fit as conceptualizations and placement criteria of the informal line of best fit. The research questions were:

What subject matter knowledge for teaching about an informal line of best fit do teacher educators have?

- What conceptualization of an informal line of best fit do teacher educators have?
- What criteria for placement for an informal line of best fit do teacher educators use?

4 Methodology

4.1 Design

In this study, we used a basic qualitative research design (Merriam & Tisdell, 2015) with the aim of identifying how participants make sense of their meanings and experiences about an informal line of best fit and how they interpret the association between variables on given tasks and questions. Accordingly, structured task-based interviews (Goldin, 2000) were used to collect data on participants' conceptions and placement criteria for an informal line of best fit.

4.2 Participants

Participants were mathematics teacher educators (TEs) working at different universities in Turkey. We used a purposeful sampling method (Merriam & Tisdell, 2015) while choosing the participants. Examining the websites of all the universities having mathematics education departments in Turkey, we searched for TEs who were experienced in teaching undergraduate and/or graduate level statistics and/or quantitative research methods courses that included the line of best fit in their course. Among eleven TEs, seven males and four females, who volunteered to attend to this study, five participants (TE1, TE3, TE5, TE6 and TE11) only taught undergraduate courses, four participants (TE2, TE8, TE9 and TE10) taught both undergraduate and graduate courses, and two participants (TE4 and TE7) taught only graduate courses. The total number of courses that participants had ever taught ranged from one to six. In addition, as a student, two participants (TE1 and TE6) had taken only undergraduate courses in statistics; seven participants (TE2, TE3, TE5, TE7, TE8, TE9 and TE11) had taken both undergraduate and graduate courses such as Statistics, Experimental Research or Quantitative Analysis, and two participants (TE4 and TE10) had taken only graduate courses, Statistics and Quantitative Analysis.

5 Data Collection

Data were collected in Spring 2018. Interviews took approximately one-hour for each participant. During the interviews, participants worked on the tasks (See Table 3) to determine a best fit line for a given dataset by means of a music wire. Music wire is chosen as a material for this process because "it is not breakable and is thin to minimize the issue of obscuring the data points when placed on the scatterplot." (Casey, 2015, p.6). Data sources were transcripts of the video recordings and the written artifacts documenting each participant's work on the tasks.

Table 3 Summary of Tasks 1–7

Task	Context	Direction	Strength	Characteristics
1	Golf ball bounce height versus drop height	+	$r_1 = 0.96$	The y-intercept of the least square regression line for **Task 1–2** is close to the origin. Multiple sets of data points on **Task 1** positioned as close to be collinear The consecutive data points on **Task 2** are collinear points
2			$r_2 = 0.97$	
3	Movie attendance versus average ticket costs	–	$r_3 = -0.76$	Relatively less strong association between variables on **Task 3**
4			$r_4 = -0.92$	Less linearly aligned data points on **Task 4**
5	Achievement motivation versus GPA scores	+	$r_5 = 0.40$	Dataset includes an outlier point
6	Students' height versus shoe sizes	None	$r_6 = 0.05$	Prior beliefs on task context might lead participants to assume positive association between variables
7	Height versus birth month	None	$r_7 = 0.03$	Non-associated data given with unrelated contextual information

5.1 Interview Protocol

The interview protocol consisted of seven task-based questions including a scatterplot of contextual data to elicit TEs' conceptualizations related to a line of best fit. The data points on the scatterplot for given tasks were situated as suitable for applying linear models. Also, an additional task-based question (Task 8) was included with hypothetical students' line placements that asked for TEs' interpretation of which line placement was better than others. The criterion for hypothetical students' line placements was determined by considering students' common placement criteria in previous studies compared with a least squares regression line. By making this comparison, TEs might conclude which criterion was more important to draw a line of best fit.

We adopted five tasks (Tasks 1–2–3–4–6) and the first and the second summary questions, asked after the task-based questions, from Casey and Wasserman (2015). We constructed Tasks 5 and 7 and modified Task 8 based on the recommendations from previous studies.

After we generated the interview protocol, two experts from the mathematics education department including one with a statistics education background examined the tasks and questions to evaluate if the tasks were appropriate to investigate research questions. Based on their feedback and suggestions, we finalized the interview protocol. Figure 1 displays the scatterplot used in Task 1. Descriptions of the tasks are provided in Table 3.

Also, following Tasks 1–7 summarized in Table 3, there were two summary questions to derive more detailed information about participants' conceptions and criteria in placement of best fit line:

> Could you tell me what you would say to a student that asked you 'What is the line of best fit?'
>
> Could you tell me what you would say to a student to help them draw the line of best fit on a scatterplot?

6 Data Analysis

For the analysis of data, we used coded analysis that:

> …focuses on observations that are assigned to predefined categories by a coder, usually from relatively small segments of a transcript. A transcript is coded when the analyst formulates criteria for recognizing a phenomenon and then lists the places where that phenomenon occurs in the transcript. The conclusions may be at the level of observation patterns alone, or, they can be used as data to support or reject theoretical hypotheses that may have been generated by other means. (Clement, 2000, p. 558)

Using the categories of conception for the line of best fit, first we examined each participants' conceptions by reading the whole interview data for all the task situations as well as the data from their responses to the first summary question "Could

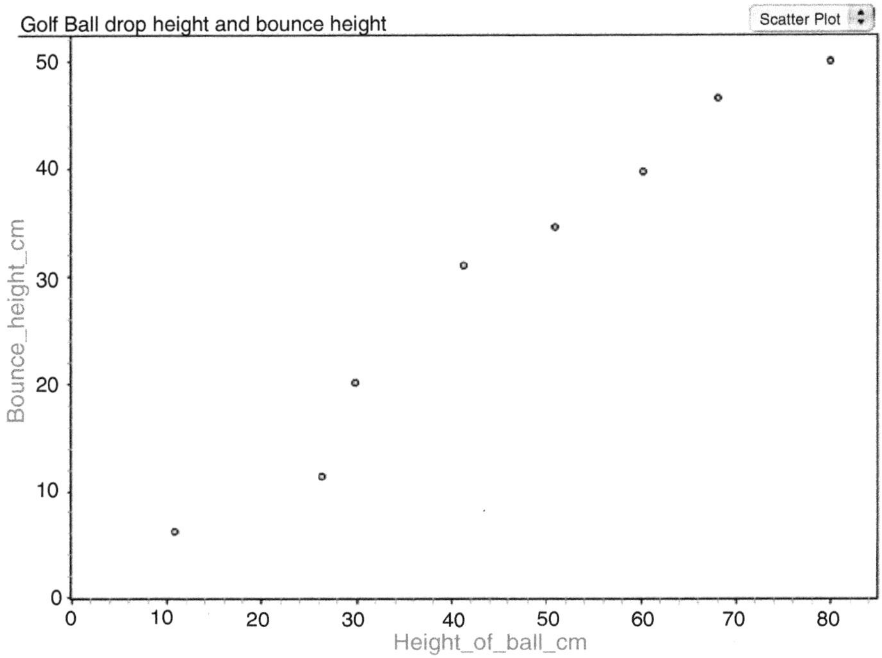

Fig. 1 Task 1 from interview protocol. (Casey & Wasserman, 2015 p.31)

you tell me what you would say to a student that asked you what is the line of best fit". The unit of analysis was the informative parts of their responses ranging from a sentence to a paragraph. We also looked for evidence of reasoning suggesting possible different conceptions not determined previously. After examining relevant data from each participant, we went back to the whole dataset to cross-compare and contrast the participants' conceptions in total. In this respect, we determined each participant's dominant conception by the frequency of the use of the relevant conception. Determining the dominant conceptions and criteria was important in explaining and elaborating on how a TE having a particular conception and criterion might reason. This way, also, the characteristics and the tendency of the reasoning with a particular conceptualization would be explicated. After determining the dominant conceptions each TE depicted, we clustered the data from different TEs showing a particular conception. Following this, for each task, we again read the data from different TEs to further conceptualize and determine the conceptions they depicted within each task. While doing that, we also looked for evidence to challenge our conjectures and modify them to cohere with the data. In this sense, we also engaged in constant comparative analysis (Strauss & Corbin, 1990).

We conducted a similar analysis for determining the criteria TEs utilized. Using the criteria categories, we read the data line-by-line and examined TEs' criteria categories on given tasks as well as their explanations for the question "what s/he would say to a student to help them draw the line of best fit on a scatter plot". Again, we identified the dominant criterion for each TE with the frequency of their reference to these criteria categories. We also analyzed their response to Task 8 as they chose the better line placement from given student answers and evaluated justifications for their responses. In this regard, Task 8 gave some further information about which criterion was more important and appropriate, according to TEs perspectives and their content knowledge, to assess students' criteria for a line of best fit.

7 Results

In this section, we explain the TEs' conceptions and criteria for placement of a line of best fit. During the interview, TEs used different conceptions and criteria based on the different task conditions. In the following subsections, we report on the results specifically focusing on each *dominant* conception and criterion TEs used and their interpretations throughout the interview in order to understand their conceptualizations of the line of best fit.

7.1 Conceptions

TEs predominantly referred to three conception categories throughout the interview, namely, *representer, predictor,* and *signal* (See Table 4).

These conceptions are elaborated below with excerpts from statements of TEs who predominantly used a particular conception in the task situations of Task 1, Task 3 and Task 5 along with the first summary question.

7.1.1 Representer

Representer was the most frequently used conception (See Table 4). For Task 1, TE8 referred to the *representer conception*:

Table 4 TE's dominant conceptions for what a line of best fit represents

Dominant conception	Participants	Frequency
Representer	TE1,TE4,TE6,TE7,TE8	5
Predictor	TE2,TE3,TE5,TE11	4
Signal	TE2,TE9,TE10	3

TE2 predominantly referred to both signal and predictor conceptions

> ...*I mean this can be the line that represents them (data points).* Otherwise a more special... I'm not affected by the context at the moment. If it was something different from the ball, it would be the same situation. What is important here is that this line cannot be vertical or horizontal. It's essential that vertical distances of these points are minimized...*I think that this is the best line that can represent this scatter of data points.*

As the excerpt suggests, TE8 described the line of best fit on Task 1 as the line that represents the data set. In particular, she focused on the data points at hand and placed the best fitting line as close as possible to all data points. In this respect, TE8 also pointed out that she was not influenced by the contextual information or general relationship between given variables; rather she focused on representing the data points while placing the best fit line.

Similarly, for Task 3, TE7, using a *representer conception,* reasoned:

> What is my purpose? Determining the best fitting curve...Therefore we always look at the distance. But my best fitting line curve, I can draw it also like this (showing with hand). But *it doesn't represent these then. In order for it to represent,* I need to look at the deviation since I cannot say equal distance or minimum distance. If it is not on the line, if you say it's as close to the line as possible, then it would be the one that passes through the most points. *But then it wouldn't represent these. It is about representation.*

TE7 used criteria for line placement with the aim of representing a given dataset. In particular, she stated that if the data points were not lined up, then the best fit line would be placed with minimum deviation from the data points in order to represent the dataset. She summarized her reasoning on the line placement process as "it is about representation" based on the *representer conception.* For Task 5, TE7 also referred to the representer conception as follows:

> After I have drawn this, I think it won't matter if I've drawn it negative or positive. *It does not represent the data... This (2,100) is too high in the deviation,* I will slide up while trying to balance this. *How can I explain this? It won't represent.* In order to equalize the deviation over there, maybe it would be something like this...

This suggests that TE7 aimed to represent a given dataset as a whole rather than focusing on any particular data point. However, there was a point (2,100) on this task that was far away from the general trend of the dataset. Therefore, she thought that the best fitting line would not represent the whole dataset since this data point (2,100) has a large deviation. In addition, TE7 explained the first summary question, "Could you tell me what you would say to a student that asked you what is the line of best fit" as follows:

> *The word represents...*I shall first think about the keywords and then I will use it in a sentence. *Represent means representing the data.* Of course, you can look at the context to see if it fits but *it's important that it represents the data. While representing, deviation...*I would say that is the balance of the distance of the points that are above the line and below the line. *So, I can represent it.* Trying to balance a deviation that will stay close to the points.

TE7's reference to the keyword "represent" indicated that she aimed to represent the data as a whole while placing the best fit line. She seemed to use the sum deviation criteria as balancing the distances of the points above and below the line in order to represent the data points at hand.

7.1.2 Predictor

Four TEs (See Table 4) predominantly referred to the *predictor conception*. Particularly, for Task 1, TE3 stated:

> There is a very strong positive relationship, if I were to predict, I would say there is a correlation close to $r = 0.90$... *This x variable – the height where the ball was dropped- means a very good predictor of the y variable.* In other words, *by knowing the value of one of the variables we can predict the second one with little error.*

As the excerpt suggested, TE3 interpreted that x variable – drop height of golf ball – could be thought as a predictor for the *y* variable, bounce height of golf ball. He further explained that he could make predictions for one of the variables with minimum error by using the value of the other variable. Therefore, he reflected the predictor conception of the best fit line on Task 1.

Similarly, for Task 3, TE11 used the predictor conception as follows: "... let's say the data are in very far points, then I can't interpret the relationship in between very well. Therefore, *I decide it to be the closest...I do that to make better predictions.*" She determined the line of best fit to be as close as possible to the data points with the aim of making better predictions with use of this line. In addition, for Task 5, TE3 depicted predictor conception of the line of best fit:

> Of course, it doesn't have a correlation coefficient as high as the previous ones, it has lower. If we were to do some more advanced work here, if we were to make statistics at the undergraduate level, we would do hypothesis testing, we would do a test to see whether the β_1 equals zero or not. What I mean by β_1 is slope here, $y' = \beta_0 + \beta_1 x$. That is to say, in this regression equation, I mean, we find β_0 and β_1 values according to the line of best fit. What we would do at the end is to look whether this β_0 and β_1 are different from 0 or not. What would it mean if it were different from zero or if it equal to zero? Are we aiming to find the motivation or GPA of the students here? We could take them all and find the average right? ...I mean we would predict the average for each student. *What we are doing with the benefit of the line of best fit is trying to predict the GPA from the line we draw and from the y values of equation we will create according to it, instead of predicting it from the average of the students' GPA scores. Therefore, these y values that we will obtain here will give better results than the y values that resulted in taking average directly.*

In the excerpt, TE3 explained the purpose of using the line of best fit to make better predictions for students' GPA scores than predictions based on taking the average of GPA scores for a given dataset. Then, she continued that since there is an association between GPA scores and achievement motivation, using the line of best fit would provide better predictions. Hence, he used a regression model for a bivariate dataset based on the *predictor conception* in this task.

Lastly, TE11 commented on the first summary question based on the predictor conception:

> I would explain to him like this: *you are trying to predict something*. For instance, I would give such a data set as an example: a child studied for 3 hours and got 10 from the exam, he studied 4 hours and got 7. Then I would ask whether there is a relationship between these or not...Let's give an example. Let's draw this, and now let's draw it like this, let's draw it like this (she is showing a positive-negative sloped line with her hand), *who will make the best prediction?* If the child gets closer to the most correct one, we can give an example and

> a data or he can say for each one that, the child studied 8 hours and got 13, will it give that result? Try it and see how close it gets. Such as, he wrote and found 13.1, he wrote the other one and found 13.5. *We want him to find the closest result* by trying on the linear equations that he has found. Because there is ready data in front of him, he can compare it with that. Therefore, an activity can be done like: *which one made it easier, so the line which the distance of these points to each other, to the line is the smallest helped my prediction.*

TE11 explained the use of the best fit line to make predictions for given variables. Then, she clarified it with an example of bivariate data that is assumed to have an association between variables. Since she believed that better predictions could be provided with a best fit line, she thought students' attention to the idea of having the closest line to data points could enable better predictions.

7.1.3 Signal

The *signal conception* was dominant for TE9 and TE10 and was frequently referenced by TE2. These TEs generally interpreted data as a combination of linear structure and variability around it. Particularly, for Task 1, TE9 made an explanation based on the signal conception. He stated:

> If the expected point were, x=10.8, let's say according to my line, it would be maybe a lower … *not 6.1 but 5.9. The n here is an error. If we add up the errors, we know that we will find zero.* For instance, I thought to have minimum value of sum of the squares. That's why *I passed it through there so the errors would balance each other.* If I hold it up, the error squares of the points below the line will be very high. Since I thought that, I did something like this…

Note that TE9 mentioned the error between the estimated value predicted by the best fit line and actual response value. In particular, he initially discussed deviation of an individual y-value on the best fit line from the actual data point and then thought of taking average of the y-values to provide balance of residuals on each side of the best fit line. He seemed to appreciate the variability in the data set while placing the line of best fit. In addition, his statement "*…If we add up the errors, we know that we will find zero…*" indicates that he seemed to be aiming for elimination of residuals while determining the line of best fit. This idea is similar to the signal-noise metaphor that defined a best fit line as a signal after eliminating the inherent noise. Similarly, for Task 3, TE10 depicted signal conception as follows:

> We are looking for linearity and *there is no linearity that is normal and social for all reasons.* No factor in bilateral relations… *let's assume we eliminate all. We wonder what kind of a relationship exists among these, it's actually an assumption.* These are totally about an assumption, *can I get a relationship among these if I rule out the other variables? If* I could, how would it be? *it would be linear, it would be a factor that affects one to one. However, since here other variables also affect, and since the point we focus on is two variables, we are trying to eliminate other variables and minimize the errors.*

TE10 noted some factors might affect the association between two variables. Therefore, irrelevant information might be included in the analysis process for bivariate data and cause some deviations from perfect linearity. Hence, he thought to eliminate inherent noise because of other factors and get an approximation for a best fit line that shows a relationship between given two variables. Then, the resulting regression line would be a *signal* for the association between the given variables.

In addition, TE2 used the signal conception while elaborating on Task 5:

> … Now, it's harder to find the best fit line. Probably, what you are looking for is this, because in the data there isn't a trend like in the others (tasks). I shall not say there is no trend but there is less correlation between two variables. *When we look at the data, the others were more linear, but the residual values in the line of best fit here will be different.* It is almost impossible to find the line of best fit from it. *While the others have more correct line of best fit, they give more or less a more correct line of best fit, error probability of the one I'll give here is really high*…But I will think in the same way and say something like this but my reliability here is quite low compared to the other ones, *there is higher deviation of this line.*

Because the dataset spreads out more than previous tasks, he noted correlation between variables would be lower, and residuals would have larger values when he determined the best fit line. According to *signal* conception, each predicted value on the best fit line could be "viewed as deviating from the actual data by a measurement error and the average of these values could be interpreted as a close approximation to the actual data" (Konold & Pollatsek, 2002, p. 269). He interpreted the Task 5 dataset with more deviations from a line of best fit than in the other tasks, using the signal meaning of the line of best fit. He also referred to the signal conception while answering the summary question:

> The notion of extracting linearity from data that is not linear is a difficult issue for them (students) to understand. What we try to do when we explain the line of best fit is to explain that we are looking for an order in the mess of the data and that one of the best ways to express this search for an order is line…

The excerpt suggested that students experience difficulty while finding a linear trend in a dataset including variations from perfect linearity. In particular, TE2 addressed this difficulty in explanations relevant to the aim of this process as extracting the line of best fit that is showing and determining the order/stability in the data varied to some extent. This indicates his use of *signal* conception. TE2 further explained his reasoning on the line of best fit that reflects signal conception as follows:

> …*when we pass a line, the line gives a value…but we want to make interpretations for actual data values from a value on the line. We have an error margin*, it might be hard to understand this…When we were making interpretations about this, we actually created the line of best fit, we had the data points in front of us. To see where each data point would be on the best fit line…I mean, if I did not see the data point, what I will predict with this value, the best fit line…But the actual value is this. Or I would predict a higher value as a lower value. Or there is a point that fits, I would predict that. I show these three situations on data and then I predict a point that I don't know. Therefore, one of these three situations is possible. I mean, *after explaining the concrete situation, predicting the abstract thing that doesn't exist. Because we don't see these (data points) anymore. Then, after extracting actual data points, my predicted values are there, but they actually might be around there as we see before. Those things which we call residual, the margin of error, show us the places where this data exist, and it shows it with some error, and we can say that it gives us that this margin of error exists.*

TE2 defined the line of best fit as a structure extracted from the data points varying in the data set. This was evident in his statement "*when we pass a line, the line gives a value…but we want to make interpretations for actual data values from a value on the line. We have an error margin…*" So, he seemed to conceptualize the line of best fit independently of the data points – the signal in the data – since he aimed to reduce the inherent variability to reach the stability within the process that varied. He also stated the need to be thinking of the line of best as the *signal* of data as well as thinking of the actual response values and the residuals in the data set. That is, from his point of view, the line of best fit seems to be a linear structure that gives predictions for variables while taking into consideration the margin of error in these predicted values and points to the three cases: that the values predicted by the line could be the same as, smaller or larger than the actual response value. This excerpt showed that he used *signal* conception.

7.2 Criteria and Methods

TEs depicted three dominant criteria categories (Table 5). Two categories, namely the closest and sum deviation, were listed in the framework, whereas the ellipse/area approach emerged from the current data.

The following shows excerpts from TEs who predominantly used the same criterion throughout the interview.

7.2.1 Closest Criterion

Closest criterion was referenced by each TE at least once during the line placement process. In addition, it was the dominant method used by seven TEs (see Table 5) to determine the best fitting line in different task situations. TE5 mentioned that the vertical distances between data points and the line should be minimized to provide the best fitting line for a given dataset in Task 1. In particular, he tried to place the line closest to all data points, applied the closest criterion for this task (Fig. 2). TE1 determined the best fit line on Task 5 by applying the closest criterion for seven data points. In particular, since there was an extreme score (2,100) far away from the general trend of the dataset, she preferred to discard this point. In this respect, she thought that if she included the extreme score in the line placement process, it would cause an increase in total distances between data points and the line. Hence, with the

Table 5 TE's dominant criteria for selecting a best fit line

Dominant criteria	Participants	Frequency
Closest	TE1,TE2,TE3,TE5,TE6,TE8,TE11	7
Sum Deviation	TE2,TE7,TE8,TE9,TE10	5
Ellipse/Area Approach	TE2,TE4	2

TE2 and TE8 used different criteria with the same frequency

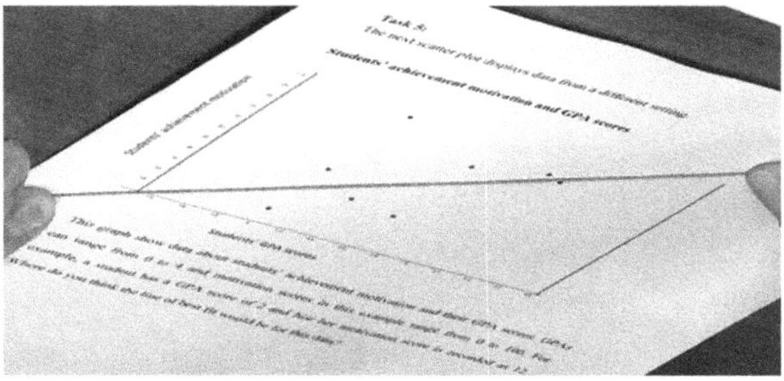

Fig. 2 TE1's line placement on Task5

aim of minimizing the sum of residuals, she used the *closest* criterion for seven data points excluding (2,100).

By the same token, while elaborating on students' line placements given on Task 8, TE2 evaluated the residuals between data points and the line for each student. For him, both students provided the elimination of residuals above and below approximately. However, what was more important was achieving the minimum distances in total.

TEs also referred to the *closest criterion* for the second summary question. TE1 summarized her dominant criterion as determining the line that is closest to all data points, stating that she focused on the whole data as opposed to any particular data point. In addition, she referred to the criterion as 'passing through two data points' which is thought as one of the most common misconceptions for students. She further elaborated that the regression line obtained with this criterion would not be useful if the dataset is varying to some extent. Therefore, she seemed to indicate that she did not determine the best fit line as passing through any data point intentionally, but instead, she aimed to place the line of best fit as closest to all data points.

7.2.2 Sum Deviation Criterion

The sum deviation criterion was obtained from five TEs throughout the interview process. TE7 focused on controlling deviations while placing a best fit line for Task 1 and 2 (Fig. 3). In this respect, she seemed to want to balance the sum of deviations above and below the line. In addition, she stated that she did not try to minimize distances between data points and the best fit line. For her, distance minimization could be provided by placing a line passing through most of the data points since she interpreted the minimization as reaching zero in distance. Hence, her reasoning while placing the line is based on *the sum deviation criterion* for these tasks (Fig. 3).

For the second summary question, TE7 referred to the sum deviation criterion when she defined the line of best fit as a representation of a given dataset. Therefore, to be able to represent the dataset, she thought to lead students to determine a line

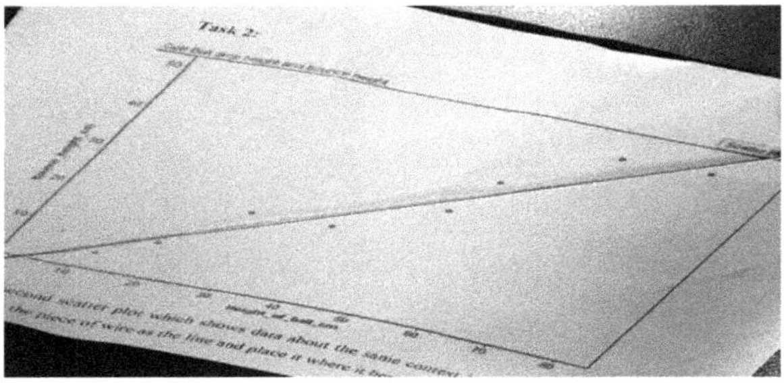

Fig. 3 TE7's line placement for Task 2

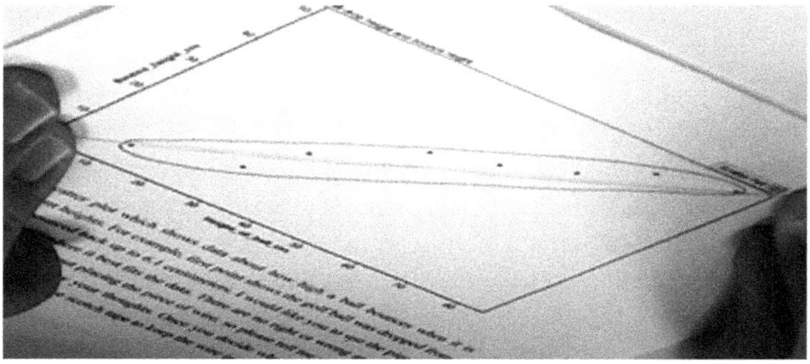

Fig. 4 TE4's line placement for Task 1

of best fit by providing a balance of distances between above points and below points with the line. Then, she further stated that the line of best fit should also be determined as close to data points as well as providing balance of distances on each side of the line. Therefore, although her dominant conception was the sum deviation criterion, she supported her perspective with the closest criterion.

7.2.3 Ellipse/Area Approach Criterion

TE2 and TE4 used different terms as "ellipse" and "area approach"; we combined them since the method and perspectives showed similarities. For Task 1, TE4 initially created an area including all data points (Fig. 4). By doing this, he believed that he could handle the dataset as a whole. Then, he found a line dividing the area into two equal parts and called this criterion the area approach. For him, this method of dividing the area into two equal pieces seemed to be easier than considering all data points at the same time in placing an informal line of best fit.

TE2 applied a similar criterion while placing a line for the tasks displaying data with smaller associated or non-associated variables. However, he used a different term –ellipse-- while describing the process. In particular, for Task 5, he thought of drawing an ellipse surrounding all the data points to determine the trend of the dataset and approximate the place of a best fitting line. In addition, he interpreted the point (2,100) as behaving differently from the general trend relating motivation and

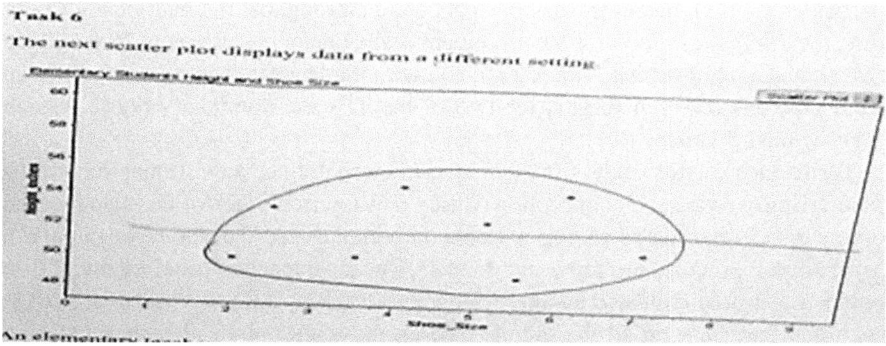

Fig. 5 TE4's line placement for Task 6

GPA scores and excluded this data point in the line placement. He stated that if he included this point, an ellipse drawn around the whole dataset would turn into a circle and cause difficulty to identify a trend in the data. In addition, he stated that this circular shape could be interpreted as an indication of weaker association. Hence, he adjusted the line of best fit applying the ellipse criterion for seven data points (except the data point with coordinates (2,100) and interpreted the line as displaying weak relationship between variables. Similarly on Task 6, TE4 used the ellipse/area criterion to create an area around the scatter of data points (Fig. 5). He placed a horizontal line with zero slope dividing the area with two equal pieces that indicated non-association between students' height and shoe sizes.

TE2 summarized the line placement process referring to the *ellipse/area criterion* for the second summary question. He stated the first step of line placement was searching for a linear trend in the dataset that was increasing or decreasing. In this respect, he mentioned drawing an ellipse including all data points to determine the correlation. Then, he interpreted that the thinner the ellipse is, the more likely the best fit line exists here. Particularly, he explained the reason for using the ellipse criterion as getting two endpoints of the ellipse that were supposed to be on the best fit line. However, when the ellipse turns into a circular shape around data, it would be difficult to identify the endpoints. In this case, the variables might have no association, and so the line of best fit could be placed horizontally.

8 Discussion and Conclusion

This paper reported on the results from research examining "What conceptualization for an informal line of best fit do teacher educators have?" and "What criteria do teacher educators use for placement of informal line of best fit?" Results in this study showed that, similar to Casey's studies of teachers and students (Casey, 2015; Casey & Wasserman, 2015), *representer conception* was the most frequent conception used by TEs although they also used *signal* and *predictor* conceptions. In particular, each TE had more than one conception throughout the interview. In addition, regarding the criteria for the placement of the line of best fit, most TEs used the *closest* and *sum deviation* criterion aligned with the teachers' criteria, but different from previous research results, for two of the TEs the dominant criteria was an *ellipse/area approach*.

Particularly, in this study, similar to teachers who defined an informal line of best fit referring to representer conception (Casey & Wasserman, 2015), TEs also focused on the data in the scatter plot as a whole and emphasized that the most important aspect of the process is to represent the data. Their perspective focusing on all data points as a whole indicated an aggregate view (Bakker, 2004; Biehler et al., 2018) such that they handled all the sample data on the scatter plot rather than focusing solely on particular data points. On the other hand, students' line placements passing *through as many points as possible* demonstrate their case-oriented view which might be an indication of having a misconception about a local conception of association (Casey, 2015). TEs with a representer conception also mostly used the *closest criterion*, the minimization of the residuals between all the data points and the best fit line. Also, they balanced deviations above and below the line suggesting that they also commonly used *the sum deviation criterion*. The only exception was TE4 with the representer conception who used *the area approach* that indicates drawing one (or more) shape surrounding all data points as closely as possible and almost equally dividing the shape to place the line of best fit. Such reasoning also reflects the common perspective of TEs with representer conception as handling data as a whole.

Regarding the results pointing to *predictor conception*, similar to the explanations provided by teachers in earlier research (Casey & Wasserman, 2015), four TEs defined the line of best fit as '*a line that gives closer predictions for values not included in the dataset.*' Also, they anticipated a strong association between variables since they thought that the stronger relationship would result in closer predictions. Similarly, they interpreted the dispersion of data points in the original data set as an indication of the variability in the predicted data. That is, the more deviated the points in the actual data, the more errors in predictions. The Conference Board of Mathematical Sciences emphasized that not only teachers' but also students' knowledge of making "…informal treatment of inference…" from the given data set with respect to the predicted values is important (CBMS, 2012, p.58). Such conceptualizations were also held by TEs while placing a line of best fit. Particularly, the TEs with *predictor conception* typically applied the *closest criterion* for line placement. Their reasoning behind the use of such criterion was to minimize the residuals to make predictions as close as possible to actual response values. The only exception was TE2 having the predictor conception by using *the term ellipse* while placing the informal line of best fit for Tasks 5–6-7 that demonstrated weak association. In particular, his approach was similar to the area approach used by TE4. TEs with the predictor conception to some extent showed similarity in criteria for placement with the TEs having the *representer conception*. However, in contrast to the TEs having the *representer conception* who evaluated the best fit line as focusing on the data points at hand, TEs having the *predictor conception* utilized the line of best fit beyond the given dataset with the aim of making predictions for future and unknown independent values.

Previous research has shown that two out of 19 teachers regarded a line of best fit as a signal for the relationship in paired data that eliminates inherent noise (e.g., the variability in the data) (Casey & Wasserman, 2015). Results in this study pointing to a *signal conception* for three out of 11 TEs, including TE2, who used this conception. They interpreted the dataset as a combination of a line of best fit and residuals around it (data = signal + noise), which is useful to construct a bridge between deterministic and stochastic approaches as well as dealing with the causation misconception. In general, they believed that perfect linear association does not occur in real life situations. For them, other factors might cause some deviations from linearity. Therefore, each individual point on a best fit line may have some deviation from the actual data. Hence, they thought to eliminate inherent noise that arose with other factors to deal with relevant and irrelevant information. This aligns with Engel et al. (2008) who stated that variation is the main reason for applying statistical methods with the aim of filtering signals from noisy data. In other words, the aim of the process is to extract a line of best fit while showing that order/stability in the data varied to some extent. For this conception, TEs simultaneously evaluated task conditions based on the data points given in scatterplots, a line of best fit that closely approximated the actual data points, and residuals that emerged with the effect of other variables in the situation. In particular, since they analyzed each individual point and the dataset as a whole concurrently to find a best fit line, evidence for the *representer conception* might also be captured from these statements. In addition, since they aimed to get close approximations to the actual response values and made reference to the functional structure as linear regression, we could say that these TEs also reflected the *predictor conception.* Therefore, *signal conception* might be thought of as embracing the other dominant conceptions of *representer* and *predictor*. Researchers advocated that mathematics teachers and TEs who are responsible for statistical instruction should have the signal meaning (Konold & Pollatsek, 2002). Regarding the line placement criterion, each participant having the *signal conception* predominantly used the *sum deviation*. In particular, the sum deviation criterion focusing on balancing deviations in the response values above and below a best fitting line seems to be congruent with the meaning of signal conception as controlling the inherent variation in the given dataset.

Considering all these conceptions in juxtaposition to each other, while the *representer conception* of an informal line of best fit seems to allow visually focusing on the best representation of the data points particular to the situation at hand, the *predictor conception* allows the utilization of a line of best fit focusing on the functional structure to make predictions beyond the given data set. In addition, the predictor conception allows for examining the dispersion of data points in the original data set to interpret the variability in the predicted data. On the other hand, the *signal conception* allows a focus simultaneously and holistically on the data points particular to the situation, the functional structure, and the residuals. Therefore, we contend that the *signal conception* might reflect and include the ideas of both *representer* and *predictor conceptions*.

As these results suggested that the signal conception seems to reflect and include other conceptions and the results came from interviews with only eleven TEs, further research needs to be conducted to support this hypothesis with a larger number of participants and use of tasks differentiating in context, scatter of data points, and strength of correlation. This is necessary as researchers (Konold & Pollatsek, 2002; Engel & Sedlmeier, 2011) point out that the signal conception is important to understand and appreciate the variation in the analysis of bivariate numerical data. Also, the data suggested a link between conceptions and the criteria used for the placement of the line of best fit. Further research focusing on how students and teachers construct such connections needs to be conducted to better support the development of their conceptualization of an informal line of best fit. Additionally, conducting research in teaching and learning environments could give more insight into understanding teachers' and also TEs' conceptualizations of and their knowledge for teaching a line of best fit.

In addition, organizing statistical education for prospective teachers aiming to reach the signal conception using a stochastic approach to linear regression might contribute to their understanding and interpretation of association and regression. TEs' reasoning process might suggest important implications for planning the teaching and learning process. Finally, using similar tasks utilized in this study to search for an informal line of best fit might promote prospective teachers' and students' understanding of the procedure involved in least squares regression.

Acknowledgement This study is a part of the first author's master's thesis, which was completed under the supervision of Dr. Gülseren Karagöz Akar at Boğaziçi University, Turkey. Opinions and conclusions in this article are those of the authors.

References

Australian Curriculum, Assessment, and Reporting Authority. (2012). *The Australian curriculum: Mathematics, Sydney, Australia: Author.*

Bakker, A. (2004). Reasoning about shape as a pattern in variability. *Statistics Education Research Journal, 3*(2), 64–83. https://doi.org/10.52041/serj.v3i2.552

Ball, D. L., Thames, M. H., & Phelps, G. (2008). Content knowledge for teaching: What makes it special? *Journal of Teacher Education, 59*(5), 389–407. https://doi.org/10.1177/0022487108324554

Batanero, C., & Díaz, C. (2010). Training teachers to teach statistics: What can we learn from research? *Statistique et Enseignement, 1*(1), 5–20.

Batanero, C., Estepa, A., Godino, J. D., & Green, D. R. (1996). Intuitive strategies and preconceptions about association in contingency tables. *Journal for Research in Mathematics Education, 27*, 151–169. https://doi.org/10.2307/749598

Batanero, C., Godino, J., & Estepa, A. (1998). Building the meaning of statistical association through data analysis activities. In A. Olivier & K. Newstead (Eds.), *Proceedings of the 22nd conference of the international group for the psychology of mathematics education* (pp. 221–236). University of Stellenbosch.

Beswick, K., & Goos, M. (2018). Mathematics teacher educator knowledge: What do we know and where to from here? *Journal of Mathematics Teacher Education, 21*(5), 417–427. https://doi.org/10.1007/s10857-018-9416-4

Biehler, R., Frischemeier, D., Reading, C., & Shaughnessy, J. M. (2018). Reasoning about data. In D. Ben-Zvi, K. Makar, & J. Garfield (Eds.), *International handbook of research in statistics education* (pp. 139–192). Springer.

Burrill, G., & Biehler, R. (2011). Fundamental statistical ideas in the school curriculum and in training teachers. In C. Batanero, G. Burrill, & C. Reading (Eds.), *Teaching statistics in school mathematics-challenges for teaching and teacher education* (pp. 57–69). Springer.

Casey, S. A. (2008). *Subject matter knowledge for teaching statistical association*. Doctoral Dissertation, Illinois State University.

Casey, S. A. (2014). Teachers' knowledge of students' conceptions and their development when learning linear regression. In K. Makar, B. de Sousa, & R. Gould (Eds.), *Sustainability in statistics education: Proceedings of the ninth international conference on teaching statistics, Flagstaff*. International Statistical Institute.

Casey, S. A. (2015). Examining student conceptions of covariation: A focus on the line of best fit. *Journal of Statistics Education, 23*(1). https://doi.org/10.1080/10691898.2015.11889722

Casey, S. A., & Wasserman, N. H. (2015). Teachers' knowledge about informal line of best fit. *Statistics Education Research Journal, 14*(1), 8–35. https://doi.org/10.52041/serj.v14i1.267

CBMS. (2012). *Conference board of mathematical sciences*.

Clement, J. (2000). Analysis of clinical interview: Foundations and model viability. In A. E. Kelly & R. A. Lesh (Eds.), *Handbook of research design in mathematics and science education* (pp. 547–589). Lawrence Erlbaum Associates Publishers.

Common Core State Standards Initiative (CCSSI). (2010). *Common core state standards for mathematics*. Author.

Engel, J., & Sedlmeier, P. (2011). Correlation and regression in the training of teachers. In *Teaching statistics in school mathematics-challenges for teaching and teacher education* (pp. 247–258). Springer.

Engel, J., Sedlmeier, P., & Wörn, C. (2008). Modeling scatterplot data and the signal-noise metaphor: Towards statistical literacy for pre-service teachers. In C. Batanero, G. Burrill, C. Reading, & A. Rossman (Eds.), *Proceedings of the ICMI study 18 and IASE round table conference*. International Commission on Mathematics Instruction and International Association for Statistical Education.

Estepa, A., & Sánchez Cobo, F. T. (2001). Empirical research on the understanding of association and implications for the training of researchers. In C. Batanero (Ed.), *Training researchers in the use of statistics* (pp. 37–51). International Association for Statistical Education and International Statistical Institute.

Franklin, C., Kader, G., Mewborn, D., Moreno, J., Peck, R., Perry, M., & Scheaffer, R. (2005). *Guidelines for assessment and instruction in statistics education (GAISE) report*. American Statistical Association.

Garfield, J., & Ben Zvi, D. (2004). Research on statistical literacy, reasoning, and thinking: Issues, challenges, and implications. In D. Ben Zvi & J. Garfield (Eds.), *The challenge of developing statistical literacy, reasoning and thinking* (pp. 397–409). Springer.

Goldin, G. A. (2000). A scientific perspective on structured, task-based interviews in mathematics education research. In A. E. Kelly & R. A. Lesh (Eds.), *Handbook of research design in mathematics and science education*. Lawrence Erlbaum Associates.

Gravetter, F. J., & Wallnau, L. B. (2013). Introduction to regression. In *Statistics for the behavioral sciences* (pp. 558–572). Wadsworth.

Groth, R. E. (2007). Toward a conceptualization of statistical knowledge for teaching. *Journal for Research in Mathematics Education, 38*(5), 427–437.

Groth, R. E. (2013). Characterizing key developmental understandings and pedagogically powerful ideas within a statistical knowledge for teaching framework. *Mathematical Thinking and Learning, 15*(2), 121–145. https://doi.org/10.1080/10986065.2013.770718

Hill, H. C., Ball, D. L., & Schilling, S. G. (2008). Unpacking pedagogical content knowledge: Conceptualizing and measuring teachers' topic-specific knowledge of students. *Journal for Research in Mathematics Education, 39*(4), 372–400. https://doi.org/10.5951/jresematheduc.39.4.0372

Kazak, S., Fujita, T., & Turmo, M. P. (2021). Students' informal statistical inferences through data modeling with a large multivariate dataset. *Mathematical Thinking and Learning, 25*, 23–43. https://doi.org/10.1080/10986065.2021.1922857

Konold, C., & Pollatsek, A. (2002). Data analysis as the search for signals in noisy processes. *Journal for Research in Mathematics Education, 33*(4), 259. https://doi.org/10.2307/749741

Konold, C., & Pollatsek, A. (2004). Conceptualizing an average as a stable feature of a noisy process. In D. Ben Zvi & J. Garfield (Eds.), *The challenge of developing statistical literacy, reasoning and thinking* (pp. 169–199). Springer.

Merriam, S. B., & Tisdell, E. J. (2015). *Qualitative research: A guide to design and implementation*. John Wiley & Sons.

Moritz, J. (2004). Reasoning about covariation. In D. Ben Zvi & J. Garfield (Eds.), *The challenge of developing statistical literacy, reasoning and thinking* (pp. 227–255). Springer.

National Governors Association Center for Best Practices & Council of Chief State School Officers (2010). *Common Core State Standards (Mathematics)*. Authors.

Sorto, M. A., White, A., & Lesser, L. M. (2011). Understanding student attempts to find a line of fit. *Teaching Statistics, 33*(2), 49–52. https://doi.org/10.1111/j.1467-9639.2010.00458.x

Strauss, A., & Corbin, J. M. (1990). *Basics of qualitative research: Grounded theory procedures and techniques*. Sage Publications, Inc.

Qualifications and Curriculum Authority (2007). *The National Curriculum 2007.* Coventry, Author.

Zieffler, A., Garfield, J., Delmas, R., & Reading, C. (2008). A framework to support research on informal inferential reasoning. *Statistics Education Research Journal, 7*(2), 40–58. https://doi.org/10.52041/serj.v7i2.469

Introducing Density Histograms to Grades 10 and 12 Students: Design and Tryout of an Intervention Inspired by Embodied Instrumentation

Lonneke Boels and Anna Shvarts

Abstract Density histograms can bridge the gap between histograms and continuous probability distributions, but research on how to learn and teach them is scarce. In this paper, we explore the learning of density histograms with the research question: How can a sequence of tasks designed from an embodied instrumentation perspective support students' understanding of density histograms? Through a sequence of tasks based on students' notions of area, students reinvented unequal bin widths and density in histograms. The results indicated that students had no difficulty choosing bin widths or using area in a histogram. Nevertheless, reinvention of the vertical density scale required intense teacher intervention suggesting that in future designs, this scale should be modified to align with students' informal notions of area. This study contributes to a new genre of tasks in statistics education based on the design heuristics of embodied instrumentation.

Keywords Density histograms · Design-based research · Embodied design · Statistics education

1 Introduction

Histograms are ubiquitous in newspapers, textbooks and research, as well as in online blogs, vlogs and television broadcasts. In education, histograms are considered as supporting the transition to continuous probability distributions (Wild, 2006), and density histograms are of key importance for this transition (Behar,

L. Boels (✉)
Utrecht University, Freudenthal Institute, Utrecht, The Netherlands

University of Applied Sciences Utrecht, Utrecht, The Netherlands
e-mail: L.b.m.m.boels@uu.nl

A. Shvarts
Utrecht University, Freudenthal Institute, Utrecht, The Netherlands

G. F. Burrill et al. (eds.), *Research on Reasoning with Data and Statistical Thinking: International Perspectives*, Advances in Mathematics Education,
https://doi.org/10.1007/978-3-031-29459-4_14

2021; Derouet & Parzysz, 2016). Research on how to teach density histograms is scarce (Boels et al., 2019a; Reading & Canada, 2011). Besides its role in statistics and probability, density is also important for other topics taught in Grades 10–12:

> The notion of density is commonly met in other matters, such as geography (population density), physics (voluminal mass), biology (cell density), and so on. In statistics, it is implicitly at play in histograms with unequal classes, but this feature remains often implicit or is not taken into account, the more so that spreadsheet [*sic*] cannot produce such histograms and hence can be no help (Parzysz, 2018, p. 70)

In density histograms, density is along the vertical axis and bin widths may vary (Fig. 1). In statistics, density is often given in percentages instead of absolute numbers (e.g., Freedman et al., 1978), but relative frequencies—a number between 0 and 1—can also be used. The latter is always used in probability density functions, such as a normal or Poisson distribution. Density can be calculated as:

$$age\ density = \frac{(relative)\ frequency\ or\ percentage\ of\ people}{age\ group\ period}$$

In density histograms, the area of a bar is proportional to the number of cases in that bar. Consider the example of the number of people per ten-year age groups (Fig. 1). The density along the vertical axis is the number of people in thousands per ten-year age group and is approximately 154,000 for the age group 15–19 or [15–20) in mathematical notation. Since this bin width is half of 10 years, 154,000 must be divided by two to get an estimated number of 77,000 people in this age group.

The aim of the present study is to explore how students can be supported in understanding density histograms. To this end, we designed a sequence of tasks inspired by ideas on embodied instrumentation—an approach that theorizes the role of artifacts such as technological tools, symbols, and graphs in teaching and

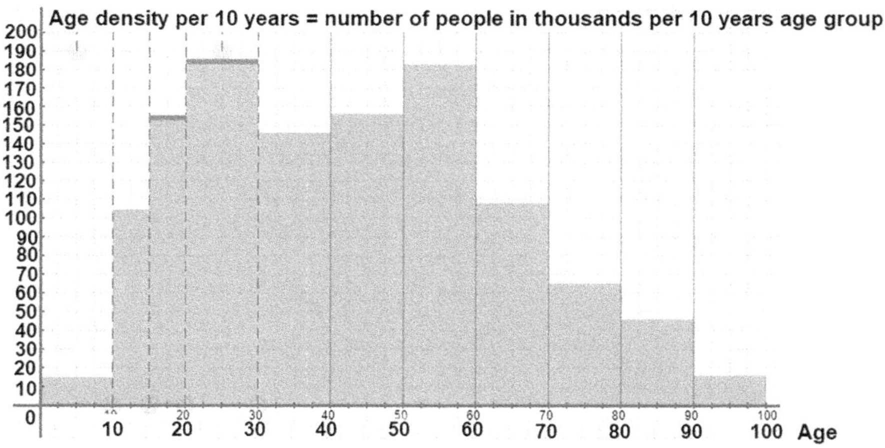

Fig. 1 Example of a density histogram with unequal bin widths. Number of people tested positive for COVID-19 in a country

learning mathematics from an embodied perspective (Drijvers, 2019; Shvarts et al., 2021). It is aligned with our intention to understand how new artifacts such as density histograms can be meaningfully introduced by taking advantage of the opportunities provided by technological environments. Therefore, we investigate the question: How can a sequence of tasks designed from an embodied instrumentation perspective support students' understanding of density histograms?

The learning trajectory is designed for Grades 10–12 pre-university track students (15–18 years old) in the Netherlands. A design research approach is used in a multiple case study with five students. Design research is an approach that aims to both test theories and contribute to further theorization of the field (Bakker, 2018), in our case, theories on how the learning of density histograms can be promoted and students' conceptions of density in histograms can be developed. Suggestions for redesign are an essential part of design research.

We discuss the results of an empirical study of a design on density histograms. Based on these results, we suggest ideas for redesign and elaborate on theoretical outcomes. With this study, we aim to inform educational practice and to add to the literature on how to teach density histograms. In addition, we hope to contribute to further usage and design of tasks from the theoretical perspective of embodied cognition and instrumentation.

When designing tasks, both form and content are relevant. In the succeeding section, we first review the literature on density histograms. Next, we address key elements of theories on embodied cognition and instrumentation.

2 Theoretical Background

2.1 Review of the Literature on Density Histograms

Histograms can support students in understanding key statistical concepts such as distribution. Density histograms are a special type of histogram. Students and teachers alike often misinterpret regular histograms (e.g., Boels et al., 2019a, b, c; Roditi, 2009). Density histograms are considered of key importance for the transition from regular histograms to continuous probability distributions (Behar, 2021; Bakker, 2004). However, the literature on how to teach density histograms is scarce, and density histograms are barely taught (e.g., Derouet & Parzysz, 2016; McGatha et al., 2002). For example, in France, teachers spend on average only 1 hour each year on histograms (Roditi, 2009) and many of them think that teaching histograms is not necessary in mathematics education.

In a French study Derouet and Parzysz (2016) analyzed Grade 10 textbooks on their usage of histograms—including density histograms—as well as how density histograms in Grade 12 textbooks were used as a preparation for the introduction of probability density curves. According to the authors, histograms are not taught in Grade 11. They found several misinterpretations regarding density histograms in the textbooks, including: (1) usage of a histogram with unequal bin widths, no scaling, and the word absolute frequency along the vertical axis instead of frequency per something; (2) using area or nothing along the vertical axis; (3) showing the frequency at the top of a bar in a histogram with unequal bin widths. In addition, French textbooks rarely give students the opportunity to summarize data in a table and construct a density histogram from this table. Furthermore, when a formula is presented to students to calculate density—height of a bar = absolute frequency/bin width—the word density never appears, and this formula is not justified. Derouet and Parzysz (2016) propose a learning trajectory from histograms to probability curves and functions to integrals. They give several suggestions for task design of which the following are taken into account in the trajectory presented here: proportional calculation such as for the heights of bars for unequal bin widths, comparing areas, and choosing and changing bin widths.

In another study aiming to inform future design of a learning trajectory on collecting and analyzing data, seventh graders were assessed on their initial understandings (McGatha et al., 2002). The task was to analyze survey data from 30 students about the number of hours per week they watched television. Students were asked to summarize the data, and most groups used a graph. One group used unequal bin widths (e.g., 11–20 and 21–25) with the number of students on the vertical axis. This seems to have gone unnoticed during class discussion. "Students are primarily concerned with school-taught conventions for drawing graphs", instead of what the graphs signify (p. 348).

University students' textbooks can also cause confusion (Huck, 2016). For example, besides the incorrect definition of histograms as bar charts for which the rectangles touch, a Gilmartin and Rex (2000) textbook concentrates on details of bin widths instead of density and area:

> Most histograms have intervals that are the same size but occasionally you may be asked to draw one with intervals of differing sizes. It is very important that you always check the size of each interval, as the width of each rectangle should correspond to its interval size. (Gilmartin & Rex, 2000, p. 21)

University students in an introductory statistics course were not able to correctly interpret areas in frequency histograms, nor were they able to compute these when a change in intervals at the tails of the graphs made it necessary (Batanero et al., 2004). In another study, graduate students taking a non-compulsory introductory statistics class constructed histograms based on a frequency table but "failed to hold [...] the intervals constant across the x-axis" (Kelly et al., 1997, p. 87).

How density histograms can best be introduced is not clear from the literature we found. Biehler (2007) compares density in a histogram with density in a boxplot. Gratzer and Carpenter (2008) explain to mathematics teachers what a density histogram is and when it should be used. Finzer (2016) had students investigate the effect of changing bin widths on the shape of a histogram with equal bins. Lee and Lee (2014) claimed that density histograms are often overlooked in curricula materials.

Based on experience in classrooms, Lee and Lee expected that many students would not be able to use area to estimate a proportion from a histogram if frequencies were not given, as this does not align with the procedures learned from textbooks.

Taken together, the literature on density histograms suggests that:

- students' understanding of density histograms needs to be developed;
- there is limited research on which to build the current study;
- tasks need to include at least choosing and changing intervals;
- interpretation of area plays a role but might be difficult in histograms;
- tasks need to focus on what a graph signifies rather than on the procedure of drawing a graph or calculating heights of the bars.

Several findings from a review study on students' difficulties with histograms (Boels et al., 2019a) were used in designing tasks for the present study. Two examples are: (1) graphs without context should be avoided (e.g., Cooper & Shore, 2010), so all our tasks have context; (2) as density histograms appear impossible to create in most professional statistical software, we used tailor-made tasks in Numworx ©.[1]

2.2 *Embodied Cognition and Instrumentation*

Our task design is inspired by an embodied view on mathematics teaching and learning. There are different variations of these theories (e.g., Abrahamson et al., 2020), but they share the idea that thinking is based on sensorimotor experiences such as movements, touch, vision, hearing. For example, most people who can ski cannot explain exactly how they do it; their bodies adapt.

As mathematical cognition is inextricably linked to artifacts usage, we also rely on an instrumental approach, which argues that artifacts are a constitutive part of instrumental actions (Artigue, 2007; Trouche, 2014; Verillon & Rabardel, 1995; Vygotsky, 1997). Learning cannot be separated from acquiring the use of a new artifact—skiing without skis makes no sense. Similarly, key concepts in statistics—such as density distribution—cannot be learned without statistical graphs such as histograms (Bakker & Hoffmann, 2005). This requires instrumental genesis, the process in which the user learns to interact with the artifact (e.g., tool, symbol, graph) by which it becomes an instrument for further actions. An instrument here is, therefore, a mixed entity that includes both the artifact and knowing when and how to use it—in order to ski, one puts down the skis, clicks the boots into the bindings, and heads down the hill. In addition, instrumental genesis also involves selecting or even designing suitable artifacts. Think of choosing snowshoes that have a large footprint for easier walking in snow or using flippers to swim faster. Similarly, histograms need to be selected or designed and further used in a mathematically appropriate manner.

Core to the *embodied* instrumentation approach is that learning is seen as the development of a body-artifacts functional system (Shvarts et al., 2021). In such a system, there is no homunculus inside of the head who regulates movements step-by-step. Instead, new forms of instrumental actions emerge when using skis to learn

[1] https://www.numworx.nl/en/

to ski down a hill: the learner barely can say how exactly the skies are used, yet a body can manage it well; the body and the skis form a body-artifact functional system. In an embodied instrumentation approach, statistical understanding emerges from *intentional* bodily actions—solving problems through multimodal sensorimotor interactions with the environment that includes artifacts—and reflection on those interactions. We briefly outline the main theoretical statements of embodied instrumentation and explicate the six design heuristics driven by those statements (e.g., Boels et al., 2022b). First, every human activity unfolds in response to some problem in the environment (Wilson & Golonka, 2013). For example, even descending a small hill on skis initially is a motor problem (Bernstein, 1940/1967) for which a new bodily functionality develops: a new functional dynamic system of skiing. Similarly, learning mathematics can be reconsidered as solving motor and perceptual problems as a learner learns to move in a mathematical way and see the world mathematically (Abrahamson, 2019; Abrahamson & Sánchez-García, 2016; Radford, 2010). The design heuristic based on this principle is to provide students with a set of problems with which they *productively struggle* while actively seeking a solution.

Second, a particularity of a functional system is that it never has a strict answer to the problem but rather exhibits an emergent, self-organized behavior that allows for flexibility across environments while still being able to achieve the target state (Bernstein, 1967). This development is fostered by presenting tasks in different environments and contexts—digital and on paper (*transfer tasks*).

Third, cognition develops in an interaction between humans and the environment in response to an underlying problem (Varela et al., 1991). This interaction can be traced as a system of sensorimotor processes and perceptual action loops, such as noticing a bump on a ski slope and going around it. Another example is projecting a height of a bar in a graph onto the vertical axis to assess its height (Boels et al., 2022a). For this reason, we provided students with an environment that allowed for sensorimotor processes and direct mathematical actions with mathematical objects, rather than having students manipulate sliders or enter numbers and let the digital environment perform the mathematical actions. Therefore, *students perform these mathematical actions*.

Fourth, while in solving a motor problem (skiing down a hill), a new movement is enough, in learning mathematics, students also need to reflect on their performance and articulate how it was done (Abrahamson et al., 2020; Alberto et al., 2022). To make sure that students go beyond manipulation and *reflect* on their enactment, we supplemented all tasks with reflective questions requiring students to explain why such a result is achieved.

Finally, mathematical artifacts are an essential part of mathematical thinking and understanding. They are part of students' perception-action loops for mathematical problem solving (Shvarts et al., 2021). We consider artifacts as crystallized histories of mathematics, hence, as efficient reification of cultural practices (Leontyev, 2009; Radford, 2003). When students encounter those artifacts in their learning, the artifacts need to be embedded into their body-functional systems. But when the artifacts are ready-made, they can be embedded without understanding, as we may never think about why skies have a particular shape or how a smartphone functions. Yet, understanding mathematical artifacts requires students to *reinvent* those artifacts by their own actions: artifacts need to become crystallizations of meaningful procedures, thus come to reify actions (Shvarts et al., 2022). For example, as students try to visualize the frequency for different outcomes, they might reinvent histograms. This last theoretical statement unfolded for us in two design heuristics. First, we reflected on what actions formed the *target artifact*—density histograms. Second, we *melted—decomposed—this artifact back into actions* (Shvarts & Alberto, 2021), the hypothetical source of its historical development.

When applying this last design heuristic to density histograms, we came to an overview of the main actions with previously acquired artifacts that led to the constitution of density histograms according to our logical-historical reconstruction (Fig. 2). We made sure that the mathematical *problems we exposed students to promote those actions as solutions*, thereby allowing *reinvention* by the students of the target artifact. In summary, our design heuristics are: (1) distinguish the target knowledge and melt this back to its constituting actions (2) create motor-control or perception problems in which students reinvent artifacts (3) create productive struggle, (4) have students perform the (digital) actions, (5) include reflection, (6) create

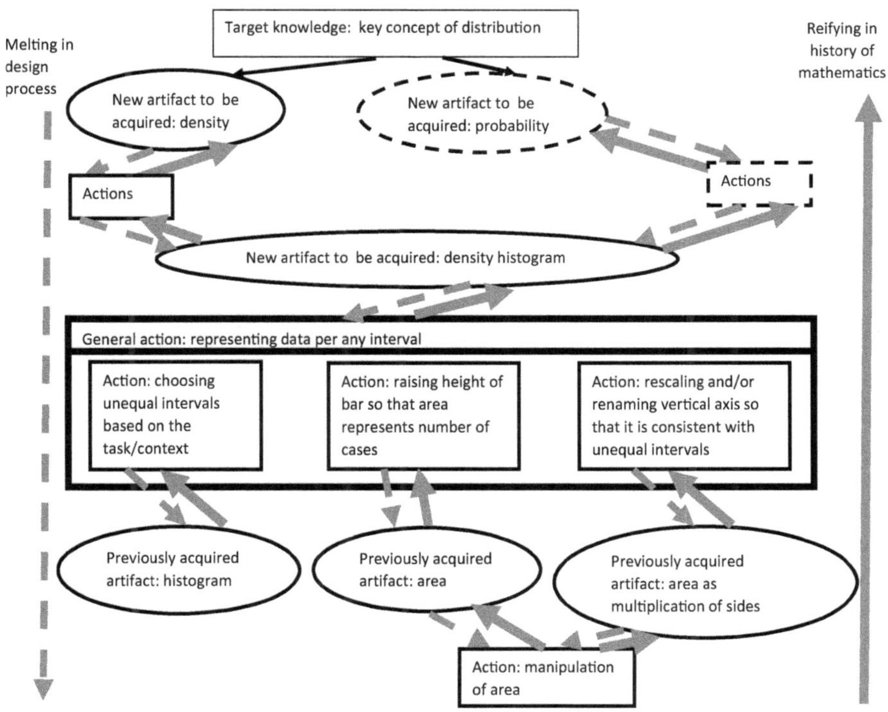

Fig. 2 Melting (decomposing) artifacts to actions that constituted them in the history of mathematics. The tasks in our design concentrate on the general action of representing data per any interval

possibilities for transfer. Those heuristics allowed us to develop a technological educational environment and a sequence of statistical problems. Aiming to solve these problems by sensory-motor interaction within the environment, the students would reinvent the target action and further target artifacts in their own sensory-motor activity of problem solving.

In this paper, we focus on the target action of *presenting data per any interval* to reinvent a density histogram artifact. Previously acquired artifacts are not part of our design since we are building on previous education. For example, in primary education, students worked with area and we assumed they acquired it as an artifact, and in the learning trajectory preceding the present one, students acquired knowledge of a histogram.

3 Materials and Method

The present study represents one part of a larger design research study, which showed that students could successfully reinvent the role of the horizontal and vertical scale in histograms and reinvent regular histograms (Boels et al., 2022b).[2] Some reflection tasks belonging to the digital tasks were presented in the lesson materials on paper. For all lesson materials (in Dutch), readers can contact the first author. Furthermore, the italic words in the task description refer to the design heuristics discussed in the theoretical background with numbering corresponding to the summary at the end of that section.

3.1 Reinventing Bin Widths

In this task, students completed a histogram (Fig. 3) by adding bars between 10 and 30 years using the given frequency table. The age group 15–19 years contains all people from the age of 15 up to the day before they turn 20. The aim of this task was that students should perceive that bin width is a choice, see Fig. 2, in line with our design heuristic to *distinguish the target knowledge and melt this back to its constituting actions (1)*. By the design of this task, students were given the choice of bin widths of 5 or 10 years for the age group 5–29 years old or [5, 30) although even bin widths of 15, 20 or 25 were possible—in line with our design heuristic to *create a perception problem in which students reinvent artifacts (2)*. When choosing bin widths of 5 years, a *productive struggle (3)* could occur because this age group is unlikely to contain all students from a 10-year age group. *Reflection (5)* was stimulated by questions posed in the lesson material on paper, such as: "How can you

[2] All digital tasks can be found here: https://embodieddesign.sites.uu.nl/activity/histograms/

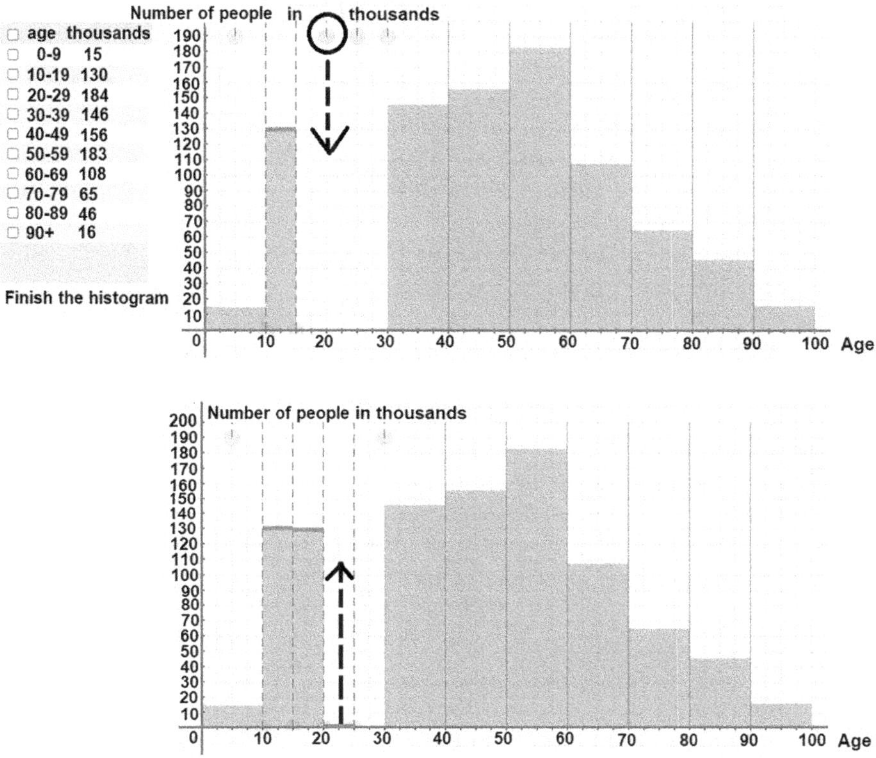

Fig. 3 Screenshots for completing a histogram from a frequency table. Students draw down sliders—indicated by the black circle (top)—after which a strip appears on the bottom of the bin. Next, they pull up this strip to the desired height (bottom). Black circle, arrows, translation, and thicker numbers added for readers convenience

check by yourself if you have performed this task correctly? Why? Perform this check."

What is new in the design of this and the next task is that students moved their hands in a specific way to *create (4)* things happening, such as raising a strip up to build a histogram instead of selecting the correct option in the software and seeing things happening or—even more common—the result of the hidden action. In a technological environment, we can constrain the possibilities and, therefore, provide better support for students so that their *struggle* remains *productive (3).*

The role of area comes into play when students compared the bars to the previous task. In this previous task, not reported here, students created a regular histogram with 10-year wide bars from the same frequency table as in the present task. In this table, the number of people per five years is not given while in the present task—by design—students were invited to make bars per five years. The total area of two bars together—for example, for the 10–19 years age group—is as large as the area of the 10-year wide bar in the previous task.

3.2 Reinventing Density in a Histogram

The aim of this task was that students *reinvent (2)* density in a histogram. Students needed to *perform two actions (4)*: raise the height of the bar so the area represents the number of cases and change what is along the vertical axis so it is consistent with the unequal bin widths—in line with the design heuristic *distinguishing the target knowledge and melt this back to actions (1)*. The finished previous task was intended to reappear for this task, but this did not technically work. Therefore, students first needed to complete the histogram once again (see Fig. 3).

Next, students were given the context of a school principal who wants to know if different COVID-protection measures should be taken for the age groups 10–14 and 15–19 years. For this, new numbers were reported for these specific age groups: 51,903 and 77,820. Students were asked to split the bar for the 10–19 year age group into two, proportional to the number of students in the 5-year age groups, in such a way that the total area of the histogram does not change (Fig. 4). The intention of this split was, in combination with the question that follows, to *create a productive struggle (3)* and *induce reflection (5)*. The first question was what now is represented along the vertical axis. Students could choose the following options that were given on paper: "Number of people in thousands, or this per five-year age group, per one-year age group, or per ten-year age group". The next question was: "Why?"

The last questions concerned what to advise the school principal about different COVID-measures per age group and what the student discovered about a histogram with unequal bin widths. Note that in this task students do not yet work with density

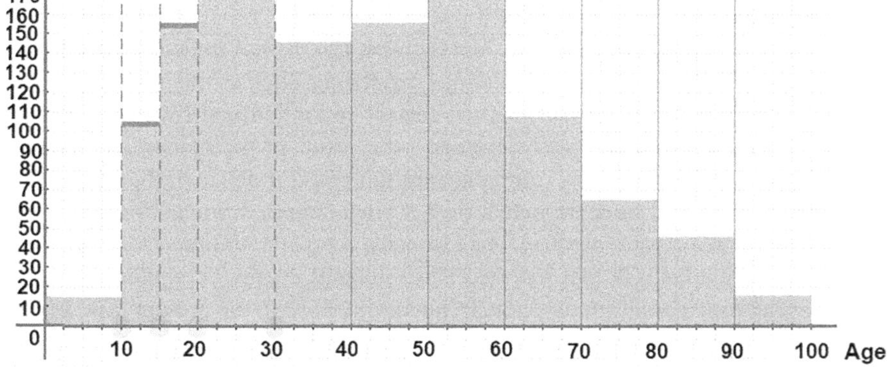

Fig. 4 Completion of the histogram after proportional splitting. Thicker numbers added for convenience of the reader

in percentages or relative frequencies but instead use people per age group to avoid possible difficulties with proportions or percentages.

3.3 *Confirming learning: Transfer*

Between the reinventing density task and the present task was a transfer task not reported here that involved the construction of a regular histogram on paper based on a frequency table in a realistic context. It *created possibilities for transfer* (6) as both the environment—paper—and the context were new. This context was reused in the present paper task, but now a histogram was provided with more fine-grained income groups. In the present task, we decided to explore students' notions of area in a histogram by providing students with a histogram without frequencies listed (Fig. 5) as suggested by Lee (1999). As there is no label on the vertical axis, students could solve this task by reasoning with only the area of the given histogram. This was new to them and, therefore, also a transfer task. The context and first question in this task was:

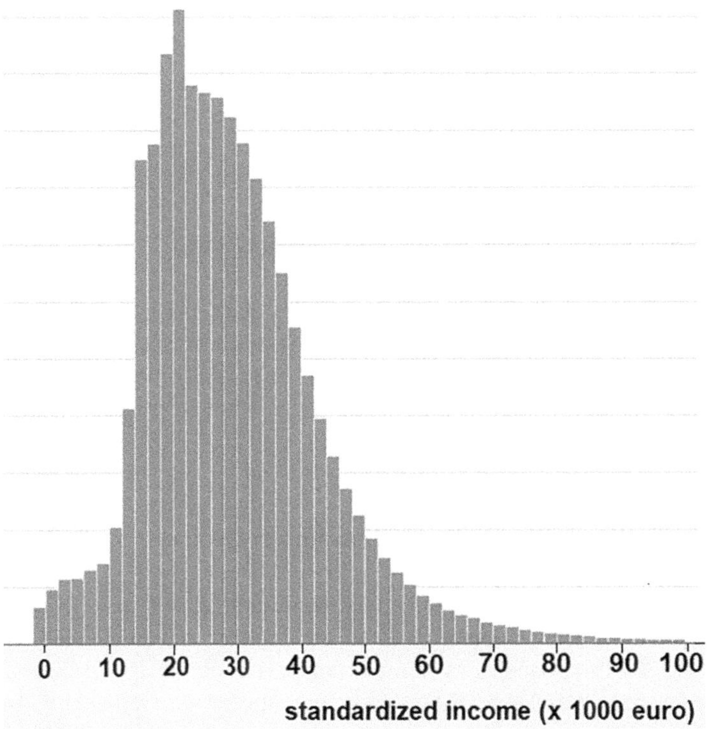

Fig. 5 Distribution of net annual income [in thousands of euros] in the Netherlands. Thick numbers and text added for convenience of the readers. (Figure source: Statistics Netherlands (CBS))

> In reality, the distribution [of annual income, Figure 5] is much more skewed than you might think given the histogram from the preceding task [task 18]. About what percentage of households has less than 27000 euros of disposable income per year? Explain how you arrived at your answer.

3.4 Method

3.4.1 Participants

The materials were tested with six pre-university Grades 10 and 12 students over two mornings. Each morning, two students worked in a pair and one student worked alone. One alone working student was excluded from the data analysis, as special educational needs made it impossible for this student to work with the materials. The five remaining students were 15–18 years old, two males and three females. They all took mathematics A, which stands for applied mathematics and includes an introduction to statistics. Students were recruited from different schools in Utrecht. The trajectory presented here took roughly 40–50 minutes. A 30-euro fee was given to the students for their participation, and approval from the Science-Geosciences Ethics Review Board and consent from the participants and legal representatives—if necessary—was obtained. Age, grade, math level and mathematics mark were collected. Furthermore, students' pre knowledge was tested with a questionnaire. Discussions between students and with the teacher-researcher (first author) and observer (second author) were audio- and videotaped. Students' grid papers and writings on lesson materials were also collected.

3.4.2 Intervention in the Teaching and Learning Lab

Students worked in the Teaching and Learning Lab (TLL) of the Freudenthal Institute at the Utrecht University. Lesson materials were partly on paper but the embodied design tasks were on a digital whiteboard, which was connected to a laptop. All students finished almost all 22 tasks of the whole learning trajectory as well as the questionnaire on their pre knowledge.

3.4.3 Data Analysis—Conjectured Learning Trajectory

For the data analysis, we used a conjectured learning trajectory (e.g., Simon & Tzur, 2004) which starts after students used messy dotplots and reinvented histograms (Boels et al., 2022b). Sequence is an important part of our design (Bakker, 2018). In the present learning trajectory (Table 1), students started with reinventing bin widths while completing a histogram from a given frequency table (task 16) and reinvented a density histogram from the same data in task 17. Furthermore, we added a task (19) on paper.

Table 1 Summary of the conjectured learning trajectory

Step	Task	Student activities	Conjecture
1 Reinventing bins	Task 16 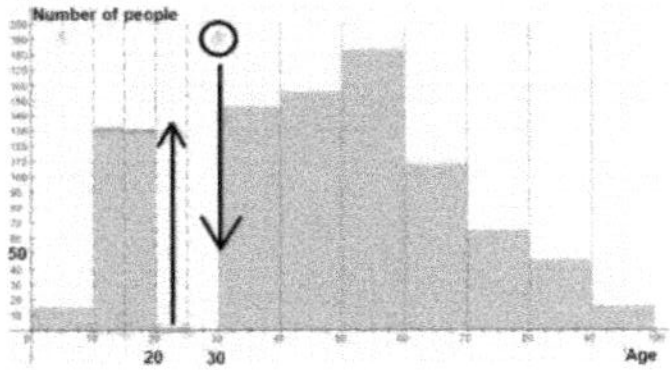	Students choose bin widths by deciding which slider to pull down.	H16: By dragging down (dotted) lines which create bins, students perceive that bin width is a choice.
2 Reinventing density histogram	Task 17 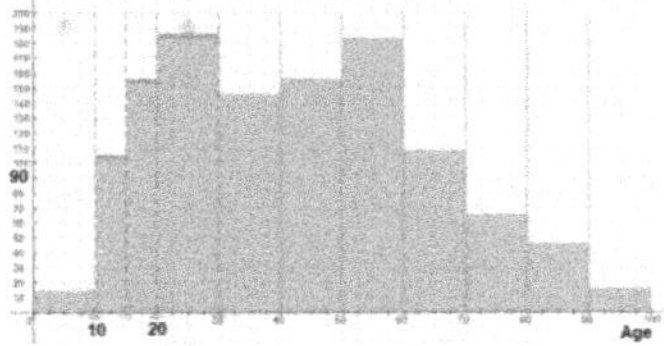	By splitting one bin into two bins, students need to rethink what the vertical axis is depicting	H17a: By keeping the total area of two smaller (5 years) bins equal to the area of the former larger (10 years) bin, students discover that the vertical axis in a density histogram does not depict frequencies. H17b: By keeping the total area of two smaller bins equal to the area of the former larger bin, students question what the vertical axis depicts and, in this way reinvent a notion of density (e.g., number of people per 5 years) in a histogram with unequal bin widths.
3 Confirming learning: transfer task	Task 19 (Data: CBS) 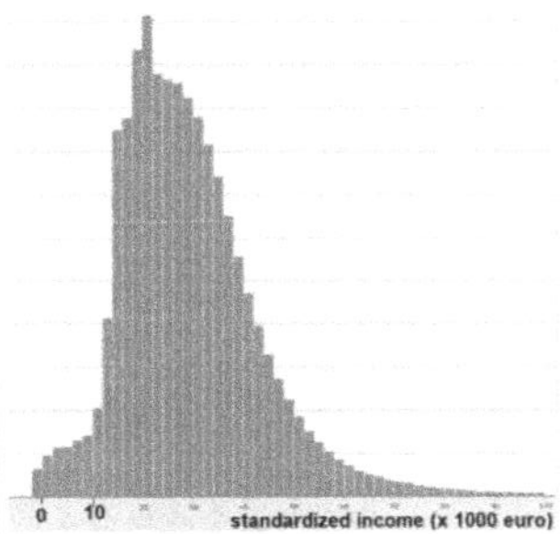	Students estimate percentages from a total area	H19a: By estimating the percentage of a population in a more fine-grained version of a previously drawn graph, students notice that the total area of all bars represents 100% of the data points and that this can be used to estimate the percentage for a proportion of the total area (i.e., the population). H19b: In addition, students perceive that it is impossible to give an exact value for the percentage or proportion of people if a measured value is cutting a bar into two portions.

4 Results

4.1 Reinventing Unequal Bin Widths Task 16

Most students had no difficulties with the task itself. However, they did not seem to reinvent bin widths. Students S1, S2, S4 and S5 worked together in two pairs and created bin widths of 5 years age groups (Fig. 6). After creating these bins by pulling all sliders down, they discussed how to pull the strips up:

S5: This is from 10 to … [age; horizontal axis]
S4: 130 [the number in thousands; vertical axis]
S5: but what is the age? 10 to 19
S4: yes
S5: but that's both of these [both bars, namely from 10–14 and from 15–19]
S4: yes, you have to slide both of those then. To 130.
S5: ok. Oh no, not this one, this one. Come on guys. 130
S4: yes
S4: [long pause]. Well! [Sounds little bit frustrated] I want to be exact [bars' heights do not align perfectly due to some technicalities in the software]
S5: No, [inaudible] don't have to look perfect ha ha [laughs a bit]

Student S3 created bins of 10 years. Although this student did not encounter any difficulties in completing the histogram, he also easily stepped over the possibility of creating smaller bins. The students that used smaller bins did not struggle with it. They all almost immediately pulled up both smaller bars for 5 years age groups to the number that belonged to the corresponding 10 years age group. Taken together, the written materials and videos do not provide evidence that supports the conjecture (H16) that students perceive bin widths as a choice. Instead, their notions of keeping the area for two bars together the same as for one wider bar seemed to be so strong that they do not question the height of the bar.

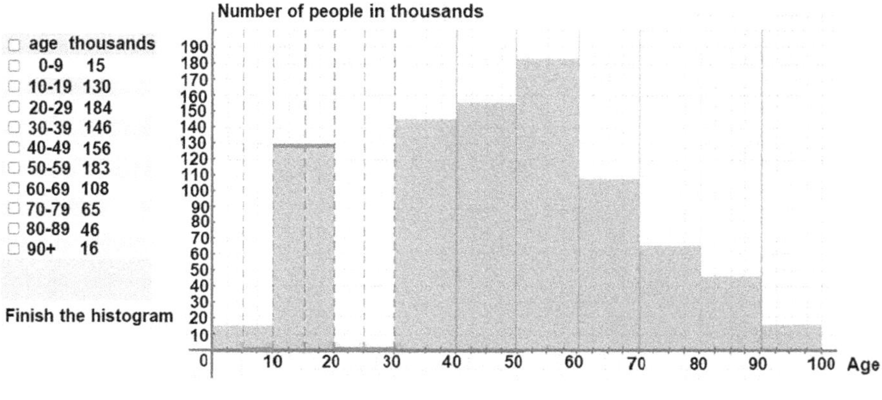

Fig. 6 Students S4 and S5 created bins of 5 year at the start of task 16

4.2 *Reinventing Density Histogram Task 17*

Again, students S1, S2, S4 and S5 dragged down all sliders, thus creating five bins each of 5-year widths. S3 used two bins of 10-year widths. Task 17 caused a lot of struggles. An example of S1 and S2 who worked together is below. A long intervention by the teacher was needed to have them reinvent the density scale along the vertical axis, although the first intuition of student S2 when questioned by the teacher was correct (Fig. 7, left):

S1: Yes. I would just set it to 50. [...] [the second bar is dragged down to approx. 78][...] [reads task aloud. Then the teacher intervenes.]

T: Yes, but you haven't quite done the task yet, because it also says split the bar for this age group into two bars, and in such a way, way that the **total area** [stresses these words] of the histogram doesn't change.

S1: hm?

S2: Um. So basically, we need to have these two... So [...] you have to double or something? [moves left and right bar up to double height ...]

S2: no, but wait. Look, look, look. A moment ago we had from here to there [moves mouse from left to right in the middle of the two bars around the height of 130] 130 and now this average is also 130, right?

S1: yes, but then the numbers don't match the number of infections.

S2: uhm.

S1: because with the rest you don't have to divide by two and with that one bar you do. That would be weird. That doesn't match anyway.

What equal area means seemed clear to students S1 and S2. When the teacher continued asking questions about what the vertical scale now represented, these students became very confused and dragged the sliders back to the position they had first chosen (see Fig. 7, right).

T: If you look at the next question that's there: what variable is along the vertical axis? On this paper there are four possibilities. What does it say now do you think? [Pause]

S1: the number of people. [Pause] Right?

T: so in the age group of 10–14 there are, 100 what is it, 103 thousand students

S2: um, no. [pause].

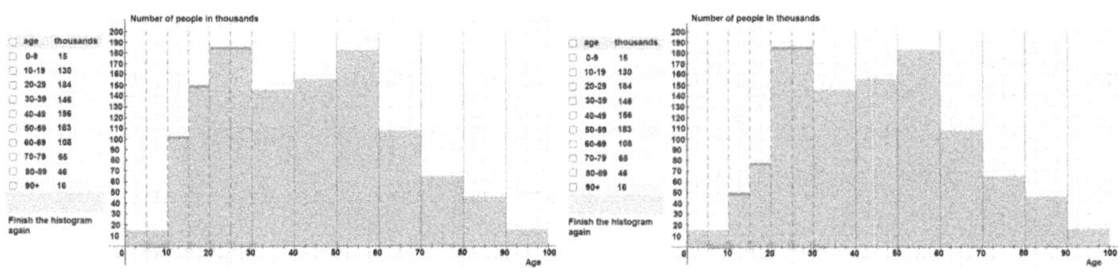

Fig. 7 Correct heights of bars for five-year age groups by students S1 and S2 (left) and their modification (halving) when asked about the scaling (right)

T: ok, what about then?
S2: Yes, even [equal] in terms of percentage but then the um… Hmm. [takes the mouse and lowers both bars back to their halves]. Then it changes, the whole thing.
T: the area.

Meanwhile, these students seemed to understand that area represented all people.

T: but we just agreed that this area wasn't right [points to the board with the two bars that were too short because halved].
S2: no.
T: and in a histogram, no matter how you make the layout, the amount of people should
S1: remain the same
T: yes [gestures with two hands and points with them to the whole area of the histogram]. The total area is what percent of the people?
S1: 100%
T: so that total area, yes we didn't change anything about the total number of people. [pause]
S1: but then shouldn't you just divide everything by two?
S2: um, no.
T: and if you were to do that, what do you get?
S1: well, then you get the same number of people, don't you? […]
S2: but then the total area of the histogram does change?

Building on the halving bars, they had another good idea that was not possible in this environment: to split all bars and then rescale the vertical axis.

S1: Oh, no, wouldn't you? [Points to the board. Sounds engaged] Eh. Wait. If you now because we've now done 1 bar, divided by 2 and split that. If you split each bar into two… if you split each bar,
T: yes
S1: then it's right.
S2: but we don't know how to split it, do we?
S1: yes, because you know the number of people, then you just can
S2: oh, you just divide it into two.

Students then discussed whether this would change the area:

S2: but then your total area changes, right?
S1: your area does but you still have the same number of people. The number of people is still 100%…. Right? […]
T: So the number of people is still 100% then. So I, I um, I would think that's a very good idea. We haven't made that possible. And, what have. If you think on that for a moment, huh, what you just said,
S1: yes
T: so you're actually going to do every class [bin] by half. […] then in one class you have the number of people per … what?
S2: five years.

After about 8 minutes of discussion, students seemed almost there. However, the system did not allow halving all bars both vertically and horizontally. Therefore, the heights of the 5 year bars needed to be adjusted.

T: Yes. But I'd like to know now what to do if I have a width of 10 for some and a width of 5 for others. [Pause] And we agree that the group of 30–40 that that's correct, right?
S2: hmhm [confirming] [...] but I had just now done this number times two.
T: yes
S2: only... But then why wasn't that correct?
T: who said it wasn't right?
S2: [indignant, louder] I had that just now!!! [...]
T: But you started adjusting it yourselves.
S2: Oh, hmpf. [moves the bars up again to the correct position, taking the ratios of the bars into account] [...] Well, now you have the same...now it's the same area again. Only...the numbers don't match. Right?
T: yes, you have the same area
S2: on average. Average that would be, but average would make 130 [...] But say what is wrong with this graph now?
S1: The area
S2: Yes [she means: no] it is [correct], the area that is correct. [...]
T: you just agreed on that. So the area is correct.[...] And then with these numbers along the vertical axis. What does it say? Because that was the point where you adjusted it. When I asked that question. Because then you said: number of people. But if it's the number of people, you have a problem. Because from 10 to 15 there's no ... 100 ... what is it? what did you put it at?.... 104 people. 104 thousand
S2: but this is just going to be uhmm, number of people per 1000 ... and then age group of ... let's see [pause] [...] of ... 10 years [louder]. C seems to me. Because in steps of ... Oh no, wait. [...] then I think five years makes the most sense because with that you can read the other ones too
T: hmhm. So from 30 to 40 you have 146 thousand every five years.
S2: [pause] No... [hesitantly]
S1: no, you have to divide those by two then anyway

Note that student S2 gave the correct answer (C) but then changed her mind. It then took another 8 minutes to get the students back on track. The teacher tried different approaches, building on the students' ideas. S2, for example, moved the mouse horizontally at 104 thousand and 130 thousand (the mean for the 10–14 and 15–19 years of age group together) but when the teacher tried to build on that idea, the students got confused. The teacher then decided to apply the same idea on a bar that had not been split: the 40–50 years of age group.

T: Then we're going to apply it to this what you just said. Here we had? [draws the horizontal line segment at the top of the bar, see Fig. 8]
S2: 156

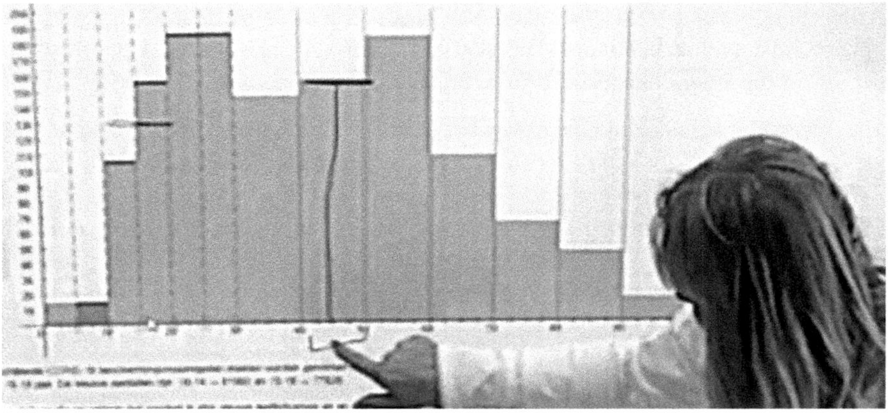

Fig. 8 The teacher halved a bar in line with an idea from a student

T: 156 thousand. And now we're going to split it in mind [draws a vertical line that splits the bar].

S1 and S2: Yes.

T: both the same amount. Then how many are in this one here? [left bar in split bar]

S1: half of 156 thousand [...]

T: only what does it say? Then what does the 156 thousand mean? Because there are no 156 thousand in here [points at left bar in the vertically split bar] you just said [waves at S1]. You said: there is only half in it.

S1: yes that's the total actually of those [...] of those ten

T: yes, of that age group of [draws a tray under 40 to 50 years, Fig. 8]

S1: [of] ten years, of five years

T: of ten years, you said. Yes. So this number [points at the top of the bar at 156 thousand] belongs to the age group of ten years. [...]

T: so what is now along the vertical axis then?

S1: the total of the 10-year age group?

T: yes, and if you have to choose from the answers that are there? [on paper] [...]

S1: the number of people by age group of ten years. [...]

T: and why is that?

S2: because you read it off by ten years.

T: yes. So even though you then make bins that are smaller, you still have to keep reading it as if it were per ten years.

In the discussion with students S1 and S2, it also became clear that they wanted to change the scale and halve all other bars, too. Therefore, on the second day, task 17 was available on paper for the students throughout the teacher-students-intervention. Moreover, students S3, S4 and S5 joined together in front of one whiteboard during this teacher intervention. Student S3 became engaged and used

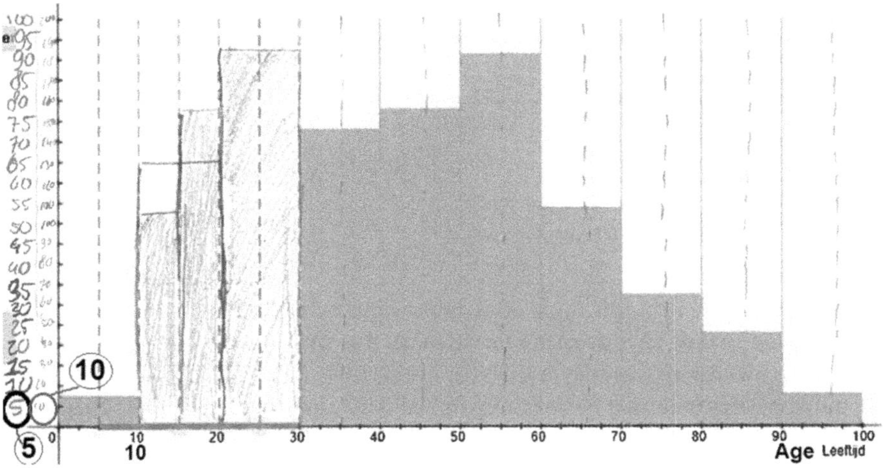

Fig. 9 Student's S3 new scaling for the vertical axis (left: 5 instead of right: 10). Circles and thicker numbers added for convenience of the reader

the paper. He first wrote down the numbers from the digital task and then halved all numbers along the vertical axis, as well as all bars in the density histogram (see vertical dotted lines in Fig. 9). He wrote about this scale:

> As a result, you are looking at groups of 5 years. With a modified scale (halve the bars), then you halve the numbers [along the vertical scale]. By also halving the groups [all other bars], the ratio is correct again. With the groups of 10–20 (etc.), everything is the same, with the groups where it is in 5 years, it also fits proportionally.

For the question regarding which variable the vertical axis is now representing, students could choose from four options given on paper (see Materials section). S3 and S4 chose option A ("per five year age groups") in line with the new scaling above. Student S5 chose an incorrect option D ("Number of people"). Students S1 and S2 chose option C ("per ten year age groups"), which is in line with the scale given in the digital environment. Taken all together, the results suggest that four students fulfilled our expectation formulated in H17a. Given their answers on the written materials, there are also some indications that students started to grasp the idea of density (H17b). Nevertheless, as can also be inferred from the excerpts of the transcripts, these ideas are still fragile and not yet well articulated.

Transfer Task 19. S1, S2 and S4 answered "50%" to the question given on paper for this task: "About what percentage of households has less than 27,000 euros of disposable income per year?" S3 drew a vertical line and braces along the horizontal axis but did not come up with a percentage. S5, who worked together with S4, answered "About 60% large part is below 27000 euros." Four out of five students did not seem to have any trouble with this task. For these four students, the results seem to support conjecture H19a. As 27,000 euros is in the middle of a bar and all students but S3 seemed to have no problems with this, H19b is likely to be supported as well. However, as we did not explicitly address this in either the task or in the discussion, we cannot be sure.

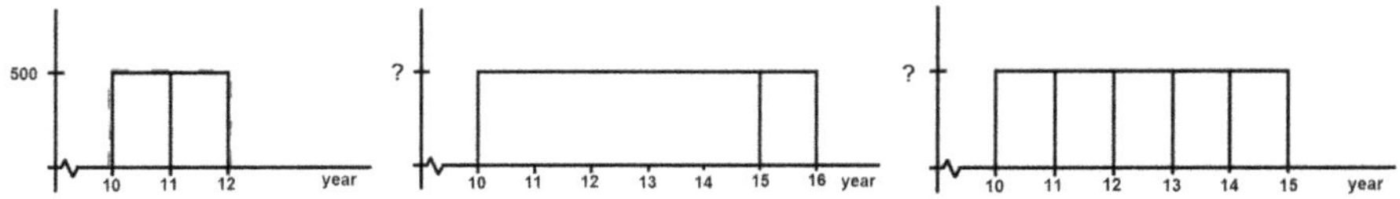

Fig. 10 Ideas for area tasks in preparation for tasks 16 or 17

5 Ideas for Redesign

In retrospect, task 17 needs to be done in previous activities. First, we could add a check button to task 16 to reassure the students that their initial construction is correct and to make sure that they use 5 year bin widths. The students' visual intuition about the histogram might be so strong they did not question the smaller bins at all. Therefore, we would ask them how many students there are in the 10–19 years age group. Next, we would ask how many there are in the 10–14 years group. We expect this would trigger the intuition that it is half in the latter, although we cannot be sure.

Additionally, we would redesign task 17 in such a way that students need to join two bars instead of splitting one. Area as a multiplication might be an easier access to density than the division we used. Moreover, the area task 19 might have been a good preparation for density histograms and histograms with unequal bin widths. In a next cycle, we would, therefore, consider either constructing a digital area task and/or giving this task earlier in the trajectory.

Furthermore, before task 16, we suggest introducing simpler tasks using area and number of people (Fig. 10). Possible information and questions are as follows: First, there are 500 students in the [10, 11) year age range (left). "What is along the vertical axis? How many people are in the [11, 12) year age range?" Both bars are now combined into one bar (dashed line). "How many students are together in that new, dashed bar? What is the 500 along the vertical axis now depicting?" Second, together there are 1000 students in the [10, 15) year age range (middle). "How many students are there in the [15, 16) year age range? What number should be along the vertical axis at the question mark?" Third, together there are 1000 students in the [10, 15) year age range, equally spread across each year (right). "How many students are in each year? What number should be along the vertical axis at the question mark? What is this number depicting?"

6 Conclusions and Discussion

Theoretical analysis of an embodied instrumentation approach led to the six design heuristics described in Sect. 2.2 that would facilitate conceptual understanding. Using these six heuristics we designed three tasks for supporting students' understanding of density histograms. We used these design heuristics to unravel the statistical key concept distribution into artifacts that need to be understood (here: density histogram). Next, we melted this artifact back to actions that historically constituted this artifact, such as choosing unequal bin widths, raising the height of a bar so that the area of that bar represents the number of cases, and (re)scaling or renaming the vertical axis so that it is consistent with the unequal bin widths.

The research question was: How can a sequence of tasks designed from an embodied instrumentation perspective support students' understanding of density histograms? We found it was important to include tasks targeting students' notions of area, and tasks in which students can reinvent unequal bin widths and density in a histogram. Both halving bin widths without changing the ratios of bins and using area of bars to estimate a proportion of the total area in a histogram without a vertical scale seemed to have worked well in our multiple-case study. Reinventing how to proportionally split a bar into two and then determine what is along the vertical scale in a density histogram required intense teacher intervention.

After the teacher's intervention, students started have a tentative sense of what density is. It might be good to introduce a density scale in simpler situations such as with only two bars and round numbers to better align with students' notions of areas. In addition, having students combine bars might align better their notions than the splitting bars approach we took. As students wanted to change the vertical scale when reinventing density, the second day task (17) was available on paper for students and used by them in their discussions.

A limitation of this study is that we worked with only five students in a laboratory setting. Therefore, further research is needed to explore how this design works for other students and in classrooms. To facilitate use in classrooms the tasks are freely available on the web and require a computer with a mouse only. Introduction in a classroom is a big step and will require (alongside resources) attending to how teachers can be trained to work with embodied pedagogies and tutoring multiple students at the same time (Abrahamson et al., 2021).

As our study is in an early stage of design, our empirical tryout revealed directions for further development rather than an end product. A contribution of our work is first that it provides early researchers insight into the step of rethinking the design based on empirical results. Further, in contrast to statements of Lee (1999), four out of five students seemed to have no trouble understanding that the whole area in a histogram is 100% and that area in a histogram can be used for estimating a proportion of a population.

In addition, reporting on not-so-successful parts of the design is important as published research tends to be biased toward successful implementations. As design research is becoming more and more accepted as a method, it is important to describe how to overcome difficulties in the design based on empirical tryouts of the design. This includes rethinking what actions students need to perform and how theory can be put into practice, paying close attention to the step of rethinking the design. In this way we hope to contribute to further application of design research methods in the field of statistics and mathematics education.

In discussing how our work fits into a chain of other works, we built on artifacts acquired in primary education such as area (e.g., constant area task, Shvarts, 2017) and artifacts acquired in the learning trajectory that preceded this learning trajectory such as histograms (Boels et al., 2022b). As a possible follow-up to these tasks, an embodied design for probability (Abrahamson, 2009; Abrahamson et al., 2020) or a design preparing for probability density (Van Dijke-Droogers, 2021) can be considered. Van Dijke-Droogers' intervention was not developed from an embodied design framework but nevertheless shares some ideas and similarities with the design of Abrahamson and colleagues.

Our six design heuristics can be applied to other key concepts and artifacts within and outside statistics education. Particularly new is that we decompose a key concept—distribution—into artifacts (e.g., density histogram) and further melt these down into actions that have constituted them in the history of mathematics (e.g., manipulate area). The six design heuristics open a new way for task design. We, therefore, hope that our study will contribute to a new genre of tasks in statistics education based on the heuristics of embodied instrumentation designs.

Our study is also important for teachers and designers of tasks and textbooks, as the literature on how to teach and learn density histograms is scarce. By describing in detail what actions the teacher took to support students, we hope to provide insight into students' thinking and how teaching of density can be done. Our description of tasks and suggestions for redesign aim to contribute to further development of tasks that advance students' learning of the key concept density as well.

Acknowledgements The authors thank Arthur Bakker and Paul Drijvers for their comments on the design of tasks and the conjectures, Dani Ben-Zvi for providing some literature, and Nathalie Kuijpers and Ciera Lamb for checking the English.

References

Abrahamson, D. (2009). Embodied design: Constructing means for constructing meaning. *Educational Studies in Mathematics, 70*, 20–47. https://doi-org.proxy.library.uu.nl/10.1007/s10649-008-9137-1

Abrahamson, D. (2019). A new world: Educational research on the sensorimotor roots of mathematical reasoning. In A. Shvarts (Ed.), *Proceedings of the annual meeting of the Russian chapter of the International Group for the Psychology of Mathematics Education (PME) & Yandex* (pp. 48–68). HSE Publishing House. https://www.igpme.org/wp-content/uploads/2020/01/PMEYandex2019Final.pdf

Abrahamson, D., & Sánchez-García, R. (2016). Learning is moving in new ways: The ecological dynamics of mathematics education. *Journal of the Learning Sciences, 25*(2), 203–239. https://doi.org/10.1080/10508406.2016.1143370

Abrahamson, D., Nathan, M. J., Williams-Pierce, C., Walkington, C., Ottmar, E. R., Soto, H., & Alibali, M. W. (2020). The future of embodied design for mathematics teaching and learning. *Frontiers in Education, 5*, 147. https://doi.org/10.3389/feduc.2020.00147

Abrahamson, D., Dutton, E., & Bakker, A. (2021). Towards an enactivist mathematics pedagogy. In S. A. Stolz (Ed.), *The body, embodiment, and education: An interdisciplinary approach*. Routledge. https://doi.org/10.4324/9781003142010

Alberto, R. A., Shvarts, A., Drijvers, P., & Bakker, A. (2022). Action-based embodied design for mathematics learning: A decade of variations on a theme. *International Journal of Child-Computer Interaction, 32*, 100419. https://doi.org/10.1016/j.ijcci.2021.100419

Artigue, M. (2007). Digital technologies: A window on theoretical issues in mathematics education. In D. Pitta-Pantazi & G. Philippou (Eds.), *Proceedings of the 5th congress of the European*

society for research in mathematics education (pp. 68–82). Cyprus University. http://erme.site/wp-content/uploads/CERME5/plenaries.pdf
Bakker, A. (2004). *Design research in statistics education: On symbolizing and computer tools*. Doctoral dissertation. Utrecht University. https://dspace.library.uu.nl/bitstream/handle/1874/893/full.pdf?sequence=2
Bakker, A. (2018). *Design research in education. A practical guide for early career researchers*. Routledge. https://doi.org/10.4324/9780203701010
Bakker, A., & Hoffmann, M. H. (2005). Diagrammatic reasoning as the basis for developing concepts: A semiotic analysis of students' learning about statistical distribution. *Educational Studies in Mathematics, 60*(3), 333–358. https://doi-org.proxy.library.uu.nl/10.1007/s10649-005-5536-8
Batanero, C., Tauber, L. M., & Sánchez, V. (2004). Students' reasoning about the normal distribution. In D. Ben-Zvi & J. Garfield (Eds.), *The challenge of developing statistical literacy, reasoning and thinking* (pp. 257–276). Springer. https://doi-org.proxy.library.uu.nl/10.1007/1-4020-2278-6_11
Behar, R. (2021). *El histograma como un instrumento para la comprensión de las funciones de densidad de probabilidad* [The histogram as a tool for understanding probability density functions]. Project description on researchgate. https://www.researchgate.net/project/El-histograma-como-un-instrumento-para-la-comprension-de-las-funciones-de-densidad-de-probabilidad
Bernstein, A. N. (1967). *The coordination and regulation of movements*. Pergamon press.
Biehler, R. (2007). Students' strategies of comparing distributions in an exploratory data analysis context. In *56th session of the International Statistical Institute*. https://www.stat.auckland.ac.nz/~iase/publications/isi56/IPM37_Biehler.pdf
Boels, L., Bakker, A., Van Dooren, W., & Drijvers, P. (2019a). Conceptual difficulties when interpreting histograms: A review. *Educational Research Review, 28*, 100291. https://doi.org/10.1016/j.edurev.2019.100291
Boels, L., Bakker, A., & Drijvers, P. (2019b). Eye tracking secondary school students' strategies when interpreting statistical graphs. In M. Graven, H. Venskat, A. A. Esien, & P. Vale (Eds.), *Proceedings of the forty-third psychology of mathematics education conference* (pp. 113–120). PME. https://www.igpme.org/wp-content/uploads/2019/07/PME43-proceedings.zip
Boels, L., Bakker, A., & Drijvers, P. (2019c). Unravelling teachers' strategies when interpreting histograms: An eye-tracking study. In U. T. Jankvist, M. Van den Heuvel-Panhuizen, & M. Veldhuis (Eds.), *Proceedings of the 11th congress of the European society for research in mathematics education* (pp. 888–895). Freudenthal Group & Freudenthal Institute, Utrecht University & ERME. https://hal.archives-ouvertes.fr/hal-02411575/document
Boels, L., Bakker, A., Van Dooren, W., & Drijvers, P. (2022a). *Secondary school students' strategies when interpreting histograms and case-value plots: An eye-tracking study*. [Manuscript submitted for publication] Freudenthal Institute, Utrecht University.
Boels, L., Bakker, A., Van Dooren, W., & Drijvers, P. (2022b). *Understanding histograms in upper-secondary school: Embodied design of a learning trajectory*. [Manuscript submitted for publication]. Freudenthal Institute, Utrecht University.
Cooper, L. L., & Shore, F. S. (2010). The effects of data and graph type on concepts and visualizations of variability. *Journal of Statistics Education, 18*(2). http://jse.amstat.org/v18n2/cooper.pdf
Derouet, C., & Parzysz, B. (2016). How can histograms be useful for introducing continuous probability distributions? *ZDM Mathematics Education, 48*(6), 757–773. https://doi-org.proxy.library.uu.nl/10.1007/s11858-016-0769-9
Drijvers, P. (2019). Embodied instrumentation: Combining different views on using digital technology in mathematics education. In U. T. Jankvist, M. van den Heuvel-Panhuizen, & M. Veldhuis (Eds.), *Proceedings of the 11th congress of the European society for research in mathematics education* (pp. 8–28). Freudenthal Group & Freudenthal Institute, Utrecht University & ERME. https://hal.archives-ouvertes.fr/hal-02436279v1

Finzer, W. (2016). What does dragging this do? The role of dynamically changing data and parameters in building a foundation for statistical understanding. In A. Rossmann & B. Chance (Eds.), *Working cooperatively in statistics education. Proceedings of the seventh international conference on teaching statistics*. Salvador, Bahia, Brazil.

Freedman, D., Pisani, R., & Purves, R. (1978). *Statistics*. W. W. Norton & Co.

Gilmartin, K., & Rex, K. (2000). *Student toolkit: More charts, graphs and tables*. Open University. https://ahpo.net/assets/more-charts-graphs-and-tables-toolkit.pdf

Gratzer, W., & Carpenter, J. E. (2008/2009). The histogram-area connection. *The Mathematics Teacher, 102*(5), 226–340.

Huck, S. W. (2016). *Statistical misconceptions. Classic edition*. Routledge.

Kelly, A. E., Sloane, F., & Whittaker, A. (1997). Simple approaches to assessing underlying understanding of statistical concepts. In I. Gal & J. B. Garfield (Eds.), *The assessment challenge in statistics education* (pp. 85–90). IOS Press.

Lee, J. T. (1999). It's all in the area. *Mathematics Teacher, 92*(8), 670–672. https://www.jstor.org/stable/27971168

Lee, J. T., & Lee H. S. (2014). Visual representations of empirical probability distributions when using the granular density metaphor. Invited paper. In K. Makar (Ed.) *Proceedings of the ninth international conference on teaching statistics, Flagstaff.*

Leontyev, A. N. (2009). Activity and consciousness. *Marxist internet archive*. http://www.marxists.org/archive/leontev/works/activity-consciousness.pdf

McGatha, M., Cobb, P., & McClain, K. (2002). An analysis of students' initial statistical understandings: Developing a conjectured learning trajectory. *The Journal of Mathematical Behavior, 21*(3), 339–355. https://doi.org/10.1016/S0732-3123(02)00133-5

Parzysz, B. (2018). Solving probabilistic problems with technologies in middle and high school: The French case. In N. Amado et al. (Eds.), *Broadening the scope of research on mathematical problem solving* (Research in mathematics education). Springer.

Radford, L. (2003). On the epistemological limits of language: Mathematical knowledge and social practice during the renaissance. *Educational Studies in Mathematics, 52*(2), 123–150. http://www.jstor.org/stable/20749442

Radford, L. (2010). The eye as a theoretician: Seeing structures in generalizing activities. *For the Learning of Mathematics, 30*(2), 2–7.

Reading, C., & Canada, D. (2011). Teachers' knowledge of distribution. In C. Batanero, G. Burrill, & C. Reading (Eds.), *Teaching statistics in school mathematics. Challenges for teaching and teacher education* (New ICMI Study Series, 14). Springer. https://doi.org/10.1007/978-94-007-1131-0_23

Roditi, E. (2009). L'histogramme : à la recherche du savoir à enseigner. [The histogram: In search of knowing how to teach it]. Spirale. *Revue de Recherches en Éducation, 43*, 129–138. https://halshs.archives-ouvertes.fr/halshs-00609704

Shvarts, A. (2017). Eye movements in emerging conceptual understanding of rectangle area. In B. Kaur, W. K. Ho, T. L. Toh, & B. H. Choy (Eds.), *Proceedings of the 41st conference of the International Group for the Psychology of Mathematics Education* (Vol. 1, p. 268). PME. https://www.igpme.org/wp-content/uploads/2019/05/PME41-2017-Singapore.zip

Shvarts, A., & Alberto, R. A. (2021). Melting cultural artifacts back to personal actions: Embodied design for a sine graph. In M. Prasitha, N. Changsri, & N. Boonsena (Eds.), *Proceedings of the 44th conference of the International Group for the Psychology of Mathematics Education* (Vol. 4, pp. 49–56). https://pme44.kku.ac.th/home/uploads/volumn/pme44_vol4.pdf#page=61

Shvarts, A., Alberto, R. A., Bakker, A., Doorman, M., & Drijvers, P. (2021). Embodied instrumentation in learning mathematics as the genesis of a body-artifact functional system. *Educational Studies in Mathematics, 107*, 447–469. https://doi.org/10.1007/s10649-021-10053-0

Shvarts, A., Bos, R., Doorman, M., & Drijvers, P. (2022). *Reifying actions into artifacts: An embodied perspective on process-object dialectics in higher-order mathematical thinking*. [submitted]. Utrecht University.

Simon, M. A., & Tzur, R. (2004). Explicating the role of mathematical tasks in conceptual learning: An elaboration of the hypothetical learning trajectory. *Mathematical Thinking and Learning, 6*(2), 91–104. https://doi.org/10.1207/s15327833mtl0602_2

Trouche, L. (2014). Instrumentation in mathematics education. In S. Lerman (Ed.), *Encyclopedia of mathematics education*. Springer. https://doi.org/10.1007/978-94-007-4978-8_80

Van Dijke-Droogers, M. (2021). *Introducing statistical inference: Design and evaluation of a learning trajectory*. Doctoral dissertation. Utrecht University. https://www.fisme.science.uu.nl/publicaties/literatuur/2021_van_dijke_introducing_statistical_inferences.pdf

Varela, F. J., Thompson, E., & Rosch, E. (1991). *The embodied mind: Cognitive science and human experience*. MIT Press.

Verillon, P., & Rabardel, P. (1995). Cognition and artifacts: A contribution to the study of thought in relation to instrumented activity. *European Journal of Psychology of Education, 10*, 77–103. https://www.jstor.org/stable/23420087

Vygotsky, L. S. (1997). *Educational psychology*. CRC Press. https://www.taylorfrancis.com/chapters/mono/10.4324/9780429273070-9/

Wild, C. J. (2006). The concept of distribution. *Statistics Education Research Journal, 5*(2), 10–26. https://iase-web.org/ojs/SERJ/article/download/497/367

Wilson, A. D., & Golonka, S. (2013). Embodied cognition is not what you think it is. *Frontiers of Psychology, 4*, 58. https://doi.org/10.3389/fpsyg.2013.00058

Margin of Error: Connecting Chance to Plausible

Gail Burrill

Abstract An intriguing question in statistics: how is it possible to know about a population using only information from a sample? This paper describes a simulation-based formula-light approach used to answer the question in a course for elementary preservice teachers. Applet-like dynamically linked documents allowed students to build "movie clips" of the features of key statistical ideas to support their developing understanding. Live visualizations of simulated sampling distributions of "chance" behavior can enable students to see that patterns of variation in the aggregate are quite predictable and can be used to reason about what might be likely or "plausible" for a given sample statistic. The discussion includes an overview of a learning trajectory leading to understanding margin of error, summarizes a key lesson in the development, describes a brief analysis of student understanding after the course, and concludes with some recommendations for future investigation.

Keywords Chance · Interactive dynamic technology · Margin of error · Sampling · Simulation

1 Introduction

"… it is crucial to focus on helping students become better educated consumers of statistical information by introducing them to the basic language and the fundamental ideas of statistics, and by emphasizing the use and interpretation of statistics in everyday life" (GAISE College Report, 2016). For example, a typical quote from the media:

> 76% of US parents are completely or somewhat satisfied with the quality of education their child is receiving, but only 43% of US adults agree based on telephone interviews conducted

G. Burrill (✉)
Program in Mathematics Education, Michigan State University, East Lansing, MI, USA
e-mail: burrill@msu.edu

G. F. Burrill et al. (eds.), *Research on Reasoning with Data and Statistical Thinking: International Perspectives*, Advances in Mathematics Education,
https://doi.org/10.1007/978-3-031-29459-4_15

> Aug. 3–7, 2016, with a random sample of 1,032 adults, aged 18 and older, living in all 50 U.S. states and the District of Columbia. The margin of sampling error is ±4%. For results based on the total sample of 254 parents of children in kindergarten through grade 12, the margin of sampling error is ±8%. (http://www.gallup.com/poll/194675/education-ratings-show-record-political-polarization.aspx?gsource=Education&gmedium=lead&gcampaign=tiles)

Reasoning from a sample to a population as in the quote is a core concept in statistics, yet ideas related to sampling have been identified as problematic for students to learn and for the public to understand (e.g., Ben-Zvi et al., 2015; Castro et al., 2007). Issues include not understanding sampling distributions; confusing the distribution of a population, the distribution of a sample from the population and the sampling distribution of a statistic (Lipson, 2002; Tversky & Kahneman, 1971), and a lack of conception of distribution to make visible the unusualness of an observation (Liu & Thompson, 2009).

Such concerns related to understanding statistical concepts as well as the rapid development of technology led to a call by many statistics educators for a different approach (e.g., Pfannkuch & Budgett, 2014; Tintle et al., 2019). As early as 1999, researchers such as del Mas et al. (del Mas et al., 1999) argued that using simulations to investigate concepts such as sampling distributions and confidence intervals allowed students to repeat a process over and over, control sample size and note patterns, and explain the behavior they observe rather than try to interpret theoretical approaches. Cobb (2007) called for simulation-based methods for teaching inference to replace the traditional approach to inference using theory-based methodology. Technology generated simulations allow students to view a process as it unfolds rather than just observing the end result. However, some researchers suggest caution in reasoning from simulations, as in some cases observations may be generalized incorrectly (e.g., Watkins et al., 2014). With appropriate guidance, experiencing these processes can help students develop the abilities and intuitive thinking that enhance powerful mental conceptualizations (Sacristan et al., 2010). Interactive apps provide students opportunities to build mental images – dynamic "movie clips"-of the features of a concept that can become the basis for understanding. Apps such as these formed the basis for a statistics course for elementary preservice teachers at a large midwestern university.

This paper examines one aspect of the course related to sampling and inference, margin of error; lays out a learning trajectory for developing the concept; describes a key lesson in the development; and concludes with an analysis of student understanding after the experience. Two things are different in this work from much of the research related to simulation-based inference: (1) the course devoted a major portion of instruction to developing conceptual understanding of core statistical ideas relying heavily on interactive dynamic technology to do so; and (2) the approach is formula light and based on identifying how chance outcomes behave in known situations. The question explored in this paper is "How does a formula-light applet-based approach support students' reasoning from a sample to the population? In particular how do students reason about margin of error?"

2 Background

A number of studies have focused on issues related to reasoning from a sample to a population (e.g., Makar & Rubin, 2009; Pfannkuch et al., 2015). Much of the work is related to understanding confidence intervals (e.g., Crooks et al., 2019; del Mas et al., 2007; Fidler, 2006; Fidler & Loftus, 2009; García-Pérez & Alcalá-Quintana, 2016; Grant & Nathan, 2008; Henriques, 2016; Hoekstra et al., 2012; Hoekstra et al., 2014; Kalinowski et al., 2018). The research clearly identifies that people from many populations - school and college students, teachers, education researchers, psychologists, graduate and undergraduate students - struggle with confidence intervals, from being able to interpret them to understanding how they relate to margin of error (Budgett & Rose, 2017; Thompson & Liu, 2005). In particular, studies show confusion about confidence levels and how these relate to confidence intervals (e.g., Kalinowski et al., 2018). However, these studies typically do not mention the words margin of error; none of the seven articles in the *Educational Studies in Mathematics* special issue on "Statistical Reasoning: Learning to Reason from Samples" (Ben-Zvi et al., 2015), for example, contain the words. One notable exception is the work by Saldanha and Thompson (2014) that directly addresses margin of error; the connection to the work in this paper is described below.

3 Building Concept Images of Margin of Error

Conceptual understanding has been identified as central in learning mathematics (Bransford et al., 1999). Conceptual understanding is related to the notion of concept images: the total cognitive structure including the mental pictures and processes associated with a concept created in students' minds through different experiences associated with the ideas (Tall & Vinner, 1981). Without a coherent mental structure, students are left to construct their understanding based on ill formed and often misguided connections and images (Oehrtman, 2008). The work of understanding subsequent topics is then built on isolated understandings specific to each topic (e.g., center as separate from variability, distribution as a set of individual outcomes, randomness as accidental or unusual). This makes it difficult for students to see and make the connections needed for deep understanding. Understanding statistical inference, for example, entails connecting and integrating a collection of emerging concepts (Saldanha & Thompson, 2014).

This paper addresses the introduction of margin of error as a concept uncoupled from degree of certainty ("I am 95% confident that …"), which is rarely done. The focus is on *interval estimate* (an interval of plausible population parameters for the truth) for the population parameter rather than a *level of confidence* (how confident we are that our interval includes the population value we want to estimate). As noted above, a margin of error is used as a measure in describing the outcomes of polls and surveys and is a common term familiar to most in the public. However, there is confusion about what the term means even in some technical documents (Saldanha, 2003; Thornton & Thornton, 2004). The word "error" suggests a mistake, which leads people to believe that the result accounts for mistakes made in the process because of strong colloquial associations with the term error. In some cases, margin of error is defined by a formula (Rumsey, 2022); in others it is defined as part of the definition of a confidence interval (Saldanha, 2003). It is sometimes defined using words familiar to only a statistician (e.g., "The margin of error expresses the maximum expected difference between the true population parameter and a sample estimate of that parameter;" Stat Trek) or defined in words that can lead to misunderstanding such as measure of reliability or degree of error. In this paper, margin of error is used to determine a set of plausible populations from which a random sample with an observed attribute could have been drawn. The margin of error is an estimate to measure the maximum amount by which the sample results are expected to differ from those of the actual population (Rumsey, 2022).

Statistical ideas rarely exist in isolation but instead are developed in a coherent learning trajectory, building from and connecting foundational ideas. Effectively reasoning about sampling and making the connection between sampling and margin of error involves understanding foundational concepts: (1) samples, populations, and sample statistics, distinguishing among these and their corresponding distributions; (2) distributional reasoning such that students develop intuitions "for a reasonable amount of variation around an expected value" (Shaughnessy, 2007, p. 982); (3) sampling representativeness including understanding that larger samples better represent the population from which they were sampled and their sample statistics are closer to the population parameters; (4) notions of sampling variability among multiple samples drawn from the same population; (5) randomness and chance and the long run stability that occurs in random behavior (Bakker, 2004; Pfannkuch, 2008).

The approach in the paper has some similarities to that used by Saldanha and Thompson (2014) who studied the results of two teaching experiments with non-Advanced Placement[1] high school statistics students. In their study, students began an experiment with a hands-on activity, and then used a microworld that simulated sampling from a population with information about samples chosen from the population represented numerically, in tables, and in histograms. The results indicated students had some success identifying relatively rare events but "students' thinking was fuzzy regarding the population's composition" (p. 24) with students unable to correctly articulate the link between a sample outcome and the population. While the approach described in the work that is the focus of this paper also begins with a similar hands-on activity and also involves simulations, the emphasis is on identifying plausible populations that might have produced the sample attribute using words such as "likely" and "happen by chance" and not on quantification. The simulated outcomes are represented in dot plots, where each sample outcome is visible in the

[1] Advanced Placement is a program that allows high school students to receive college credit for successfully taking a college course in high school.

distribution as opposed to histograms used in the Saldanha and Thompson (2014) experiment, where the individual outcomes are collapsed into bars.

A study by Budgett and Rose (2017) also used a simulation approach; however, only the teacher had access to the technology for simulating samples, and the simulated results were displayed in the bars of histograms. The focus was not on chance outcomes; margin of error was defined as half of a bootstrap confidence interval and the approach led directly to confidence levels. Another study by Van Dijke-Droogers et al. (2020) involved repeated sampling with a black box, quite similar to the idea of a mystery bag, to introduce ninth-grade students to the concepts of sample, frequency distribution, and simulated sampling distribution but did not extend the work to include margin of error.

4 The Course

The project was carried out in several iterations of a semester long statistics course for elementary preservice students who were underclassmen in an elementary teacher preparation program at a large midwestern public university. The semester-long class met twice a week for 15 weeks in 110-minute sessions. The goals of the course were to enable students to interpret and make sense of data, in particular data related to education, and to give them tools and strategies for their own teaching. The first phase of the design research process took place with Cohort 1, and changes across the course were made before the course was implemented with Cohort 2. Thus, results are reported here for students who gave permission for their work to be used from Cohorts 2 (29 students), 3 (12 students), and 4 (17 students). Overall, 67% of the students had never taken a statistics course; 10% had taken an introductory university course and the others had taken some statistics in high school. The fourth cohort was forced by the pandemic to go online for the last six weeks of class, which included the lessons on margin of error.

In a pre-course survey, students were asked to rate their familiarity with margin of error on a five-point Likert scale that ranged from 1 "not at all familiar" to 5 "very familiar". The results were surprising as in every cohort the majority of students selected 2 or higher (in Cohort 2, 45% selected 3 or higher), and these were often students who had no prior work in statistics. Their explanations, if they gave one, consisted of "used it in chemistry or physics or psychology or calculus", "learned in middle school".

The project was designed using cycles of design, test and improvement (Bakker, 2018) such that learnings from one iteration of the course informed the next. Changes in the approach from the first iteration included dropping the use of a textbook, taking time at the beginning of the course to consider ways in which mathematics and statistics are different, modifications to the activities to make them shorter and more focused, and weekly homework specifically designed to probe both conceptual and procedural understanding of the prior week's work. Students had their own computers and used TI© Nspire software to access files from *Building*

Concepts: Statistics and Probability (https://education.ti.com/en/building-concepts) and later StatKey (www.lock5stat.com/StatKey/). The question explored in this paper is "How does a formula-light applet-based approach support students' reasoning from a sample to the population? In particular how do students reason about margin of error?"

A typical class began with a brief overview of key ideas or considerations from the prior class, then a short introduction to a new concept followed by working in randomly generated small groups through a series of carefully designed questions using the applets to investigate the concept, with guidance from the instructor to ensure that students don't go wrong with the technology (Drijvers, 2015). Weekly homework assignments consisted of short tasks or questions focused on the key ideas covered the prior week. Assignments were ungraded with students receiving full marks for turning in a completed assignment, but problematic answers were highlighted by the instructor and discussed in class. Students often presented their thinking about the tasks, with discussion focused on making connections across the content they had been learning as well as on considerations that would be important when teaching the concepts to students. Ideas were deliberately revisited in different contexts to ensure that students developed robust images for the concepts (Oehrtman, 2008). Students took part in frequent formative assessments such as a quick quiz, thumbs up/thumbs down on answers, exit tickets or small group discussions on "What am I [the instructor] worried about" with statements taken from student work.

5 Learning Trajectory for Margin of Error

Interactive apps were carefully designed considering specific content objectives mindful that even a well-designed simulation is unlikely to be an effective teaching tool unless students' interaction with it is carefully structured (Lane & Peres, 2006). The apps were used early in the course to introduce the prerequisite knowledge necessary to understand sampling distributions including ideas related to distribution, randomness, variability, chance, and sampling as described above. Distributions were introduced in analyzing univariate data at the beginning of the term, with an emphasis on measures of variability including mean absolute deviation.

5.1 Random Behavior and Long Run Stability

Typically, students explored concepts by hand before using technology (Chance & Rossman, 2006; GAISE College Report, 2016). For example, students were given small bags containing an unknown number of blue chips, similar to the growing samples activity in Ben-Zvi et al. (2015). They drew one chip, recorded its color as blue or not blue, returned the chip to the bag and repeated the process, keeping track of the probability of drawing a blue chip both numerically and graphically by plotting the probabilities versus the number of draws. After five draws they made a conjecture about the actual probability of selecting a blue chip from the bag, continued the process, refining their conjectures, until they felt fairly comfortable with their claim about the probability of getting a blue chip in randomly drawing a chip from the bag. They checked their conjecture with the actual number of blue chips in their bags and then reflected in small groups on the process, considering the accuracy of their estimates, the number of draws before they were ready to make a claim about the probability of randomly selecting a blue chip from their bag, and how their graphs were alike and how they were different.

This activity was followed by using an app to model the physical simulation, where students could generate many samples from the same bag and observe the fluctuations in the probability of getting a blue chip in the first few draws and the tendency for the probability to stabilize after a certain number of draws. They could change the proportion of blue chips in the bag and notice whether their observations were the same or different. The simulations enabled students to build their understanding by seeing the results grow one repetition at a time before moving to groups of repetitions and by repeating this progression many times (Figs. 1 and 2).

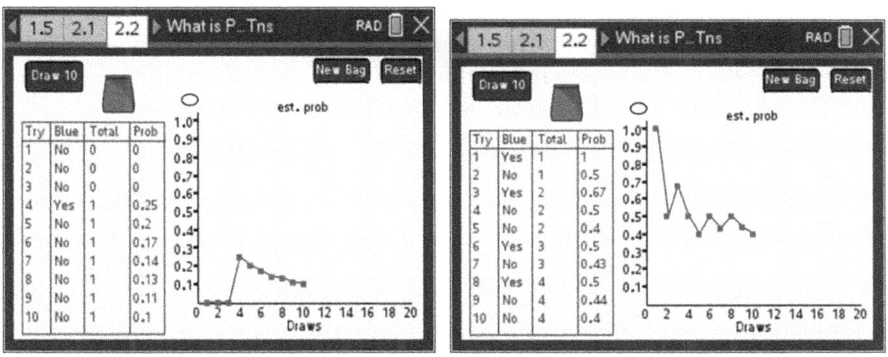

Fig. 1 Probability of a blue chip after 10 draws

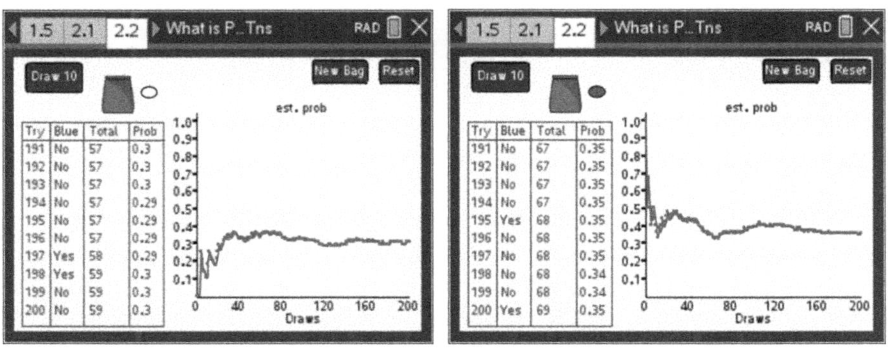

Fig. 2 Probability of drawing a blue chip stabilizes as the number of draws increases

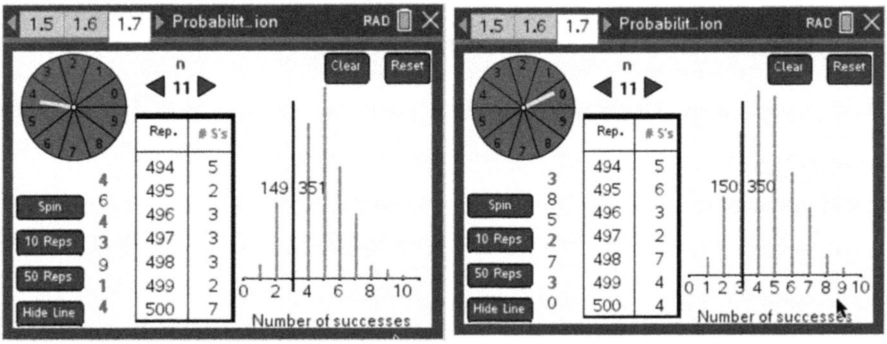

Fig. 3 Students simulated 500 sets of 11 games for a player with a 40% shooting average to estimate the probability the player would have a game with only a 30% success rate

5.2 *The Role of Chance*

In the unit on probability, students worked through activities like those in *The Art and Techniques of Simulation* (Gnanadesikan and Schaeffer Gnanadesikan & Scheaffer, 1987) continuing the emphasis on the long run stability in random behavior and explored the typical variability around expected outcomes in simulated sampling distributions. They explored the role of chance in answering questions by comparing an outcome to simulated sampling distributions to get a sense of how surprising the outcome seemed to be; for example, the technology allowed students to quickly reproduce a simulation to get a sense of the validity of their estimate for the probability related to the number of baskets a player with a 40% shooting average could be expected to make (Fig. 3). It is critical in developing intuitions that each student generate many sampling distributions, comparing their results to others and discuss what is the same and what is different (Van Dijke-Droogers et al., 2020; Wild et al., 2011).

5.3 *Sampling*

Students engaged in a variety of tasks collectively designed to investigate the relationship between sampling distributions of a sample statistic and the population from which the sample was drawn. The tasks were framed to develop the notions that samples will vary from expected outcomes and that this variation will be different from sample to sample, that small samples typically have more variability from sample to sample than larger ones, and that usually random samples will be somewhat representative of the population. For example, in Fig. 4, students compared the mean and the interval defined by the mean plus/minus the mean absolute deviation

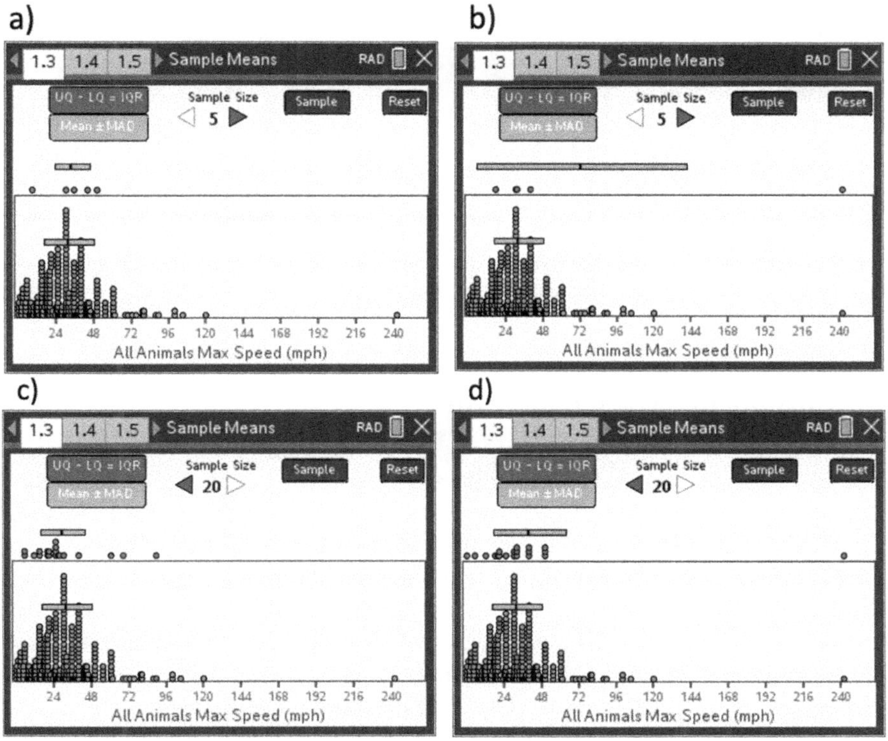

Fig. 4 Investigating the variability, represented by the horizontal bars in the figures, in simulated distributions of sample (top dot plot) of the same size (**a** to **b** for $n = 5$, and **c** to **d** for $n = 20$) from the population of maximum animal speeds (bottom dot plot) and of different size samples (**a** and **b** to **c** and **d**)

for the samples drawn from a population of maximum speeds for various animal species and noted how these changed from sample to sample and as the sample size changes. This relates to the discussion in Wild et al. (2011) of looking through rippled glass where students learn that small samples may have distortions (like very rippled glass) but big samples typically have only small distortions.

Students explored the notion that random samples are typically representative of the population using activities such as the Random Rectangle Activity (Scheaffer et al., 1996). To ensure they understood that characteristics of the population should be present in about the same proportion as they are in the population, they used an app that allowed them to visualize representativeness by contrasting sampling distributions from a population where each outcome is equally likely to sampling distributions where the outcomes are not all equally likely (Fig. 5).

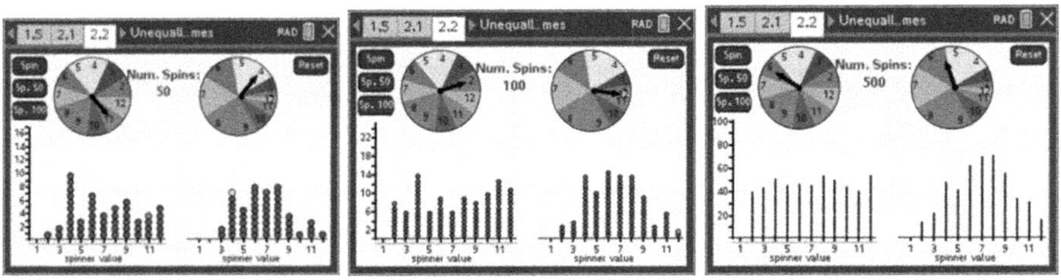

Fig. 5 Generating simulated distributions where the outcomes on the spinners are and are not equally likely, noting that as the number of repetitions grows the distributions of the outcomes stabilize

6 Introducing Margin of Error

The lesson introducing the concept of margin of error was based on the Quantitative Literacy Project module, *Exploring Surveys and Information from Samples* (Landwehr et al., 1987), motivated by the poll described at the start of this paper. A sample of size 30 is randomly drawn with replacement from a "mystery bag" containing an unknown proportion of blue chips, and the question is posed: how can we use the information from this sample (say nine blue chips) to learn something about the actual proportion of the blue chips in the bag. Note that unlike the preceding experiment drawing a chip from a bag, here the sample size is clearly defined, and only one sample of that size can be drawn from the mystery bag. Students used known population proportions to investigate the question. They worked in pairs by drawing samples of size 30 from bags each clearly marked with the percentage of blue chips in the bag with percentages ranging from 10% to 90%. After drawing about five samples by hand and recording the number of blue chips, they used an app to simulate drawing the samples from their bags and created a simulated sampling distribution of the number of blue chips for each population proportion. Students identified plausible populations for the mystery bag by comparing the observed sample value (the nine blue chips referred to above) to their simulated distributions. At first, students individually responded to the question "could the sample from the mystery bag have come from your bag?", and their yesses and corresponding populations were recorded. Then the work was summarized in a chart (Fig. 6) with each group displaying the interval representing the number of blue chips in the collection of samples they had drawn from their individual bags with known proportions of blue chips. A vertical line representing the observed number (in this case, nine) of blue chips in the mystery bag was drawn on the chart.

Thus, any population having from 20% to 50% blue chips might by chance produce nine blue chips in a sample of 30; a plausible population for such a sample would be 35% with a 15% "margin of error". Students reflected on the activity by responding to probing questions and finding margins of error for a variety of situations using the chart, emphasizing "likely", "plausible" and "chance". The discussion specifically addressed two obstacles to understanding noted by Saldanha and

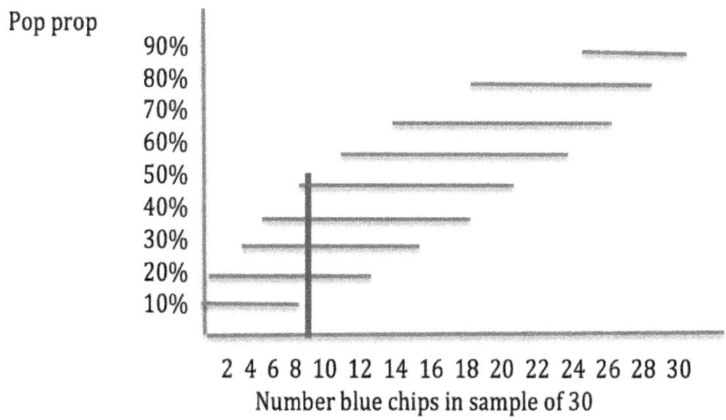

Fig. 6 Number of blue chips in samples of 30 chips drawn from known population proportions. The horizontal bars represent the range of blue chips obtained in simulating 500 samples of size 30 drawn with replacement for nine populations where the probability of a success ranged from 10% to 90%

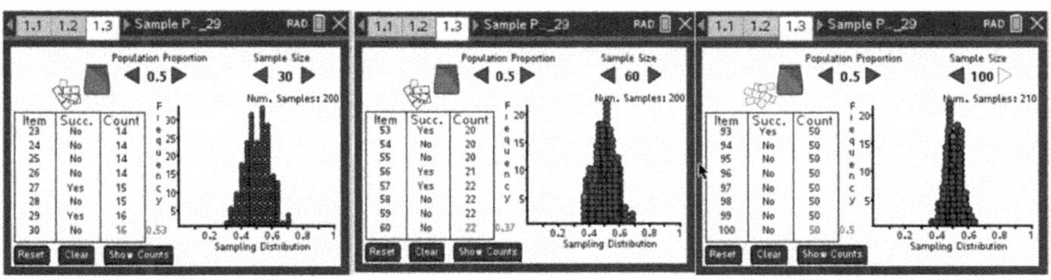

Fig. 7 Simulated sampling distributions illustrating decreasing variability (spread) in the number of successes as the sample size increases

Thompson (2014): (1) making explicit that in real life a single sample is typically drawn in order to infer a population's unknown parameter and the simulations to develop the chart were drawing many samples from populations having known parameters and (2) the context itself was not important; for the given sample size of 30, the same chart could be used for any context.

The class returned to the two margins of error (4% and 8%) in the original poll. They conjectured this had to do with sample size and verified their conjectures using an app that allowed them to set the sample size, the population proportion and to "grow" the sampling distribution (Fig. 7). Note that because the apps were designed to develop conceptual understanding, the choices were limited, the method of sampling was visible, and the last sample could be recalled by scrolling.

After working through several examples, students revisited their original bags. Each group drew a random sample of 30 chips from their bags and used the result to find a set of plausible population proportions for that outcome and the corresponding margin of error. Because the proportion of blue chips was known for each of their bags, they could easily check to see if their interval of plausible population proportions contained the true proportion of blue chips. If the class is large enough, typically at least one group will by chance have an interval that does not contain the actual proportion for their bag. This could be revisited in the context of the probability app from Fig. 2, where students generated samples of size 30, created a set of plausible population proportions for the true proportion of blue chips in the bag and then used the long run relative frequency to check the validity of their claim. Engaging students in these activities was important for consolidating understanding that a set of plausible populations is not guaranteed to contain the truth, anticipating a typical misinterpretation of margin of error.

Activities and class discussions focused on nuances (why "error"; why not use counts to compare sample sizes; how many samples before the distribution begins to "stabilize"; how can we estimate the standard deviation from the sampling distribution?). In response to "What do you wonder about?" students raised questions such as "Is it possible that the "truth" will ever be outside of the margin of error?"; "If you want a more specific margin of error, would you have to test by 0.01 and so on?" Typically one question that emerged after several classes was "How do you get the known populations without doing it for each percent?" This led to making the connection from simulated sampling distributions to normal distributions and the use of standard deviation to describe what is likely or plausible. Eventually, students used StatKey to simulate the sampling distributions, revisiting the original poll to verify and consolidate their earlier thinking.

7 Data Analysis

The data consist of written records of student comments and descriptions of their approaches as they worked through instructional materials, reflections of the instructor after each class, quizzes, exams, student projects, and a final survey related to the use of the apps. The first phase of the design research process took place with Cohort 1, and major changes across the course were made before the course was implemented with Cohort 2. Thus, results are reported here for Cohorts 2, 3, and 4. Note again that Cohort 4 was in class during the onset of the pandemic, which made it necessary to modify what had been done with the preceding cohorts including the ommission of one test. Because of exam security, the questions on tests were not identical across the three cohorts, with several exceptions noted below.

Questions from tests, quizzes and information from student projects related to margin of error were categorized as: (1) recognizing why margin of error is important; (2) finding a margin of error (either from a set of sampling distributions drawn from knowing populations or by calculations); (3) interpreting margin of error in a context; and (4) applying margin of error. The results are displayed in Table 1. On similar test questions about why a margin of error is important, about 60% to 70% of the students in Cohort 2 and Cohort 3 identified sampling variability as the

Table 1 Percent of students correctly answering questions related to margin of error[a]

Skill	Situation	Cohort 2 (n = 29)	Cohort 3 (n = 12)	Cohort 4 (n = 17)
Identify importance of margin of error	Test 2	67%	62%	—
Find/calculate margin of error	Given set of simulated sampling distributions for proportions[b]	96%	92%	82%
Interpret margin of error in context	Given set of simulated sampling distributions for proportions[b]	37%	46%	53%
	Mean, final	45%	54%	35%
Apply in context	Mean, final	69%	92%	65%
	Find confidence interval for mean, final	—	46%	—
	Interpret confidence interval for mean, final	—	62%	—
	Used correctly in projects - of those who chose to use margin of error	100%	100%	75%

[a] Assessments for students in Cohort 4 were adjusted because of constraints imposed by the pandemic
[b] Test 2 for Cohort 2 and Cohort 3; Final exam for Cohort 4

reason, whereas about 15% to 20% identified choices related to thinking of error as making a mistake or misinterpretation, and about 5% to 25% identified all of the choices. Students in all three cohorts were better at finding a margin of error than correctly interpreting the result in a context. Misinterpretations of margin of error, whether for means or proportions, included an "absolute" statement: a margin of error defines an interval *that contains* the population parameter or the interval repesented *the* range of possible population parameters, common errors noted by other researchers (e.g., Kalinowski et al., 2018; Saldanha & Thompson, 2014). On the final exam for Cohort 2, about 60% correctly interpreted margin of error in the context of confidence intervals. Of the 10 groups across the three cohorts that used margin of error in their final projects, only one had an incorrect interpretation claiming the population parameter was in the interval defined by the margin of error: "the margin of error says commercials [from a survey about the length of commercials for a given TV station] are between 22.13–28.25 sec."

The two questions addressed in this study were: "How does a formula-light applet-based approach support students' reasoning from a sample to the population? In particular how do students reason about margin of error?" To better understand how students were thinking, the data were analyzed by classifying the responses using a hierarchical performance level based on the SOLO taxonomy (Structure of Observed Learning Outcomes, Biggs & Collis, 1982). Table 2 is adapted from Reading and Reid's Interpretation of the SOLO taxonomy for statistical reasoning (2006) to focus on the development and consolidation of a concept image (Burrill, 2018).

Table 2 SOLO taxonomy and concept images adapted from Reading and Reid (2006)

Description of application to reasoning process
Does not refer to key elements of the concept.
Focuses on one key element of the concept.
Focuses on more than one key element of the concept.
Develops relational links between various key elements of the concept.

Table 3 Features associated with concept image for margin of error

SOLO taxonomy level	Students' reasoning about margin of error
Prestructural (P)	Thinks of margin of error as accounting for mistakes in the process; associates margin of error with the number of repetitions of a simulation
Unistructural (U)	Recognizes variability in the collection of samples, uses counts in making statement about population; finds numerical value for margin of error without interpretation; correct but adds an unacceptable comment
Multistructural (M)	Recognizes margin of error is making a statement about a population parameter; selects appropriate method for finding margin of error; finds an interval for margin of error; recognizes sample size changes outcome but not connected to margin of error
Relational (R)	Links margin of error to sample size; connects margin of error to visual image of interval for plausible populations; correctly interprets margin of error in a context; applies concept to both means and proportions; connects margin of error to repeated sampling from the same population

The analysis with respect to a specific concept was done in three parts: First, identifying elements or features of a concept that could be associated with the levels in the SOLO taxonomy and linking these to possible misinterpretations. Second, categorizing examples used during class and student responses with respect to the elements in the taxonomy and, third, summarizing the SOLO levels attained by the students with respect to the concept. Table 3 illustrates some features that might be associated with reasoning about margin of error, and the discussion that follows describes examples of student work related to margin of error. Note that these results do not directly relate to the percentage correct in Table 1 because the responses were analyzed according to the features in the taxonomy and not necessarily correctness.

For each cohort, the final exam included a question related to a diagram from the Michigan Department of Education (MDE) reporting student achievement results in different content areas (Fig. 8). The margin of error, as labelled by the MDE, is represented in the grey bar around the black vertical segment. The diagrams for the other cohorts were similar.

The exam questions about the report for all three cohorts were:

(a) How would you explain to a parent what the margin of error represents in terms of their child's language arts performance score on the M-Step?

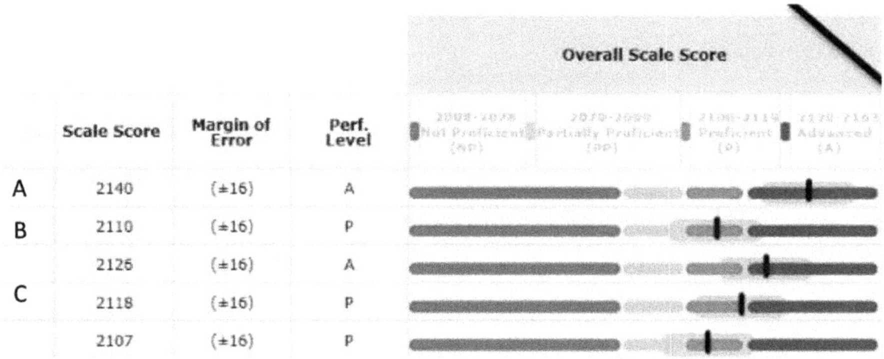

Fig. 8 MDE Student Achievement Report (Michigan Department of Education, 2016)

(b) The report to the teacher looks like the one above. For which, if any, of the students whose scale scores are shown is the margin of error most problematic in terms of the student's level of proficiency? Explain why.

Correct responses to the first question were coded relational because those responses involved connecting variability to margin of error from a visual image and interpreting margin of error in a context. A typical correct response was "The margin of error is most problematic for student B with a scale score of 2110. His scale score places him within the proficient group. But the margin of error means that he could actually be in the partially proficient group or the advanced group. This is extremely problematic because the student could be placed in a group that he does not actually fit in." An incorrect response was often something like "Your son's score on the M-Step was 1294, with a margin of error of 6. This means that although his score was 1296, his scores on all components could have ranged from anywhere between 1288 and 1300." This response would suggest the student understood how to find a margin of error, that a margin of error is an interval and that a margin of error is about the population, all multistructural level knowledge.

To answer question (b) correctly, a student had to connect margin of error and the visual image to a context, which would be at the relational level. Students at the multistructural level typically responded with the wrong student or wrong category. Responses that identified the students with the lowest score rather than taking into account the margin of error were coded unistructural. Tables 4 and 5 show the results of coding the responses for both questions (a) and (b) using the Solo taxonomy.

Overall, on these questions, at least 70% of the students across the three cohorts was able to identify students for whom a designated margin of error was problematic and to explain to a parent what the margin of error represents in terms of their child's math performance score on the state achievement test at the multistructural level ("I would explain to a parent that the margin of error represents the area that if the student were to keep retaking the M-Step their scores would be likely to

Table 4 Categorization of responses with respect to explaining margin of error

SOLO taxonomy level	Reasoning about margin of error: Explaining margin of error to parents			
	Cohort 2 (*n* = 29)	Cohort 3 (*n* = 12)	Cohort 4 (*n* = 17)	Overall (*n* = 58)
Prestructural (P)	14%	8%	17%	12%
Unistructural (U)	8%	17%	24%	14%
Multistructural (M)	67%	25%	24%	46%
Relational (R)	11%*	50%	35%	27%

*Results for this cohort reported elsewhere (Burrill, 2018) contained an error

Table 5 Categorization of responses with respect to applying concept of margin of error in context

SOLO taxonomy level	Reasoning about margin of error: Identifying students for whom designated margin of error was problematic			
	Cohort 2 (*n* = 29)	Cohort 3 (*n* = 12)	Cohort 4 (*n* = 17)	Overall (*n* = 58)
Prestructural (P)	14%	0%	6%	9%
Unistructural (U)	10%	8%	24%	14%
Multistructural (M)	28%	8%	17%	21%
Relational (R)	48%	84%	53%	56%

continue to fall within that range of numbers.") or higher. Fourteen percent gave an explanation that focused accurately on one aspect (unistructural) but added something extra such as a reference to a normal distribution. For example:

> Within the students who scored partially proficient, the margin of error represents the likely interval that 95% of the students scored in. This student score right about in the middle of where 95% of the students scored. So this student scored in a range that about 68% of students scored in.

On a major test given only to students in Cohorts 2 and 3 (not given to Cohort 4 due to the pandemic), two thirds of the students identified a plausible interval for a margin of error from graphical representations of simulated sampling distributions of proportions from known populations (relational). Those at the multistructural level typically made absolute statements, such as: "the likely interval for the actual population proportion has to be from a% to b% because the other simulated sampling distributions never contained the sample proportion." On the test questions related to interpreting margin of error, at least two thirds of the students in these cohorts (2 and 3) were at the multistructural level, with 38% at the relational level while the others had no focus on relevant aspects (prestructural), with most making typical errors such as not connecting the interval to the population (e.g., "in a random sample of 40 Reese's Pieces, between 45% and 75% of them will be brown"). Unistructural level responses made statements using "likely" and an interval but described the sample such as "it is 45%-75% likely to have 25 Reese's Pieces that were brown."

On questions about the connection between sample size and margin of error, 78% of Cohort 2 was able to correctly explain how the margin of error would change if the sample size changed from size 50 to size 30. When the question was stated more openly on the final exam for Cohort 4 (e.g., why were three margins of error reported in describing the results of a poll), only 31% of the students were able to make the connection (relational), with 46% of the students noting that the sample size changed but not explicitly stating the relation between the change to margin of error (multistructural).

Responses to the end of course survey provides some insight into the first question, "How does a formula-light applet-based approach support students' reasoning from a sample to the population?" Although some students found using the apps difficult at first or wanted more lecture, overall the consensus of student responses suggested they liked the structure of the class and found the technology very much supported their learning. For example:

- "The ti-nspire files [apps] really helped give me a true understanding of the concepts and what different statistical measures actual meant. I had something to visualize when I heard a concept like IQR or significance or normal distribution."
- "I found the combination of the files and worksheets helpful to get a general understanding of the concepts."

8 Discussion

The focus question for this paper was "How does a formula-light applet-based approach support students' reasoning from a sample to the population? In particular how do students reason about margin of error?" Analyzing work from two tests, several quizzes, and the final exam as well as 23 student projects suggests that, overall, nearly three fourths of the preservice students had at least a multistructural understanding of margin of error, according to the SOLO taxonomy. They seemed to be strongest in deciding whether a sampling distribution from a known population suggests a plausible population for an observed result, calculating a margin of error for a sample proportion, and articulating the impact of sample size on a margin of error. Nearly half of the students struggled with interpretations in one or more of the contexts, with typical errors such as stating the margin of error specifies the range for the population parameter or the margin of error gives the probability of getting a sample such as the one observed.

The results are encouraging in terms of student learning and consistent with analyses of other content objectives for the course; however, the findings are observational and limited by sample size. The results could be confounded by the instruction, the approach, the activities, and other classroom factors. They highlight the need for more attention to the interpretation of margin of error as well as other places where instruction needs to be revisited. For example, using counts versus relative frequencies/percentages was confusing for some students as evidenced by statements such as, "The margin of error is the interval that is likely to contain the true proportion of the population. In this case, 1307–1339 is the interval that is likely to contain the proportion of students who are not proficient in science." In general, this points to a weakness that is typical in statistical communication: how to better help students be careful with language. Some of the responses to the question about their image of statistical significance also indicated the need for a better transition to using the normal distribution and standard deviation to generalize the process of finding a margin of error. One possible avenue for this is to investigate the effect on student thinking in moving from what could be seen as a Bayesian approach using simulated sampling distributions to determine what is likely by chance to a more frequentist approach using the StatKey randomization simulation. Another possible avenue for research is to study student thinking about whether every plausible population identified by the margin of error is equally likely (Kalinowski et al., 2018).

The work raises interesting questions such as the link between student understanding of the precursor knowledge for successfully learning what margin of error is and means, and successfully being able to interpret margin of error in a variety of situations. Exploring exactly what elements of instruction were treated more lightly or the effect on student thinking of the activities that were eliminated for Cohort 4 due to the pandemic would give a better understanding of the development trajectory. For example, Cohort 4 did not go through the follow up activity revisiting the individual bags of colored chips to make the point that for random samples, it is possible that the interval defined by the margin of error calculated from a sample will not contain the true value of the parameter. This may have been connected to the number of students who made "absolute" errors in interpreting margin of error in different contexts, which seemed to be greater for students in Cohort 4. From another perspective, noting the difference in responses to the questions on the relationship between margin of error and sample size, exploring the nuances in the way tasks are stated might be useful in making sure that students have a sufficiently robust understanding of the concept to navigate slightly different versions of questions related to the concept. And an in-depth exploration of what students meant by their responses to the initial survey about their level of comfort with margin of error would be insightful (given their performance in the course) and might help instructors better anticipate prior knowledge and beliefs students bring to their classes.

In conclusion, introducing students to margin of error from a formula-light, simulation-based approach and the use of the apps to support the development of understanding seems promising but opens the door for many other questions that can be important in fulfilling that promise.

References

Bakker, A. (2004). Reasoning about shape as a pattern in variability. *Statistics Education Research Journal, 3*(2), 64–83. https://doi.org/10.52041/serj.v3i2.552

Bakker, A. (2018). Design research in education. In *A practical guide for early career researchers*. Routledge. https://doi.org/10.4324/9780203701010

Ben-Zvi, D., Bakker, A., & Makar, K. (2015). Learning to reason from samples. *Educational Studies in Mathematics, 88*(3), 291–303. https://doi.org/10.1007/s10649-015-9593-3

Biggs, J. B., & Collis, K. F. (1982). *Evaluating the quality of learning: The Solo taxonomy*. Academic.

Bransford, J. D., & Brown, A. L. (1999). In R. R. Cocking (Ed.), *How people learn: Brain, mind, experience, and school*. National Academy Press.

Budgett, S., & Rose, D. (2017). Developing statistical literacy in the final school year. *Statistics Education Research Journal, 16*(1), 139–162. https://doi.org/10.52041/serj.v16i1.221

Burrill, G. (2018). Concept images and statistical thinking: The role of interactive dynamic technology. In M. A. Sorto (Ed.). *Proceedings of the tenth international congress on teaching statistics*. Kyoto, Japan. https://iaseweb.org/Conference_Proceedings.php?p=ICOTS_10_2018

Castro, A., Vanhoof, S., Van den Noortgate, W., & Onghena, P. (2007). Students' misconceptions of statistical inference: A review of the empirical evidence from research on statistics education. *Educational Research Review, 2*(2), 98–113. https://doi.org/10.1016/j.edurev.2007.04.001

Chance, B., & Rossman, A. (2006). Using simulation to teach and learn statistics. In A. Rossman & B. Chance (Eds.), *Working cooperatively in statistics education: Proceedings of the seventh international conference on teaching statistics* (pp. 1–6). Salvador. https://iase-web.org/Conference_Proceedings.php?p=ICOTS_7_2006

Cobb, G. (2007). The introductory statistics course: A Ptolemaic curriculum? *Technology Innovations in Statistics Education, 1*(1), 1–15. https://doi.org/10.5070/T511000028

Crooks, N., Bartel, A., & Albali, M. (2019). Conceptual knowledge of confidence intervals in psychology undergraduate and graduate students. *Statistics Education Research Journal, 18*(1), 46–62. https://doi.org/10.52041/serj.v18i1.149

del Mas, R., Garfield, J., & Chance, B. (1999). A model of classroom research in action: Developing simulation activities to improve students' statistical reasoning. *Journal of Statistics Education, 7*(3). https://doi.org/10.1080/10691898.1999.12131279

del Mas, R., Garfield, J., Ooms, A., & Chance, B. (2007). Assessing students' conceptual understanding after a first year course in statistics. *Statistics Education Research Journal, 6*(2), 28–58. https://doi.org/10.52041/serj.v6i2.483

Drijvers, P. (2015). Digital technology in mathematics education: Why it works (or doesn't). In S. J. Cho (Ed.), *Selected regular lectures from the 12th international congress on mathematical education* (pp. 135–151). Springer. https://doi.org/10.1007/978-3-319-17187-6_8

Fidler, F. (2006). Should psychology abandon *p*-values and teach CIs instead? Evidence-based reforms in statistics education. In A. Rossman & B. Chance (Eds.), *Working cooperatively in statistics education: Proceedings of the seventh international conference on teaching statistics (ICOTS-7)*. Salvador. https://iase-web.org/documents/papers/icots7/5E4_FIDL.pdf

Fidler, F., & Loftus, G. R. (2009). Why figures with error bars should replace *p*-values: Some conceptual arguments and empirical demonstrations. *Zeitschrift für Psychologie/Journal of Psychology, 217*(1), 27–37.

GAISE College Report ASA Revision Committee. (2016). *Guidelines for assessment and instruction in statistics education college report*. http://www.amstat.org/education/gaise

García-Pérez, M. A., & Alcalá-Quintana, R. (2016). The interpretation of scholars' interpretations of confidence intervals: Criticism, replication, and extension of Hoekstra et al. (2014). *Frontiers in Psychology, 7*, 1–12. https://doi.org/10.3389/fpsyg.2016.01042

Gnanadesikan, M., & Scheaffer, R. (1987). The art and technique of simulation. Pearson Learning.

Grant, T., & Nathan, M. (2008). *Students' conceptual metaphors influence their statistical reasoning about confidence intervals*. WCER Working Paper No. 2008-5. Wisconsin: Wisconsin Center for Education Research. [Online: https://wcer.wisc.edu/docs/working-papers/Working_Paper_No_2008_05.pdf]

Henriques, A. (2016). Students' difficulties in understanding of confidence intervals. In D. Ben-Zvi & K. Makar (Eds.), *The teaching and learning of statistics* (pp. 129–138). Springer. https://doi.org/10.1007/978-3-319-23470-0_18

Hoekstra, R., Kiers, H., & Johnson, A. (2012). Are assumptions of well-known statistical techniques checked, and why (not)? *Frontiers in Psychology, 3*. https://doi.org/10.3389/fpsyg.2012.00137

Hoekstra, R., Morey, R., Rouder, J., & Wagenmakers, E. (2014). Robust misinterpretation of confidence intervals. *Psychonomic Bulletin & Review, 21*(5), 1157–1164. https://doi.org/10.3758/s13423-013-0572-3

Kalinowski, P., Lai, J., & Cumming, G. (2018). A cross-sectional analysis of students' intuitions when interpreting CIs. *Frontiers in Psychology, 9*(112). https://doi.org/10.3389/fpsyg.2018.00112

Landwehr, J., Swift, J., & Watkins, A. (1987). *Exploring surveys and information from samples*. Pearson Learning.

Lane, D., & Peres, S. (2006). Interactive simulations in the teaching of statistics: Promise and pitfalls. In A. Rossman & B. Chance (Eds.), *Working cooperatively in statistics education: Proceedings of the seventh international conference on the teaching of statistics*. Salvador. https://iase-web.org/Conference_Proceedings.php?p=ICOTS_7_2006

Lipson, K. (2002). The role of computer-based technology in developing understanding of the concept of sampling distribution. In *Proceedings of the 6th International Conference on Teaching Statistics*.

Liu, Y., & Thompson, P. W. (2009). Mathematics teachers' understandings of proto-hypothesis testing. *Pedagogies, 4*(2), 126–138. https://doi.org/10.1080/15544800902741564

Makar, K., & Rubin, A. (2009). A framework for thinking about informal statistical inference. *Statistics Education Research Journal, 8*(1), 82–105. https://doi.org/10.52041/serj.v8i1.457

Michigan Department of Education. (2016). *M-STEP Final Reports Webcast*.

Oehrtman, M. (2008). Layers of abstraction: Theory and design for the instruction of limit concepts. In M. Carlson & C. Rasmussen (Eds.), *Making the connection: Research and teaching in undergraduate mathematics education*. http://hub.mspnet.org//index.cfm/19688

Pfannkuch, M. (2008). Building sampling concepts for statistical inference: A case study. In *Proceedings of the eleventh international congress of mathematics education* (ICOTS 11). Monterrey, Mexico. Online: http://tsg.icme11.org/tsg/show/15

Pfannkuch, M., & Budgett, S. (2014). Constructing inferential concepts through bootstrap and randomization-test simulations: A case study. In K. Makar, B. de Sousa, & R. Gould (Eds.). Sustainability in statistics education. *Proceedings of the ninth international conference on teaching statistics* (ICOTS9) Flagstaff, Arizona, USA.

Pfannkuch, M., Arnold, P., & Wild, C. (2015). What I see is not quite the way it really is: Students' emergent reasoning about sampling variability. *Educational Studies in Mathematics., 88*, 343–360. https://doi.org/10.1007/s10649-014-9539-1

Reading, C., & Reid, J. (2006). An emerging hierarchy of reasoning about distribution from a variation perspective. *Statistics Education Research Journal, 5*(2), 46–68. https://doi.org/10.52041/serj.v5i2.500

Rumsey, D. (2022). *Statistics II. For dummies* (2nd ed.). John Wiley & Sons.

Sacristan, A., Calder, N., Rojano, T., Santos-Trigo, M., Friedlander, A., & Meissner, H. (2010). The influence and shaping of digital technologies on the learning – And learning trajectories - of mathematical concepts. In C. Hoyles & J. Lagrange (Eds.), *Mathematics education and technology - rethinking the terrain* (The 17th ICMI Study) (pp. 179–226). Springer. https://doi.org/10.1007/978-1-4419-0146-0_9

Saldanha, L. A. (2003). "Is this sample unusual?" *An investigation of students exploring connections between sampling distributions and statistical inference*. Unpublished doctoral dissertation, Vanderbilt University, Nashville, TN.

Saldanha, L., & Thompson, P. (2014). Conceptual issues in understanding the inner logic of statistical inference: Insights from two teaching experiments. *Journal of Mathematical Behavior, 35*, 1–30. https://doi.org/10.1016/j.jmathb.2014.03.001

Scheaffer, R., Gnanadesikan, M., Watkins, A., & Witmer, J. (1996). *Activity-based statistics*. Springer.

Shaughnessy, J. M. (2007). Research on statistics learning and reasoning. In F. K. Lester Jr. (Ed.), *Second handbook of research on mathematics teaching and learning, 2* (pp. 957–1009). Information Age.

Stat Trek. Statistics dictionary. Teach yourself statistics. Accessed 6/30/2019. https://stattrek.com/statistics/dictionary.aspx?definition=margin%20of%20error

Tall, D., & Vinner, S. (1981). Concept image and concept definition in mathematics with particular reference to limits and continuity. *Educational Studies in Mathematics, 12*, 151–169. https://doi.org/10.1007/BF00305619

Thompson, P. W., & Liu, Y. (2005). Understandings of margin of error. In S. Wilson (Ed.), *Proceedings of the twenty-seventh annual meeting of the International Group for the Psychology of mathematics education*. Virginia Tech.

Thornton, R., & Thornton, J. (2004). Erring on the margin of error. *Southern Economic Journal, 71*(1), 130–135. https://doi.org/10.1002/j.2325-8012.2004.tb00628.x

Tintle, N., Carver, R., Chance, B., Cobb, G., Rossman, A., Roy, S., Swanson, T., & Vander Stoep, J. (2019). *Introduction to statistical investigations*. Wiley.

Tversky, A., & Kahneman, D. (1971). Belief in the law of small numbers. *Psychological Bulletin, 76*, 105–110. https://doi.org/10.1037/h0031322

Van Dijke-Droogers, M. J. S., Drijvers, P. H. M., & Bakker, A. (2020). Repeated sampling with a black box to make informal statistical inference accessible. *Mathematical Thinking and Learning, 22*(2), 116–138. https://doi.org/10.1080/10986065.2019.1617025

Watkins, A., Bargagliotti, A., & Franklin, C. (2014). Simulation of the sampling distribution of the mean can mislead. *Journal of Statistics Education, 22*(3), 1–21. https://doi.org/10.1080/10691898.2014.11889716

Wild, C. J., Pfannkuch, M., Regan, M., & Horton, N. J. (2011). Towards more accessible conceptions of statistical inference. *Journal of the Royal Statistical Society: Series A (Statistics in Society), 174*(2), 247–295. https://doi.org/10.1111/j.1467-985X.2010.00678.x

The Mystery of the Black Box: An Experience of Informal Inferential Reasoning

Soledad Estrella, Maritza Méndez-Reina, Rodrigo Salinas, and Tamara Rojas

Abstract The development of statistical thinking is essential for citizens to make decisions in situations of uncertainty, as well as to overcome the fallibility of intuition, analyze the behavior of data with skepticism, and make inferences with foundations. In this chapter, we present the analysis of two lessons designed with tasks adapted from the statistical education literature by a lesson study group that seeks to promote the informal inferential reasoning of seventh-grade students (ages 12 and 13 years). The responses of the students to the tasks were classified using a hierarchical model, and the results show that they coordinate their informal knowledge about the context and the problem data, and they make reasonable statements beyond the data they possess. Although the implemented learning sequence allows the progressive integration of some reported components of informal statistical inference, several students show evidence of difficulties in expressing inferences that reflect the uncertainty and variability involved.

Keywords Informal statistical inference · Inferential reasoning · Lesson study · Statistical reasoning

1 Introduction

Improving learning in mathematics, and in statistics in particular, is a challenge for many educational systems. For more than two decades, the statistical discipline has been incorporated into the mathematics curricula of several countries with the aim of developing statistical thinking in citizens. Different studies and international

S. Estrella (✉) · M. Méndez-Reina · R. Salinas · T. Rojas
Pontificia Universidad Católica de Valparaíso, Valparaíso, Chile
e-mail: soledad.estrella@pucv.cl; maritza.mendez.r@mail.pucv.cl; rodrigo.salinas@pucv.cl; tamara.rojas@pucv.cl

G. F. Burrill et al. (eds.), *Research on Reasoning with Data and Statistical Thinking: International Perspectives*, Advances in Mathematics Education,
https://doi.org/10.1007/978-3-031-29459-4_16

reports, through investigative approaches, have highlighted the need to encourage early statistical thinking in students (Ben-Zvi, 2016; Ben-Zvi & Makar, 2016; Burrill, 2020; Cobb, 2015; English, 2012; Franklin et al., 2007; Estrella et al., 2022; Estrella, 2018; Franklin & Mewborn, 2006; National Council of Teachers of Mathematics, 2000, 2009). Statistical thinking has long been exhorted as a primary goal of statistics education, yet we are still learning about how best to teach and assess statistical thinking (Garfield et al., 2015).

Statistical inference, a means to establish estimates under uncertain conditions using available data, constitutes the focus of much research. Thus, statistical inference is considered a cornerstone of statistics; hence, it is desirable that at the school level, students have access to experiences that allow them to make statements about the unknown based on samples (Makar et al., 2011).

Moore (1991, p. 330) defines statistical inference as "formal methods for drawing conclusions from data taking into account the effects of randomization and other chance variation" and notes that most school curricula exclude the ideas of formal methods, randomization, and chance variation. Informal statistical inference (ISI) has often been perceived as a way to make the transition to formal inference easier and more successful (Zieffler et al., 2008).

Makar and Rubin (2009) consider that ISI is a reasoned but informal process of creating or testing generalizations from data, that is, not necessarily through standard statistical procedures, and that informal inferential reasoning (IIR) in statistics is the process of making probabilistic generalizations from (evidenced with) data that extend beyond the data collected. Garfield and Ben-Zvi (2008) regard IIR as the reasoning that leads to making an ISI. Given this scenario, as a theoretical and pedagogical approach (Ben-Zvi, 2016), ISI advocates the development of new ways of reasoning by students to support them in making decisions in situations of uncertainty, understanding that the most important function of inference is the production of authentic new knowledge.

Some authors suggest that prior to making formal inferences, it is necessary to understand the meaning of samples and sampling (Bakker, 2004) as well as variability and distributions (Castro Sotos et al., 2007; Konold et al., 2015; Van Dijke-Droogers et al., 2020, 2021; Wild, 2006) to develop students' reasoning about samples as a foundation for statistical thinking. Garfield et al. (2015) indicate that students may develop ideas of sampling variability that enable them to think statistically and make better predictions when making an inference. Offering informal inference activities at an early age, before more formal school activities, facilitates the understanding of key concepts and the probabilistic reasoning required for statistical inferences (Paparistodemou & Meletiou-Mavrotheris, 2008; Van Dijke-Droogers et al., 2020, 2021).

Our research sought to characterize the informal inferential reasoning of seventh-grade students through a learning sequence that promotes generalization beyond the data, using data as evidence, and probabilistic language (C.f., Makar & Rubin, 2009) through lessons that promote learning in an online education environment. Elements of the situations proposed in the learning sequence are described, and some characteristics of the progression of IIR are discussed based on responses by seventh-grade students, particularly analyzing the verbal and graphical reasoning that arises in two lessons out of a total of four constructed by a lesson study group.

2 Conceptual Framework

Some researchers maintain that IIR has the potential to create more coherence in the statistics curriculum as a constant line that becomes more complex as students' progress (Bakker & Derry, 2011). Bringing this reasoning to the classroom requires professional development for teachers—for example, the lesson study methodology—providing teachers with opportunities to learn and strengthen skills to help their students develop as statistical thinkers (Estrella et al., 2020).

2.1 *Statistical Inference and ISI*

The field of statistics has a key role in decision-making in various fields and sciences. In situations where data are needed, the data can be used for different purposes, either to summarize the information of a sample when determining descriptive measures or to analyze the distribution of samples and estimate unknown parameters. Estimation of population parameters (e.g., construction of confidence intervals) is one use of inferential statistics. Another use is to check claims made regarding the values of population parameters (i.e., perform hypothesis testing). For the latter purpose, statistical inference is relevant because, through this type of reasoning, conclusions or statements are generated under conditions of uncertainty when only partial data are available (Makar & Rubin, 2018).

Some studies have proposed advancing education regarding the treatment of statistical inference to work in an articulated way with statistical and probabilistic notions (Makar, 2014; Meletiou-Mavrotheris & Paparistodemou, 2015; Watson & English, 2016) to gradually reach the formal and standard treatment of inference (confidence intervals, hypothesis tests, and regression models). This progression would allow students to draw conclusions from the observation and comparison of data distributions, which corresponds to informal inference (Pfannkuch, 2006).

Statistics curricula with a descriptive approach can be transformed to a more inferential approach, which could support the advancement of statistical learning in students (Ben-Zvi, 2016; Van Dijke-Droogers et al., 2021). The initiation of inference through a pedagogy is given through the ISI approach, which allows the teaching of content associated with uncertainty prior to the use of formal inference techniques.

Various frameworks have been proposed when characterizing ISI and the reasoning that supports it. However, the common components of these frameworks require assessing and considering the available data to establish arguments associated with a question or problem, and investigators must consider the evidence that the data provide over their own experiences or personal opinions (e.g., Makar & Rubin, 2009; Pfannkuch, 2011).

Generalizing beyond the data provides the ability to communicate conclusions derived from particular data, generating inferences that are applied to a broader set of data (e.g., Zieffler et al., 2008). Additionally, expressing uncertainty implies manifesting the uncertainty in generalizations, being aware that statements cannot be stated in absolute terms (e.g., Ben-Zvi et al., 2012), coordinating aggregate views on data (e.g., Konold et al., 2015), and integrating context (e.g., Langrall et al., 2011) since the interpretation of patterns in data requires a recognition of contextual knowledge. These studies support the tendency to place informal inference at the center of the school curriculum and require rethinking with regard to how to build this reasoning on the concept of inference and how to teach it (Garfield et al., 2015).

2.2 Lesson Study

Lesson study (LS) is a form of school improvement that can positively impact student learning and teacher professional development (Estrella et al., 2018; Lewis et al., 2006; Murata, 2011). LS positions action research as the key to improving the capabilities of teachers based on the evidence they gather about student learning and development (Isoda et al., 2022; Isoda & Olfos, 2009; Lee & Tan, 2020). LS focuses on teamwork and shared responsibility around a lesson plan and the implementation and improvement of that lesson plan. One or more teachers prepare the lesson, selecting the necessary materials to achieve the objective stipulated in the lesson plan. Subsequently, one of the teachers involved in the planning implements the lesson, which in some cases is observed by other teachers or researchers. Once the lesson is over, teachers and observers meet in a session to review and analyze the implemented lesson; this collaborative process allows improvements to be made to the lesson plan before reteaching and its subsequent dissemination (Isoda & Olfos, 2009, 2021). This process has an impact on the self-efficacy of teachers (Estrella et al., 2022), a result that has been positively and empirically linked to better student performance (Darling-Hammond et al., 2017).

Considering the above, a learning sequence focused on the development of IIR in seventh-grade students was planned, implemented and analyzed through the LS process to investigate the inferences produced by the students after extracting population characteristics using samples. The following research question was used to structure the current study and the analysis of the data collected: What are the characteristics of informal inferences produced by seventh-grade students when analyzing the frequency distributions of repeated samples?

3 Methodology

This research involves a qualitative study that employs an action research methodology. This methodology entails carrying out a systematic intervention on a practice and collecting, interpreting and analyzing information about the practice, with the intention of solving problems specific to this practice and improving it, but without the intention of making a theoretical generalization (Bakker & van Eerde, 2015; Kelly & Lesh, 2000).

3.1 Participants

A total of 36 seventh-grade students, aged between 12 and 13 years, who had no experience with sampling participated (i.e., they had not yet considered the difference between populations and samples). All participants were students at two similar urban schools, both in a central region of Chile with a low socioeconomic status. The two schools were selected for convenience since the preservice teachers were working there. In School A, the course was made up of 20 male students, and at School B, it was made up of 16 students (boys and girls). Ethical consent was obtained from the students for the collection of data and digital recordings.

3.2 Data Collection

The described synchronous online lessons occurred at the end of the first semester of the 2020 school year and were carried out during the regular mathematics lessons over a period of two weeks.

For the design and implementation of the learning sequence, a lesson study group (LSG) was formed that worked collaboratively for two months. This LSG included a specialist in statistical education and LS, an in-service teacher, and two preservice teachers with experience in LS (who implemented the lessons). A total of 13 LSG meetings were held, and four lessons were implemented. Each lesson was always taught in one school before the other and the LSG held a meeting after the first implementation.

3.3 Tasks

The learning sequence seeks to familiarize students with the concepts that support inference and take a probabilistic view of statistics so that they can imagine populations and their relationship to the concepts of samples and sampling variability (Pfannkuch et al., 2012).

The LSG designed a learning sequence for the development of IIR that consisted of four lessons. The central task of lesson 1 addressed iconic data that were slightly modified from Watson and Callingham (2003) and sought to enable students to predict the variable's category in which missing data were found, beyond the data provided in the data representation. The task of lesson 2 focused on tabulated temperature data, with the values and context modified based on Oslington et al. (2020), and sought to equip students to predict the possible values of the variable to complete a table that had missing data. The tasks of lessons 3 and 4 were associated with the black box experiment proposed by Van Dijke-Droogers et al. (2020, 2021).

Considering studies on developing students' informal ideas of inference (e.g., Ben-Zvi, 2006), the experiment by Van Dijke-Droogers et al. (2020, 2021) was employed in this study because these authors argue that the ideas of repeated sampling with a black box seem to be useful for the introduction of ideas of statistical inference.

The lessons were conducted through videoconferencing platforms (Zoom or Meet), which allowed the session to be recorded and narrative data to be collected, for example, verbatim phrases from the lesson, student responses in digital forms (Google Forms), photographs of graphics or written productions by the students, and field notes of two of the members of the LSG (see Appendix 1).

Because most of the students at both schools did not have stable Internet access and/or computer equipment at home, they could not simulate the sampling distributions by themselves and explore the sampling distribution application. However, the teacher interacted with the distribution sampling app, and the students visualized on the zoom screen, the population size, the population proportion, the sample size, the population displayed by coloured balls, and the changes in the distributions simultaneously.

Considering that IIR develops progressively, in this chapter, these last two lessons of the sequence of four lessons on IIR designed in the LSG are described, both focused on repeated sampling, see Table 1. This table describes the teaching activity, IIR concepts, components learning activity, and resources used in Lessons 3 and 4.

Next, the tasks of lesson 3 and lesson 4 are described. In both lessons, students were asked to justify their arguments.

3.3.1 Description of the Task of Lesson 3: Black Box Experiment

For lesson 3, the black box experiment with 1000 balls of two different colors (yellow and pink) was considered. The 1000 balls in the box were in a proportion of 300 pink and 700 yellow. In this experiment, the students observed a video in which a box was shaken, and the interior was viewed through a small window to investigate the content of the black box by collecting and comparing sample values. The key concepts were sample, sampling variability, repeated sampling and sample size.

The task focuses on making inferences about unknown subsequent samples based on observed samples without using formal techniques and consists of predicting the number of pink balls in the black box despite only "seeing" through the window. The prediction was based on the knowledge obtained from five given samples (with n = 100), and the students were asked to make a prediction individually regarding the values of the subsequent three samples, answering questions (a and b) and completing the table (Fig. 1).

Table 1 Learning sequence of lessons 3 and 4

	Teaching activity	Concepts	IIR components	Learning activity	Resources
Lesson 3	Conduct a black box experiment (with a small window). A total of 100 balls of two colors (70 yellow and 30 pink) are inside the box, and the sample is observed through the window.	Random variable; sample; population; random experiment; frequency distribution; sampling variability; sample size; repeated sampling; estimation of the sample proportion corresponding to a variable category	Argue with evidence based on samples obtained in a random experiment; generalize possible proportions in samples (of one of the variable categories); express themselves with uncertainty	Predict the possible proportions of balls of one color in three random samples	
Lesson 4	Draw a frequency distribution of repeated samples. Subsequently, increase the number of samples, and using simulation software (sample size of 100), present a frequency distribution for the black box experiment; request predictions of the proportion of balls of a given color, considering repeated sampling and the variability in the sampling distribution.	Random variable; sample; population; random experiment; frequency distribution; sampling distribution; different number of repeated samples; variability; estimation of proportions	Argue with evidence based on 20 samples; present the variability in a frequency distribution; estimate the proportion associated with the population, with a certain degree of certainty	Draw a frequency distribution of repeated samples; observe the concentration of data in the sample distribution simulations; predict the proportion of balls of a given color in the population	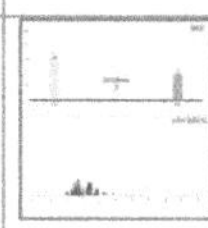 Statistical education app https://www.vustat.eu/apps/yesno/index.html

3.3.2 Description of the Task of Lesson 4: Black Box Experiment

In the central task for lesson 4, students were asked to write a list of 20 samples with possible values of pink balls and a description of each sample. Based on predictions from the black box experiment, the students drew a sampling distribution of 20 samples; the points in the dot plot represented the recognized aspects of a sampling distribution so that each of the estimates from the sample was visualized.

a. How many balls will appear in the subsequent samples?

Sample	Yellow balls	Pink balls
Sample 1	68	32
Sample 2	73	27
Sample 3	69	31
Sample 4	76	24
Sample 5	70	30
Sample 6		
Sample 7		
Sample 8		

b. What did you base your answer on?

Fig. 1 Table for lesson 3 to be completed individually by the student

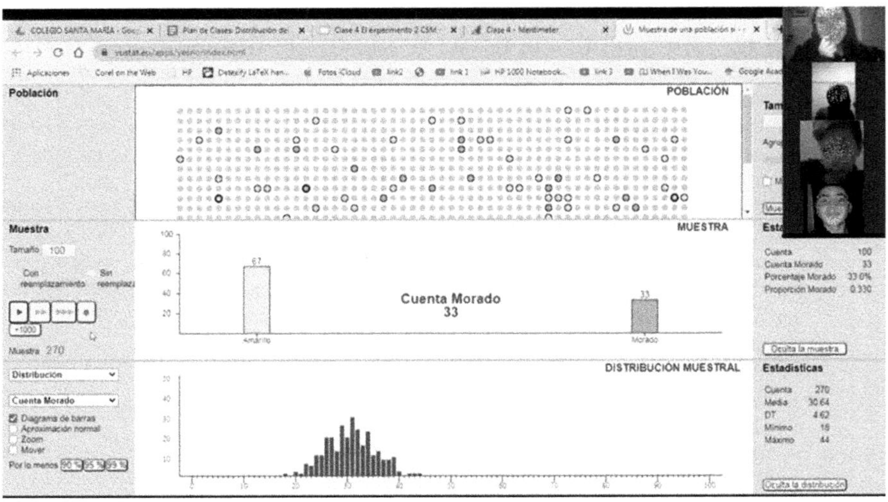

Fig. 2 Collective discussion in lesson 4 about sample size and sampling distributions

After the lesson task, the teacher simulated the sampling in the experiment, and the students observed the software application of Van Blokland and Van de Giessen (2020): the sampling distribution, the size of the population of balls, the proportion of pink balls in the population and the sample size (Fig. 2).

Table 2 General description of the learning outcomes with levels of IIR

Level	General description
5	Predicts values of variables based on the given data and argues with some degree of uncertainty.
4	Predicts values of variables and argues with some degree of uncertainty, without specifying evidence-based data.
3	Predicts values of variables based on the given data without expressing any degree of uncertainty.
2	Estimates plausible values and argues based on quantities, approximations or numerical patterns without explaining the evidence-based data or expressing some degree of uncertainty.
1	Establishes plausible values but argues without using evidence or by providing incoherent arguments and does not express any degree of uncertainty.
0	Establishes values outside the range of variability of the data provided, argues without using evidence or by providing incoherent arguments and does not express any degree of uncertainty.

3.4 Data Analysis

The analysis was qualitative and interpretive and focused on the arguments distinguishing components of the IIR of the students for each task. Rojas and Salinas (2020) proposed six levels for IIR (as a hierarchical sequence, see Table 2) that reflect the theoretical frameworks introduced by Biggs and Collis (2014). The levels described allowed us to classify, analyze and describe the reasoning of the students from greater to lesser degrees of sophistication and complexity in relation to the achievement of the desired result.

Two of the authors reviewed all the transcripts and marked and distinguished the arguments given in response to the tasks. Subsequently, the other two authors compared the arguments in light of the hierarchical model of IIR (generalization beyond the data; evidence based on the data; and use of language with uncertainty). The six levels relate to the three components of IIR in terms of observable learning outcomes that place them in a continuum at levels of ascending cognitive complexity. The analysis included repeated reviews of the data sources in several rounds of consensus.

4 Results

4.1 Answers to the Task of Lesson 3: Black Box Experiment

The black box allowed students to explore the variability involved in repeated sampling. Thus, through the question "How many balls will appear in the subsequent samples?", the students proposed values (number of pink balls); the question "What did you base your answer on?" allowed the students to argue the reason for their choices. Appendix 2 presents the level assigned, the sample values, and the arguments for each of the 36 students.

Sixty-four percent of the students (23) provided arguments that were classified as level 0, 1 or 2, the lowest in the hierarchy, indicating a lack of appreciation of uncertainty, perhaps because the experience of sampling was new to them. Levels 1 and 2 reflect intuitive statistical beliefs by giving plausible values. It seems that the arguments classified as level 2 were not statistical, and in level 1, there were no arguments referring to the data provided, or they were inconsistent.

Responses classified as level 0 (8.33% of students) included values outside the range of variability of the data provided and without arguments about the reason for their choice or with incoherent arguments; these students' predictions do not consider the given samples. For example, one of the students with a level 0 response answered *73–56-62* and did not answer the question "What did you base your answer on?" Another student proposed several values outside the range of variability and argued based on the logical fallacy of past events influencing future events. For example, *50–44-60; "I simply relied on the percentage because there were more yellow [balls]. Now, the majority should be pink [balls]"*.

The responses of ten students (27.78%) were classified as level 1. Although these students were able to estimate plausible values, they did not present arguments about the chosen values, nor did they express themselves using language with degrees of uncertainty. For example, two of these students responded *25–31-31; "I don't know" and 28–34-33; "I do not know what I based it on"*.

Regarding the ten students with responses classified as level 2 (27.78%), although they estimated plausible values, they considered the data as numbers without context, omitting the scenario of the random experiment of the black box. Thus, their arguments were based on considering quantities or numerical patterns, without some degree of uncertainty about the estimated values. The following are examples of the responses of these students: *21–28-18; "I based my answer on the numerical sequence for each one"* and *21–29-18; "There was a sequence of numbers in which you subtract 1 from the first number and 3 and so on"*.

Thirty percent of the students (11) provided arguments classified as level 3. They made predictions estimating possible values for the variables based on the data for the repeated random samples presented in the table. In their arguments, they explained that evidence was provided by the given samples considering the context of the random experiment; however, they did not use expressions that denoted uncertainty about these predictions. The following responses provide examples: *29–25-30; "In the previous samples of pink balls"; 27–25-32; "In the data of the first five samples"; and 31–26-30; "I took into account the other samples and estimated"*.

Only 6% of the students (2) provided arguments that were classified as level 4. The students appreciated the uncertainty present in the situation, and they predicted the possible values corresponding to the proportion of the population using expressions such as "could be" and "more probable", language that indicates levels of possibilities. However, their arguments did not include evidence-based values based on the repeated samplings given. The following are examples of such responses: *26–24-30; "Predicting [from the samples] what the result could be" and 37–29-40; "I thought that every time I shook the bottle, it would be random. So, I thought that for each turn, [the sample] could be more likely to go up or down [the number of pink balls]".*

Regarding the values to be predicted for the three samples, the majority of the 36 students (92%) generalized about plausible values for the content of the physical black box; that is, they produced sample values for the pink balls corresponding to the proportion of the population. The median and mode of the estimates by the students were 28 and 29 (the expected value was 30), respectively, and the values given oscillated between 14 and 70 pink balls (see Table B.1, Appendix 2).

Given that no student responses were classified as level 5 and that the responses of only two students were classified as level 4, the learning sequence requires spaces for dialog in which the certainty of the inferences that are constructed from data and subsequent generalizations can be discussed to support students' understanding of uncertainty.

4.2 Answers to the Task of Lesson 4: Drawing a Frequency Distribution

Next, some student productions associated with the central task of lesson 4 are presented. The students had to draw a hypothetical frequency distribution of pink balls using 20 samples from the black box experiment studied in the previous lesson. Table 3 summarizes some student productions related to the frequency distribution of the black box experiment for a small number of samples (20). The productions of six students from both schools were analyzed by the authors; the plots were chosen as representative of the whole due to their readability, the differences among them, and the use of a scale axis and the concentration of points (sample data). Since the lessons were online, several students sent their plots, but they did not explain them; therefore, we make assumptions about their understanding from their dot plots.

The six examples of frequency distribution drawings considered data close to 30 (mostly without showing maximum frequencies at 30), with minimum values starting at 18 and some values greater than 40 (although with frequency 1). Additionally, the drawings suggested that the students understood that the samples varied. Through this variability and given a specific sample size, the students were able to make a generalization of the behavior of the sample distribution beyond the data provided in the previous samples (observed in the simulation), and the construction of their graphs realized this informal reasoning about samples and sampling variability.

Before the teacher showed the simulation of a sample size of 100, some students predicted that if the number of samples increased, the number of pink balls would increase, for example, E6, "from approximately 70 to 80". To discuss this

Table 3 Analysis of six frequency distribution drawings of 20 repeated samples

Frequency distribution plotted for 20 repeated samples	Production characteristics
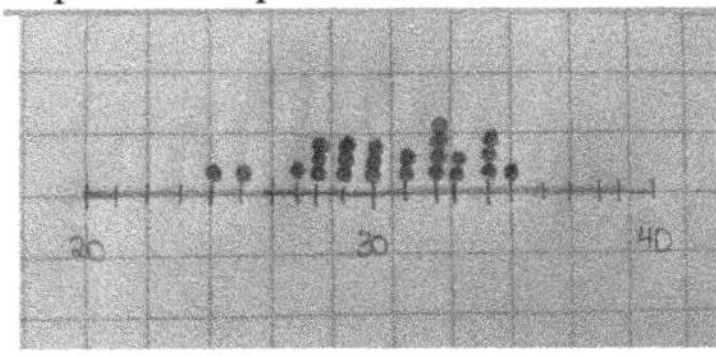	The horizontal axis is graduated from 20 to 40. This student represented the predicted data with values close to the value 30 (although he/she presents 24 samples instead of 20). The sampling distribution has a mode of 32, with frequency 4, close to the expected value (30). The student may have considered the variability because the predictions are concentrated between 24 and 35.
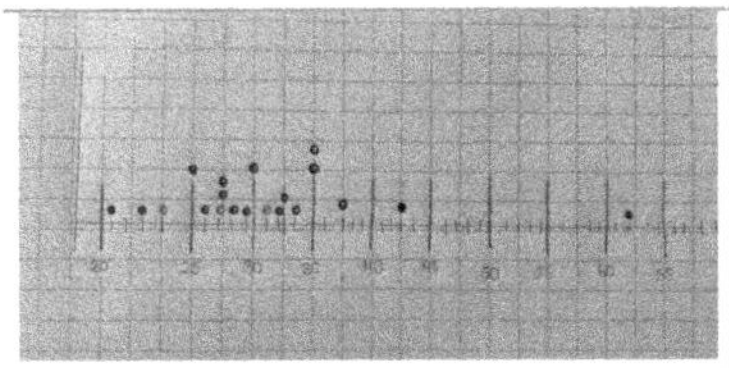	The horizontal axis was graduated from 20 to 65 (although it presented segments for which possible data were positioned that could not allow the distribution to be visualized correctly). In this case, the student centered most of the data between 25 and 35, with only a mode of 27 and frequency of 3. The student may have considered high sampling variability because he/she drew possible distant values, noting values such as 21 and 62.
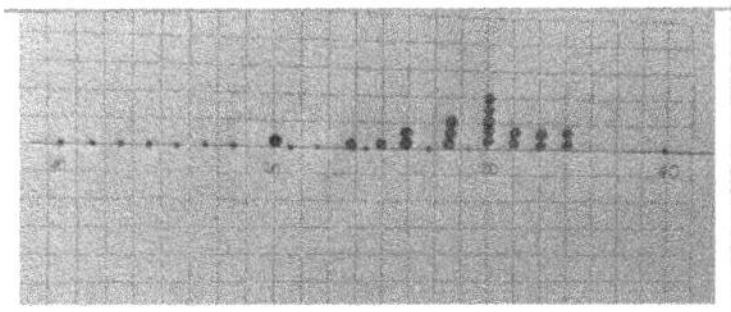	There is a horizontal axis graduated from 10 to 40 (although the distances between gradations were not consistent). The student centered the values close to 30, with a mode of 30 and frequency of 5. The student may have considered the variability when drawing values between 20 and 33, with a value of 20 for one of the samples.
	The horizontal axis was graduated from 10 to 40. The data were concentrated between 26 and 35; a central value was not identified; there were seven values with a maximum frequency of 2. The student considered the variability in the distribution of the values between 18 and 37.
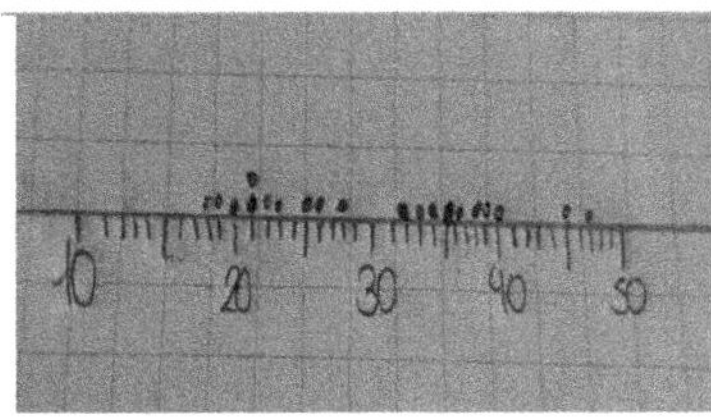	The horizontal axis was graduated between 10 and 50. The data were not concentrated around a value because the values ranged from 17 to 47, with 21 being the only value with a frequency greater than 1 (2); there were no data points between 29 and 31. The student may have considered the concept of variability but not of a central value close to the expected sample value.
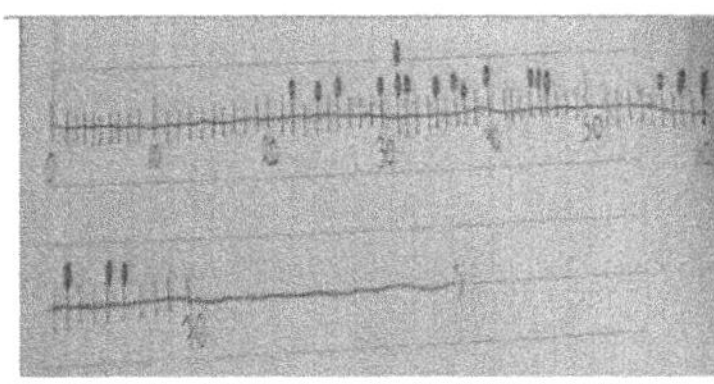	The axis was graduated from 0 to 70, with 14 data points between 22 and 46; 31 was the only value with a frequency greater than 1 (2). Six data points (56, 58, 60, 62, 65, 66) were outside the range of the expected sample value. The student may have considered high sampling variability.

misconception, the teacher proposed that the students observe the behavior of the data in the dot plot for $n = 100$.

Teacher: Let's see [in the App] if, instead of 50 samples, I took 100 samples. What is happening?

E7: The numbers are repeated a lot… the same numbers show up; they come out between the limits, between 20 and 50. And when [the teacher] included more samples, the numbers began to be repeated.

Teacher: (…) But you said that if I included more samples, then I should get larger numbers. Did that happen?

Students: No… no… [several students in the class express this answer].

Teacher: What happened?

E2: It went to 40 and 50 [referring to the maximum values], but not much larger.

Teacher: We increased the samples to 200 [in the app]. Did what E4 and E6 predicted happen?

E4: Yes, more numbers appeared; more appeared closer to 45.

Teacher: What is happening with this graph?

E4: It is changing; it is growing.

Teacher: Between what values is the graph growing?

E4: […] between the random numbers that are coming out.

E8: Between 20 and 40.

Through the collective discussion (E2, E4, E7 and E8 of School B), the students were able to confront their previous ideas about what could happen when considering a greater sample size of 100, visually identifying some attributes of the distributions, such as maximum and minimum values in the samples and the concentration of the data as a first approximation of the expected number of pink balls. Subsequently, the students observed the sampling distribution graphs on the software application (resource indicated in Table 1) when the number of samples increased beyond 200 and 900. The following dialog between the teacher and some students (E3, E4, and E5 of School B) illustrates how the lesson plan anticipated the incipient transition from the value in which the data are concentrated (regarding the behavior of the pink balls) to the population value.

Teacher: Think about the following. In the sample, we have 100, […] and if we talk about 1000 balls, what will happen then?

E3: It will be concentrated between 40 and 60.

E4: It is between 30 and 40 all the time. For example, when shaking [the black box], a certain number of balls will come out; so, 100 will not come out immediately; it will always be rounded to a number.

Teacher: What is the total number of pink balls?

E5: 500.

E3: 310… I don't know.

Teacher: How?

E3: If there are 31 and you want to do 1000, that should be multiplied, I do not know…

E4: We must multiply the largest number of those that are left by the number of attempts.

E3: 310, because I tried to multiply… because if I want to see ten times more, I added zero.

The excerpt from the last lesson shows the need to expand the number of lessons together with the manipulation of the black box and the exploration of repeated sampling in the application. Additionally, the students manifested incipient statistical reasoning when identifying the concentration of and variability in the data in context. As Noll and Shaughnessy (2012) suggest, focusing on the variability aspect of the distribution would improve students' reasoning about sampling distributions.

However, students also indicated mathematical procedures such as multiplication or rounding and moving away from the context of the data and the construction of inferences, apparently reacting in a manner that is usual in a mathematics lesson and moving away from IIR.

5 Conclusions and Discussion

This chapter reports on a part of the design of a four-lesson learning sequence to introduce informal statistical inference in seventh grade in an online environment. The characteristics of informal inferences produced by seventh-grade students when analyzing frequency distributions of repeated samples were observed. Some characteristics of IIR demonstrated by the students were analyzed based on verbal and graphic reasoning that emerged in the last two lessons, and the IIR of 36 Chilean students was characterized in two tasks related to samples in the black box experiment. Although some students overestimated the values and proposed only low local maximum frequencies, our study indicates that in a short period of time and using the black box experiment as a central task, some students can draw and recognize the distribution of frequencies (expected) and possibly begin to develop an understanding of concepts such as sample, variability and distributions.

The results suggested that seventh-grade students could coordinate their informal knowledge about the context and the data about the problem and draw inferences beyond the data they possess. The first two unreported lessons attempted to sensitize the subjects to inference (Zieffler et al., 2008) using interesting and realistic material and simple descriptive analyses (e.g., de Vetten et al., 2018; Paparistodemou & Meletiou-Mavrotheris, 2008). Although the learning sequence implemented could allow the progressive integration of some reported components of ISI, there are several cases that give evidence of difficulties in expressing inferences (probabilistic language) that reflect the uncertainty and variability involved. On the other hand, the representations produced by the students demonstrate an understanding of the frequency distribution from sampling predictions in the context of a black box and therefore establish a path to understanding sampling distribution (C.f., Van Dijke-Droogers et al., 2021).

Among the limitations of the present study, although the lesson plan indicated that the students would use the application to observe the simulations, the teacher actually presented the application that simulated the repeated sampling; this occurred as a result of the conditions of online education, in particular that the majority of students did not have access to high-speed internet and/or quality computer equipment. Even though it was key for the students to simultaneously observe the behavior of the data in the simulation and evaluate what was happening with respect to the variability of the sample size and the number of samples considered, this was not achieved.

Another limitation due to restricted access to schools and students caused by the pandemic was the lack of personal interviews that would have allowed a more complete analysis, making it possible to detect the development of IIR after the learning sequence. The above made it difficult to assess the sense of the data (Estrella, 2018; Estrella et al., 2021) that the students had developed in their schooling, for example, their way of explaining and representing variability when reporting data (samples), and thus of the students' understanding of the different representations (those drawn by the students and those delivered by the application) and their reasoning when making predictions.

Several researchers point out that an inference is uncertain and that it requires expressions that denote degrees of uncertainty regarding the situation. De Vetten et al. (2018) described probabilistic language as language that expresses the certainty of inferences and that considers sampling variability (understanding that the results for samples are similar and that, therefore, in certain circumstances, a sample can be used to make inferences) and the degree of representativeness of the sample (considering the sampling method and sample size). However, in this study, few students were able to express uncertainty in their inferences; this requires further investigation of the design of lessons that provoke dialogic conversations about aspects of sampling, such as variability and representativeness, and that confront students with the impossibility of making inferences with absolute certainties.

It is also necessary to design and implement a broader learning sequence (e.g., Van Dijke-Droogers et al., 2021) that considers the simulation of sampling distributions with different sample sizes (so that students understand that a larger sample size leads to variability in the estimation of the population and better inferences) and considers a variable number of repetitions (so that students understand that this leads to less variability in the sample mean and therefore to a better estimate of the population).

Dvir and Ben-Zvi (2021) indicated a need for students to make informal inferences but perform an in-depth examination of the mechanism or random sources and understand the difference between the population and the sample. Considering the stochastic components, we plan to investigate a learning sequence with lessons that promote dialog and the use of simulation software so that students can develop IIR by comparing and using distributions such as a statistical model to interpret variability and uncertainty.

Acknowledgments Support from ANID Fondecyt 1200346, FONDEF ID20i10070, VRIEA PUCV 039.439/2020, National Doctorate Scholarship ANID 21210862, and ANID/PIA/Basal Funds for Centers of Excellence FB0003 is gratefully acknowledged.

Appendix 1

Table A.1 shows the technological platforms that were necessary in each session of the learning sequence. The resource column lists the technologies used by the teacher to implement the lesson. In the "other records" column, the platforms or means necessary for the students to respond to the central activity of the lesson can be observed.

Table A.1 Technologies implemented online for each lesson of the learning sequence

Lesson	School	Video conference	Resources	Other records
1	B	Zoom	Video – Chat	Google Forms
	A	Google Meet	PPT – Video – Chat	Google Forms
2	B	Zoom	PPT – Chat	Google Forms
	A	Google Meet	PPT – Chat	Google Forms
3	B	Zoom	PPT – Video – App – Chat -Mentimeter	Google Forms
	A	Google Meet	PPT – Video – App – Chat -Mentimeter	Google Forms
4	B	Zoom	PPT – Video – App – Chat – Mentimeter	Photographs
	A	Google Meet	PPT – Video – App – Chat – Mentimeter	Photographs

Appendix 2 (Table B.1)

Table B.1 IIR level based on the values for three samples (S6, S7, S8) and on the students' arguments provided for the task of lesson 3 *(n* = 36)

IIR Level	Student	S6	S7	S8	Students' arguments given to the question, What did you base your answer on?
4	E4B	26	24	30	"Predicting [from the samples] what the result could be"
4	E7B	37	29	40	"I thought that every time I shook the bottle, it would be random. So, I thought that for each turn, [the sample] could be more likely to go up or down [the number of pink balls]"
3	E10A	31	26	30	"I considered the other samples and estimated"
3	E13A	33	30	29	"I relied on the numbers of balls that appeared in the previous samples to give an approximate value"
3	E1A	27	23	28	"An approximation of the number of pink balls"
3	E6A	29	25	30	"The previous samples of pink balls"
3	E9B	27	25	32	"The data of the first five samples"
3	E14B	31	27	34	"Because these data [the given samples] are close to the temperatures, and I looked at that"
3	E4A	28	26	29	"The table [of data]"
3	E7A	28	27	31	"The number of pink balls varied within seven balls with the other samples"

(continued)

Table B.1 (continued)

IIR Level	Student	S6	S7	S8	Students' arguments given to the question, What did you base your answer on?
3	E11A	31	70	32	"It was based on the average of the previous samples"
3	E12A	25	31	26	"I made an approximation of the balls that were going to come out"
3	E18A	33	25	23	"Statistics"
2	E20A	40	37	42	"Estimated other responses"
2	E1B	26	28	30	"[I based] it on the results and got an approximation"
2	E5A	21	29	19	"On how the figures were"
2	E16A	21	29	18	"There was a sequence of numbers in which the first number subtracted 1 and the second subtracted 3 and so on"
2	E17A	28	26	29	"I based my answer on the fact that since they could not leave a defined number, I put those that had not come out yet"
2	E2B	21	28	18	"It was based on the numerical sequence that each one had"
2	E8B	31	33	24	"On previous issues"
2	E16B	28	33	26	"I relied on the previous numbers"
2	E3B	29	28	31	"I rounded the numbers"
2	E12B	28	29	27	"I did it by approximating"
1	E3A	26	33	14	Did not answer
1	E8A	39	40	26	Did not answer
1	E9A	34	29	38	"On people's algorithm"
1	E15A	23	29	22	Did not answer
1	E19A	25	31	31	"I don't know"
1	E5B	28	34	33	"I do not know what I based it on"
1	E6B	29	26	33	"I will explain it in the next class"
1	E10B	25	23	16	"I did it randomly"
1	E13B	51	24	43	Did not answer
1	E15B	27	28	26	Did not answer
0	E11B	39	60	58	"I based my answer on the data in the table"
0	E14A	50	44	60	"I simply relied on the percentage since there were more yellow ones. Now, the majority should be pink ones."
0	E2A	73	56	62	Did not answer

References

Bakker, A. (2004). Reasoning about shape as a pattern in variability. *Statistics Education Research Journal, 3*(2), 64–83.

Bakker, A., & Derry, J. (2011). Lessons from inferentialism for statistics education. *Mathematical Thinking and Learning, 13*(1–2), 5–26. https://doi.org/10.1080/10986065.2011.538293

Bakker, A., & van Eerde, D. (2015). An introduction to design-based research with and example from statistics education. In A. Bikner-Ahsbahs, C. Knipping, & N. Presmeg (Eds.), *Approaches to qualitative research in mathematics education. Examples of methodology and methods* (pp. 429–466). Springer. https://doi.org/10.1007/978-94-017-9181-6_16

Ben-Zvi, D. (2006, July 2–7). Scaffolding students' informal inference and argumentation. In A. Rossman. & B. Chance (Eds.), *Proceedings of the 7th international conference on teaching of statistics (CD-ROM), Salvador.*

Ben-Zvi, D. (2016). Tres paradigmas en el Desarrollo del razonamiento estadístico de los estudiantes [Three paradigms in developing students' statistical reasoning]. In S. Estrella et al. (Eds.), *XX Actas de las Jornadas Nacionales de Educación Matemática* (pp. 13–22). SOCHIEM.

Ben-Zvi, D., & Makar, K. (2016). International perspectives on the teaching and learning of statistics. In D. Ben-Zvi & K. Makar (Eds.), *The teaching and learning of statistics* (pp. 1–10). Springer. https://doi.org/10.1007/978-3-319-23470-0_1

Ben-Zvi, D., Aridor, K., Makar, K., & Bakker, A. (2012). Students' emergent articulations of uncertainty while making informal statistical inferences. *ZDM, 44*(7), 913–925. https://doi.org/10.1007/s11858-012-0420-3

Biggs, J. B., & Collis, K. F. (2014). *Evaluating the quality of learning: The SOLO taxonomy (structure of the observed learning outcome)*. Academic.

Burrill, G. (2020, July 6–12). Statistical literacy and quantitative reasoning: Rethinking the curriculum. In P. Arnold (Ed.), *New skills in the changing world of statistics education: Proceedings of the roundtable conference of the International Association for Statistical Education (IASE)*, Held online.

Castro Sotos, A. E. C., Vanhoof, S., Van den Noortgate, W., & Onghena, P. (2007). Students' misconceptions of statistical inference: A review of the empirical evidence from research on statistics education. *Educational Research Review, 2*(2), 98–113. https://doi.org/10.1016/j.edurev.2007.04.001

Cobb, G. W. (2015). Mere renovation is too little, too late: We need to rethink the undergraduate curriculum from the ground up. *The American Statistician, 69*(4), 266–282. https://doi.org/10.1080/00031305.2015.1093029

Darling-Hammond, L., Hyler, M., & Gardner, M. (2017). *Effective teacher professional development*. Learning Policy Institute. https://files.eric.ed.gov/fulltext/ED606743.pdf

de Vetten, A., Schoonenboom, J., Keijzer, R., & van Oers, B. (2018). The development of informal statistical inference content knowledge of pre-service primary school teachers during a teacher college intervention. *Educational Studies in Mathematics, 99*(2), 217–234. https://doi.org/10.1007/s10649-018-9823-6

Dvir, M., & Ben-Zvi, D. (2021). Informal statistical models and modeling. *Mathematical Thinking and Learning, 1–21*, 79–99. https://doi.org/10.1080/10986065.2021.1925842

English, L. (2012). Data modeling with first-grade students. *Educational Studies in Mathematics, 81*(1), 15–30. https://doi.org/10.1007/s10649-011-9377-3

Estrella, S. (2018). Data representations in early statistics: Data sense, meta-representational competence and transnumeration. In A. Leavy, M. Meletiou-Mavrotheris, & E. Paparistodemou (Eds.), *Statistics in early childhood and primary education – Supporting early statistical and probabilistic thinking* (pp. 239–256). Springer. https://doi.org/10.1007/978-981-13-1044-7_14

Estrella, S., Mena, A., & Olfos, R. (2018). Lesson study in Chile: A very promising but still uncertain path. In M. Quaresma, C. Winsløw, S. Clivaz, J. da Ponte, A. Ní Shúilleabháin, & A. Takahashi (Eds.), *Mathematics lesson study around the world: Theoretical and methodological issues* (pp. 105–122). Springer. https://doi.org/10.1007/978-3-319-75696-7

Estrella, S., Zakaryan, D., Olfos, R., & Espinoza, G. (2020). How teachers learn to maintain the cognitive demand of tasks through lesson study. *Journal of Mathematics Teacher Education, 23*, 293–310. https://doi.org/10.1007/s10857-018-09423-y

Estrella, S., Vergara, A., & González, O. (2021). Developing data sense: Making inferences from variability in tsunamis at primary school. *Statistics Education Research Journal, 20*(2), 16. https://doi.org/10.52041/serj.v20i2.413

Estrella, S., Méndez-Reina, M., Olfos, R., & Aguilera, J. (2022). Early statistics in kindergarten: Analysis of an educator's pedagogical content knowledge in lessons promoting informal inferential reasoning. *International Journal for Lesson and Learning Studies, 11*(1), 1–13. https://doi.org/10.1108/IJLLS-07-2021-0061

Franklin, C., & Mewborn, D. (2006). The statistical education of grades pre-K-2 teachers: A shared responsibility. In G. Burrill (Ed.), *NCTM 2006 yearbook: Thinking and reasoning with data and Chance* (pp. 335–344). NCTM.

Franklin, C., Kader, G., Mewborn, D., Moreno, J., Peck, R., Perry, M., et al. (2007). *Guidelines for assessment and instruction in statistics education (GAISE) report: A preK-12 curriculum framework*. American Statistical Association.

Garfield, J., & Ben-Zvi, D. (2008). *Developing students' statistical reasoning: Connecting research and teaching practice*. Springer. https://doi.org/10.1007/978-1-4020-8383-9

Garfield, J., Le, L., Zieffler, A., & Ben-Zvi, D. (2015). Developing students' reasoning about samples and sampling variability as a path to expert statistical thinking. *Educational Studies in Mathematics, 88*(3), 327–342. https://doi.org/10.1007/s10649-014-9541-7

Isoda, M., & Olfos, R. (2009). *El enfoque de resolución de problemas en la enseñanza de la matemática a partir del estudio de clases* (The problem-solving approach in the teaching of mathematics from the lesson study). Ediciones Universitarias de Valparaíso.

Isoda, M., & Olfos, R. (2021). *Teaching multiplication with lesson study*. Springer. https://doi.org/10.1007/978-3-030-28561-6

Isoda, M., Olfos, R., Estrella, S., & Baldin, Y. (2022). Two contributions of japanese lesson study for the mathematics teacher education: The effective terminology for designing lessons and as a driving force to promote sustainable study groups. *Educação Matemática Em Revista, 1*(23), 98–112. https://doi.org/10.37001/EMR-RS.v.2.n.23.2022.p.98-112

Kelly, A., & Lesh, R. (2000). *Handbook of research design in mathematics and science education*. Routledge. https://doi.org/10.4324/9781410602725

Konold, C., Higgins, T., Russell, S. J., & Khalil, K. (2015). Data seen through different lenses. *Educational Studies in Mathematics, 88*(3), 305–325. https://doi.org/10.1007/s10649-013-9529-8

Langrall, C., Nisbet, S., Mooney, E., & Jansem, S. (2011). The role of context expertise when comparing groups. *Mathematical Thinking and Learning, 13*(1–2), 47–67. https://doi.org/10.1080/10986065.2011.538620

Lee, L., & Tan, S. (2020). Teacher learning in lesson study: Affordances, disturbances, contradictions, and implications. *Teaching and Teacher Education, 89*, 102986. https://doi.org/10.1016/j.tate.2019.102986

Lewis, C., Perry, R., & Murata, A. (2006). How should research contribute to instructional improvement? The case of lesson study. *Educational Researcher, 35*(3), 3–14. https://doi.org/10.3102/0013189X035003003

Makar, K. (2014). Young children's explorations of average through informal inferential reasoning. *Educational Studies in Mathematics, 86*(1), 61–78. https://doi.org/10.1007/s10649-013-9526-y

Makar, K., & Rubin, A. (2009). A framework for thinking about informal statistical inference. *Statistics Education Research Journal, 8*(1), 82–105.

Makar, K., & Rubin, A. (2018). Learning about statistical inference. In D. Ben-Zvi, K. Makar, & J. Garfield (Eds.), *International handbook of research in statistics education* (pp. 261–294). Springer. https://doi.org/10.1007/978-3-319-66195-7_8

Makar, K., Bakker, A., & Ben-Zvi, D. (2011). The reasoning behind informal statistical inference. *Mathematical Thinking and Learning, 13*(1), 152–173. https://doi.org/10.1080/10986065.2011.538301

Meletiou-Mavrotheris, M., & Paparistodemou, E. (2015). Developing students' reasoning about samples and sampling in the context of informal inferences. *Educational Studies in Mathematics, 88*(3), 385–404. https://doi.org/10.1007/s10649-014-9551-5

Moore, D. S. (1991). *Statistics: Concepts and controversies* (3rd ed.). W. H. Freeman.

Murata, A. (2011). Introduction: Conceptual overview of lesson study. In L. Hart, A. Alston, & A. Murata (Eds.), *Lesson study research and practice in mathematics education* (pp. 1–12). Springer. https://doi.org/10.1007/978-90-481-9941-9

National Council of Teachers of Mathematics. (2000). *Principles and standards for school mathematics*. Author.

National Council of Teachers of Mathematics. (2009). *Navigating through data analysis and probability in prekindergarten-grade 2* (Vol. 1). Author.

Noll, J., & Shaughnessy, M. (2012). Aspects of students' reasoning about variation in empirical sampling distribution. *Journal for Research in Mathematics Education, 43*(5), 509–556. https://doi.org/10.5951/jresematheduc.43.5.0509

Oslington, G., Mulligan, J., & Van Bergen, P. (2020). Third-graders' predictive reasoning strategies. *Educational Studies in Mathematics, 104*(1), 5–24. https://doi.org/10.1007/s10649-020-09949-0

Paparistodemou, E., & Meletiou-Mavrotheris, M. (2008). Developing young students' informal inference skills in data analysis. *Statistics Education Research Journal, 7*(2), 83–106.

Pfannkuch, M. (2006, July 2–7). Informal inferential reasoning. In A. Rossman & B. Chance (Eds.), *Proceedings of the 7th international conference on teaching of statistics (CD-ROM), Salvador*.

Pfannkuch, M. (2011). The role of context in developing informal statistical inferential reasoning: A classroom study. *Mathematical Thinking and Learning, 13*(1–2), 27–46. https://doi.org/10.1080/10986065.2011.538302

Pfannkuch, M., Wild, C., & Parsonage, R. (2012). A conceptual pathway to confidence intervals. *ZDM, 44*(7), 899–911. https://doi.org/10.1007/s11858-012-0446-6

Rojas, T., & Salinas, R. (2020). *Una secuencia de aprendizaje que desarrolla el razonamiento inferencial estadístico informal, diseñada en un estudio de clases para una enseñanza escolar online.* [A learning sequence that develops informal statistical inferential reasoning, designed in a lesson study for online school teaching]. [Unpublished undergraduate thesis, Pontificia Universidad Católica de Valparaíso].

Van Blokland, P., & Van de Giessen, C. (2020). *VUSTAT* [computer software]. Amsterdam, the Netherlands: VUSOFT. https://www.vustat.eu/apps/yesno/index.html

Van Dijke-Droogers, M., Drijvers, P., & Bakker, A. (2020). Repeated sampling with a black box to make informal statistical inference accessible. *Mathematical Thinking and Learning, 22*(2), 116–138. https://doi.org/10.1080/10986065.2019.1617025

Van Dijke-Droogers, M., Drijvers, P., & Bakker, A. (2021). Introducing statistical inference: Design of a theoretically and empirically based learning trajectory. *International Journal of Science and Mathematics Education, 1–24*, 1743–1766. https://doi.org/10.1007/s10763-021-10208-8

Watson, J., & Callingham, R. (2003). Statistical literacy: A complex hierarchical construct. *Statistics Education Research Journal, 2*(2), 3–46.

Watson, J., & English, L. D. (2016). Repeated random sampling in year 5. *Journal of Statistics Education, 24*(1), 27–37. https://doi.org/10.1080/10691898.2016.1158026

Wild, C. J. (2006). The concept of distribution. *Statistics Education Research Journal, 5*(2), 10–26. https://doi.org/10.52041/serj.v5i2.497

Zieffler, A., Garfield, J., delMas, R., & Reading, C. (2008). A framework to support research on informal inferential reasoning. *Statistics Education Research Journal, 7*(2), 40–58.

Part IV
Data and Society

Critical Citizenship in Statistics Teacher Education

Cindy Alejandra Martínez-Castro, Lucía Zapata-Cardona, and Gloria Lynn Jones

Abstract This research shows evidence of critical citizenship when prospective teachers engage in statistical investigations. Critical citizenship is a quality of thought that promotes environmentally, socially, politically and economically conscious citizens and develops critical dispositions towards the world in which they live. The participants were 10 prospective teachers pursuing a degree in Mathematics Education at a well-known public university in northwestern Colombia. The prospective teachers voluntarily participated in the research in which they carried out four classroom statistical investigations related to social crises. The main sources of information came from discussions in eight lessons where prospective teachers studied empirically crises of society using statistical tools. Supplementary information included ideograms, autobiographies, and narratives. The results show that statistical investigations promote statistical thinking and help to make evident the sense of agency.

Keywords Statistics education · Teacher education · Critical citizenship · Statistical thinking · Sense of agency

1 Introduction

Statistics teacher education represents a challenge in a world in which statistics permeates multiple settings of modern life. Today's citizen needs to use statistical concepts and tools to make sense of the socio-political problems present in their contexts (Weiland, 2019). To meet this need, teachers are required to help their

C. A. Martínez-Castro · L. Zapata-Cardona (✉)
Universidad de Antioquia, Medellín, Colombia
e-mail: cindy.martinez@udea.edu.co; lucia.zapata1@udea.edu.co

G. L. Jones
Wadsworth Magnet School for High Achievers, Decatur, GA, USA

G. F. Burrill et al. (eds.), *Research on Reasoning with Data and Statistical Thinking: International Perspectives*, Advances in Mathematics Education,
https://doi.org/10.1007/978-3-031-29459-4_17

students make that sense. However, research in statistics teacher education (Zapata-Cardona & González-Gómez, 2017; Souza et al., 2015) has recognized the limitations of some programs that focus solely on the technical component (Loya, 2008) of disciplinary knowledge, and very timidly on the development of the critical ability (Zapata-Cardona & González-Gómez, 2017). The mastery of statistical knowledge is not enough to develop critical citizenship (Zapata-Cardona & Marrugo, 2019). Teacher education should make efforts to prepare teachers to be capable of participating in their communities by critically reading statistics and their uses in sociocultural environments (Ernest, 2015).

Statistics teacher education requires to focus on the academic skills needed for teachers' professional life and, simultaneously, on the critical skills to read and write the world (Gutstein, 2007). Reading the world means using statistics as a lens to understand the problems present in sociocultural contexts, and writing the world means acting to contribute to the solution of such problems (Weiland, 2017) as well as to contribute to the transformation of society.

One possibility of overcoming this limitation is to link teacher education programs to the promotion of critical citizenship through strategies such as *statistical investigations*. This is a way to overcome the teaching that traditionally exalts the learning of statistical concepts and procedures isolated from the context in which they have occurred (Zapata-Cardona, 2016a, p. 76). Moreover, the focus is not exclusively on the acquisition of knowledge but takes into account the social dimension of the human beings (Zapata-Cardona, 2016b). The goal of this study is to answer the research question: What evidence of critical citizenship do preservice teacher exhibit while engaging in statistical investigations?

2 Theoretical Background

2.1 *Critical Citizenship*

Critical citizenship is related to the active participation of people in their communities and governments, where they interrogate the structures that produce conditions of injustice, and work to change them (Weiland, 2019, 2017). Critical citizens can use statistics to influence, shape and transform socially constructed structures and discourses around them (Weiland, 2017). Statistics is a social construction that allows studying social justice issues (Lesser, 2014, 2007) and promoting reflections on the role it plays in society.

Contributions such as those of Zapata-Cardona and Marrugo (2019) have emphasized that statistics should be a tool to help citizens understand and transform their world beyond the domain of concepts and procedures. These authors define critical citizenship as "an intellectual tool oriented to educate critical and aware citizens who have the responsibility to participate in society and contribute to its transformation" (p. 375). Critical citizenship has to do with the active participation of subjects in their communities and societies, oriented towards the construction of social justice and democracy. Such participation involves a social empowerment of teachers to critically understand the uses of statistics in society (Ernest, 2015).

In the field of teacher education, Gutiérrez (2009) has suggested that prospective teachers need to think critically about the subject matter and its teaching to become critical citizens who overcome the mere condition of uncritical consumers in a capitalist society. Prospective teachers need to develop a political knowledge of the subject matter to (1) deconstruct the discourses of power that perpetuate inequities and to (2) recognize their position in society through the identification and visibility of existing injustices in the world (Gutiérrez, 2013).

2.2 Statistical Investigations

In this research, statistical investigations are conceived as ways of organizing the teaching of statistics. They are inspired by the investigative cycle suggested by the empirical enquiry (Wild & Pfannkuch, 1999) and by the statistical problem-solving process proposed in the GAISE—*Guidelines for Assessment and Instruction in Statistics Education*—(Bargagliotti et al., 2020), but they go beyond that structure of formulating questions, collecting and analyzing data, and interpreting results. The statistical investigations support the development of subjects as critical citizens, by studying critical social, political, environmental and economic phenomena present in their contexts (Zapata-Cardona, 2016a, b). They are holistic approaches that seeks to develop both: technical knowledge and skills in the management of statistical tools and, a social conscience (Zapata-Cardona, 2016b). In other words, the statistical investigations seek to develop statistical knowledge as well as social awareness.

A statistical investigation begins with a social crisis and uses statistical tools to study it empirically, understand it, and react to it (Zapata-Cardona, 2016b). According to Skovsmose (1999), the social crises refer to phenomena of repression, conflict, contradiction, misery, inequality, ecological devastation and exploitation. Some of them are related to the unequal distribution of goods, the differences in social and economic opportunities, and the social repression caused by some ways of managing the structures of power in society. The crises are part of reality. A statistical investigation does not end when the subjects —prospective teachers— appropriate the statistical procedures and tools, but when they use them to address social crises.

Statistical investigations can be used to communicate statistical information and offer arguments to question the structures of injustice (Weiland, 2019) that allow the construction of more just, democratic and humanized societies (Campos, 2016b). That is, statistical investigations in the classroom can be tools to identify and question the uses of statistics in society and their tensions; by asking statistical questions, using data collection and analysis methods, those tensions are highlighted (Weiland, 2017).

3 Method

To show the evidence of critical citizenship while prospective teachers engage in statistical investigations, a qualitative research paradigm (Denzin & Lincoln, 2012) and a critical-dialectical approach (Sánchez, 1998) were followed. The participants were ten prospective statistics teachers (two men and eight women) who were taking a seminar on statistics teaching methods in a Mathematics Education program at a public university in a Northwestern city of Colombia. The participation was voluntary, the privacy and confidentiality of the information was kept and anonymity was guaranteed using pseudonyms for the report.

The participants carried out four statistical investigations in 8 two-hour meetings. Each statistical investigation was related to a social crisis (global warming, minimum wage in Colombia, malnutrition and obesity in Colombia and gender inequality in Colombia). In each investigation, participants were invited to study a statistical question: Is our city (planet) warming up? Has the minimum wage in Colombia lost purchasing value? What factors are associated with malnutrition and obesity in Colombia? And, is there gender inequality in our country?

The statistical investigations began by reading and discussing news articles that addressed the social crises in question. Each statistical investigation ended when the prospective teachers were able to answer the statistical question and present to the class the process followed and the empirical inquiry tools used.

When implementing the statistical investigations, participants were expected to use real information (data from meteorological archives, reports from statistical agencies, World Bank statistics, etc.) and statistical techniques (exploratory data analysis, analysis of trends, difference of means, hypotheses test) to study and empirically understand each critical situation. It was also sought that this understanding of the crises would contribute to undertake changes to overcome them.

The data gathered included instruments such as autobiographies, ideograms, narratives, and video recordings of the meetings that were transcribed word by word. An autobiography is a retrospective and reflective writing in which the participants give an account of their trajectories. In it, crucial events that take place in a certain temporality and have an impact in the process of becoming are highlighted (Ramos & Gonçalves, 1996). An ideogram is a pictorial organization produced by each participant about a theme; it allows to highlight concepts, ideas and points of view on the subject under question (Jaramillo, 2003). The narrative is a tool (written or verbal) through which participants reveal the meanings they are producing about the events they are experiencing and how those events influence their formative process (Jaramillo, 2003).

The information produced was organized and reduced using the software, *Atlas.ti* version 7.0. The transcripts and other data sources were annotated and coded independently by each researcher and then discussed in the light of the theoretical framework to reach a consensus. The units of analysis were episodes of verbal and written statements of the participants in which indications of critical citizenship while engaging in the statistical investigations were evident. An inductive strategy

—a cyclical process based on the systematic observation from the empirical data to propose theories— was followed. In a process of exhaustive, reflective and constant review of the data, two categories emerged and they are reported in the results.

4 Results and Discussion

To answer the research question, two categories emerged in the analysis process which show evidence of critical citizenship when prospective teachers engage in statistical investigations. In the first section, some representative episodes from prospective teachers' participation suggest evidence of statistical thinking. In the second section, some representative episodes from prospective teachers' participation suggest signs of the sense of agency.

4.1 Statistical Thinking

To participate as informed citizens in today's society, it is important that subjects can learn to think statistically (Ben-Zvi & Garfield, 2004) to understand the different social crises. Statistical thinking is related to the idea of globally assessing a real-world problem, solving it using data, and understanding how and why statistical analysis is important (Campos, 2016a). Statistical thinking includes "mastery of concepts and procedures, model building, reasoning, inference, development of dispositions, which are not isolated but in relation to a process" (Zapata-Cardona, 2016a, p. 74) in the empirical study of real-world problems. The study of social crises presents in the statistical investigations urged prospective teachers to get involved in a cycle of empirical inquiry. The following fragments of speech of some participants give an account of these actions.

Valeria: The teaching of statistics should aim for us to interpret reality beyond personal experience (Valeria's narrative, April 25, 2019, lines 34–36).

Felipe: The purpose of these activities [statistical investigations] is to link the concepts of statistics in a way that allows both to generate an interest and perform analyzes that will have direct implications in the reading of a social issue (Felipe's narrative, April 25, 2019, lines 58–61).

Valeria noted the problem of anecdotal evidence to make judgments and make decisions (in the sense indicated by Wild & Pfannkuch, 1999). Felipe, in contrast, realized the possibility of connecting statistics with real scenarios where the data is meaningful. Consistent with participants reflections, the essence of statistics is "recognizing that the main purpose of collecting and investigating data is to learn more about real situations and that data-based evidence is needed for making decisions and evaluating information" (Pfannkuch & Ben-Zvi, 2011, p. 325).

The development of statistical investigations supported by a cycle of empirical inquiry also made it possible to link some constitutive elements of statistical thinking, suggested by Wild and Pfannkuch (1999), such as *transnumeration*. This element refers to the change of data representation to facilitate its understanding (Wild & Pfannkuch, 1999). In the process of trying to answer the statistical questions, the participants found it necessary to investigate data and transform their representation to make sense of them (Pfannkuch & Wild, 2004). For example, when addressing the statistical question: Is there gender inequality in our country? Why or why not? Teresa and Claudia decided to search data from official statistics that would allow them to make a comparison of some variables between the population of men and the population of women in Colombia. The participants described it in the following way:

Claudia: We took into account four important conditions to determine whether or not there is gender inequality in the country. One of them was unemployment, the other was [...] boys and girls out of school [graduation rates], the other was services and industry employment, and income.

Teresa: We always made the comparison between men and women.

Claudia and to make these comparisons we used the mean in the information (Meeting 8, April 2, 2019, lines 9–16).

To analyze the information, the participants used some transnumeration techniques —such as those suggested by Estrella (2018)— that would allow them to organize, describe and give meaning to the data investigated. For example, in one of the selected variables, the percentage of unemployment in Colombia between men and women in the period 1991–2018, the technique used by Claudia and Teresa was to change the representation of the raw data to a visual representation through the construction of a line graph (see Fig. 1).

The illustration of the graph presented in Fig. 1 allowed the participants to organize the data to observe their behavior over time and to compare the percentage of unemployment between the chosen groups: men and women. Two transnumeration techniques emerged that consist of organizing the data and forming groups to obtain a greater understanding of the global behavior. The transnumeration process provided the prospective teachers insight about the behavior of the investigated problem depicted in the following fragment:
Claudia: So, the first one is this, the percentage of unemployment in the last 27 years in relation to the active population of the country for each year. This was like the tracking we did from 1991 to 2018, we got this information […] from the World Bank database […]. So, over time it is maintained as a proportion in which there are always more unemployed women than men […]. The one for men is orange and the one for women is blue [see Fig. 1] (Meeting 8, April 2, 2019, lines 16–28).

The graph constructed by the participants is instrumental in revealing data that supports the findings that the percentage of unemployment of women is higher than that of men during the period studied. Some contributions in the literature suggest that "This changing of data representation in order to trigger new understandings

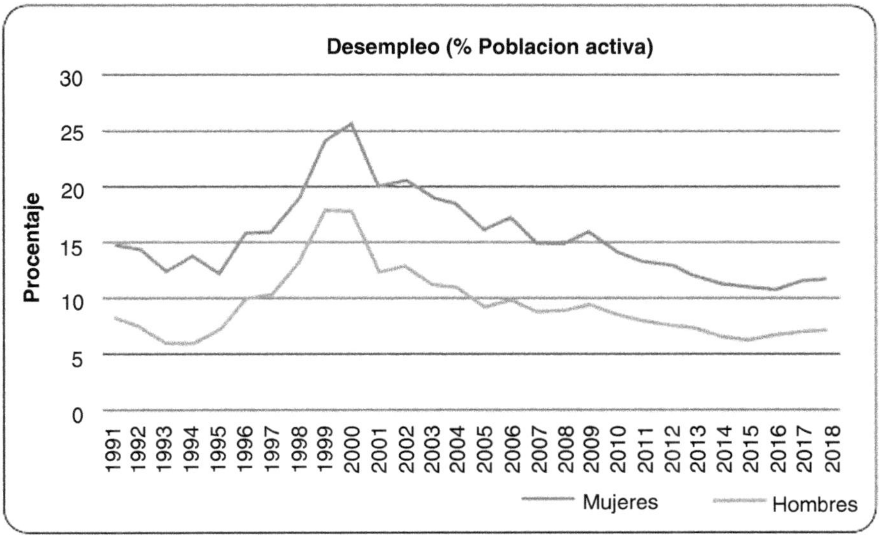

Fig. 1 Graphical representation created by Claudia and Teresa to compare the percentage of unemployed men (orange, inferior line) and woman (blue, superior line) in Colombia from 1991 to 2018

from the data or to communicate the messages in the data illustrates some fundamental statistical thinking" (Pfannkuch & Wild, 2004, p. 25) in the empirical research. Facing statistical investigations led preservice teachers to recognize the need to transform data to make sense of it and communicate it in a meaningful way as it relates to the actual situation.

The graph displayed in Fig. 1, prepared by Claudia and Teresa, as part of their investigation process to statistically study gender inequality in Colombia, became one of the statistical models from which the participants guided their reasoning and inferences. This is related to another constitutive element of statistical thinking called thinking with statistical models (Wild & Pfannkuch, 1999). Some authors (Pfannkuch & Wild, 2004) consider that statistical models are statistical ways of representing and thinking about reality, and, therefore, allow interpreting and studying social phenomena (Campos, 2016b) present in the world. In this sense, a simple tool such as the graphical representation of the data, or a specific way of structuring the data such as tables (Zapata-Cardona, 2018), can be considered statistical models.

Statistical thinking involves making graphical representations and structuring data in tables that lead to understandings and inferences (Campos et al., 2013). The research participants were able to think with their own statistical models and thus establish inferences such as those shared by Teresa in the following excerpt:

Teresa: If we take, let us say, all the tables [and graphs], all the data that we took, we could see that, yes, there is inequity with respect to women. [...] In unemployment, in the monthly earnings of women, it can also be seen. [...] So, there would be, for example, in industry there would be inequity with respect to women [...] and also, for example, in earnings and also in this part that has to do with unemployment. In conclusion, we could say that there is inequity (Meeting 8, April 2, 2019, lines 122–132).

The participants managed to infer, through the use of a probabilistic language (Makar & Rubin, 2009), that there *may be* gender inequity for the variables studied in the context of the problem, and this was thanks to the statistical analysis carried out with the help of the models and their reasoning about them. The reasoning from the statistical models built is evidence that statistic was used to study disparate opportunities between different social groups (Gutstein, 2006).

Prospective teachers undertook actions related to the construction of models (collection and analysis of data, use of tools and statistical reasoning to obtain conclusions based on data) and used the results provided by the models to understand and react to the problems under study. This means that the models became a tool to deepen the statistical knowledge of prospective teachers, but at the same time allowed them to show evidence of their awareness as critical citizens (Zapata-Cardona & Marrugo, 2019).

4.2 Sense of Agency

In today's society, subjects need opportunities to address complex socio-political problems along with learning statistical concepts and practices in an effort to read and write the world through statistics as critical citizens (Weiland, 2017). Read the world by using statistics to understand social, cultural and political contexts in society; and write the world by using statistics to try to transform those contexts (Weiland, 2019). In this dialogic process of reading and writing the world, subjects need to develop a sense of agency, that is, "a belief in themselves as people who can make a difference in the world, as ones who are makers of history" (Gutstein, 2003, p. 40). The sense of agency is associated with the ideas of empowerment and liberation. It is a transformative capacity to make a difference in the world by understanding and combating social inequalities (Porciúncula et al., 2019).

The work with statistical investigations allowed participants to reflect on the importance of statistics as a tool to study and understand complex socio-political problems and thus contribute to the transformation of such issues (Lesser, 2007). This was evident in some fragments of participants' speech:

Valeria: There are no decent and well-paid working conditions for the working class. The capitalist production model is unsustainable and generates only profit for a few. There are losses for humanity and the planet Earth. How do we generate predictions that allow us to make decisions and to change towards the results we want if it is not with true data and with concrete and consistent analyzes of that data? (Valeria's narrative, April 25, 2019, lines 38–44).

The sense of agency was linked to the increased awareness of participants about social justice issues. This awareness reveals committed citizens who can learn about the importance and power of statistics, to assess, understand, and change the inequities that persist in society (Gutstein, 2007; Lesser, 2014).

The teacher education program designed from a socio-critical perspective allowed prospective teachers to critically reflect on the non-neutral character of statistics (Campos, 2016b; Giroux, 2006). As statistics is a science that allows people to approach the analysis of phenomena from their sociopolitical contexts, it cannot be separated from the interests of the subjects when they carry out statistical studies (Weiland, 2019). This idea arose in the discourse of the participants when some of the problems present in the statistical investigations were addressed:

Bruno: with statistics you can control the scope of what you want. How far you want the person to read or be informed. [...] So, it is how that information is controlled, directed, or limited (Meeting 1, March 5, 2019, lines 544–553).

Claudia: for one to determine that there is inequity with respect to any of the genders, one has to determine very well the variable under study. [...] I insist that it has to do with what we choose (Meeting 8, April 2, 2019, lines 107–115).

Participants' reflections reveal that the way in which people approach and interpret a set of data depends on their interests. That is, the interests of the subjects play a role both in the production of arguments based on data and in the interpretation of those arguments. Statistics has a political character (Weiland, 2019), since it is influenced by the decisions and choices that people make when they use it to build arguments based on data and make decisions.

The fact that prospective teachers have pointed out the political nature of statistics when studying social crises is something that challenges the neutral and objective vision of statistical knowledge (Campos, 2016b). This points to the importance of people making their interests visible and explicit when they collect and analyze statistical information, and reflecting on how these interests could influence their arguments (Weiland, 2019). This is related to the call made by Skovsmose (1999) that both teachers and students must maintain a "critical distance" to question aspects such as the applicability of science, who uses it, where it is used and what are the implicit interests in its use.

The sense of agency can also be evidenced when the participants recognize the world in which they are protagonists. That is, they are not just passive agents who receive information, but question and criticize the current state of things and see the social, economic and political forces reflected in their reality. The method course enabled the participants to reflect on the importance of statistics as a tool to study and understand complex sociopolitical problems, and thus help achieve social change in the world (Lesser, 2007). This can be evidenced in the following excerpts:

Laura: News such as the ones mentioned, that were raised from other Latin American and European countries, managed to generate concern among the audience. They are familiar problems for Colombians. […] This type of research helps in the teaching of statistics, since it summons the pre-service teacher to understand the world through statistics and to develop human conscience, change and truth. For this, data is needed (Laura Narrative Writing, April 25, 2019, lines 23–30).

Laura's reflections show how being able to interpret and produce information related to critical issues in her society, considering quantitative data, can be an opportunity to make a lasting appreciation or even a commitment to statistics as "a tool to help understand (and maybe improve) some of our society's most profound or pressing matters" (Lesser, 2007, p. 1).

The methods course also allowed the participants to carry out a critical reflection to question discourses and actions that, on many occasions, perpetuate the current state of the existing crises in society. The fragments of Claudia and Valeria's speech give an account of this:

Claudia: what practices are we going to teach to children so that they become aware that the increase in temperatures in the country does not depend strictly on chance, but on other factors. And how, through my actions, I improve or worsen the ecological situation of the country, and in a very particular case of the city […]. Also, how to carry out and make those comparisons with children at the statistical level of all those variables, economy, population growth, sea level rise. We need to do that right now from a statistical point of view to be able to establish those variables (Meeting 2, March 7, 2019, lines 335–356).

Valeria: An important part is not reproducing it [the discourse of gender inequality]. […] through discourse we also […] legitimize things (Meeting 8, April 2, 2019, lines 874–893).

The participants revealed their sense of agency by questioning different actions and discourses that can contribute to perpetuate or counteract practices such as gender inequality and environmental deterioration. Reflecting critically has to do with "understanding and questioning our actions from the ethical and moral dimensions [that] can directly or indirectly influence others or us" (Guerrero, 2008, p. 73). The participants, in their roles as citizens and as future teachers, were able to question actions and discourses that might, directly or indirectly, affect the lives of other people and themselves in the present or future. Based on critical reflection, the participants tried to interrogate, problematize and reconstitute dehumanizing and unjust discourses (Weiland, 2019) in a struggle to contribute, as agents of social change, to the construction of fairer, more democratic and more equitable and humanized societies.

Preparing prospective teachers as agents of change for their participation in democratic societies requires the development of an ethical and political awareness (Weiland, 2019). Ethics has to do with a set of moral or philosophical values that people use to make decisions in relation to others, their communities and the world in general (Weiland, 2019), with a view to humanizing the society of which they are part of. The ethical and political awareness was evidenced in some statements of the participants, such as those of Ana, Teresa and Valeria:

Ana: According to what I found; I believe that there is [gender] inequity in Colombia. In all the variables [...] [the statistics of] women were below [than those of] men, except in education. But even with higher education for women, unemployment was lower [for men]. That does not make sense, because society tells us that the more education, the more possibilities, the more opportunities. [...] I would like to start a campaign or something, ¡women do something! (Meeting 8, April 2, 2019, lines 329–335).

Teresa: We are people, we are human beings, we have some rights [...]. Being a man or a woman should not be a disadvantage to develop our life in any aspect: academic, work, economic (Meeting 8, April 2, 2019, lines 864–868).

Valeria: We were surprised by how numbers and statistics specifically can serve [to] understand these phenomena and [to] corroborate consequences and relationships [...], not to remain in a discourse about how bad things are, as always, but to propose solutions and put them in motion (Valeria's Narrative Writing, April 25, 2019, lines 30–33).

In Ana, Teresa and Valeria's fragments of speech was evident how the participants had the opportunity to use statistics to analyze situations that seem to be unfair in their society —comparing by gender information on various economic and social variables from the city—, results that generated impotence (in Ana's case). It was also evident a feeling about being able to do something in relation to the situation. As Giroux (2006) proposes, ethical discourse expresses a concern for the human suffering of the oppressed and can allow the rejection of practices that perpetuate such suffering. Feelings of helplessness and of wanting to do something about it, upon hearing about different situations of injustice, are important challenges to encourage prospective teachers "to see themselves as capable of acting on their environment and making positive social change toward a more just world" (Gutstein, 2006, p. 90).

The opportunity for prospective teachers to reflect on dilemmas and socio-political issues allow them to combine the reading and writing of the world through statistics to develop their sense of personal and social agency. Here, it was evident what Zapata-Cardona and Marrugo (2019) propose when they argue that including statistical tasks-problems related to critical issues in society (environmental problems, social inequalities, gender bias, social indicators, among others) offer elements for subjects to become agents of change who have the responsibility to participate in society and contribute to its transformation, as they apply contents, procedures and statistical tools.

5 Conclusions

The main findings of this study reveal that by engaging prospective teachers in statistical investigations, traces of critical citizenship could be observed in two aspects: the statistical thinking and the sense of agency. These aspects became what could be called "critical citizenship tools" to study, understand, challenge and think of ways to contribute to the transformation of social crises.

Offering opportunities for prospective teachers to think statistically by empirically studying social crises is a contribution to statistical thinking and critical citizenship. Using knowledge and statistical tools is an important prerequisite for understanding the world, therefore, people who have these tools and know how to use them are more likely to participate in society. By involving prospective teachers in authentic experiences of facing real problems in sociopolitical contexts, their statistical thinking was stimulated as a contextual tool for understanding, knowing and participating in the surrounding world.

The results reveal that prospective statistics teachers from this research had the opportunity to exhibit their sense of personal and social agency as a tool to uncover and challenge some of the inequities present in their society. The sense of agency was evidenced with the evolution of the initial vision of statistics as a non-neutral science, with the transformation of the participants from spectators to protagonists of history and with the concretion of transformative actions to change the current state of things. This means that the participants had the opportunity to think about how the world could be a fairer and more equitable place, and to believe that their actions and those of others count and represent the world; and therefore, they can influence change and contribute to social justice and democracy. Prospective teachers from this study challenged through their voices some social crises, and relied on statistics to argue through quantitative data the need to transform actions and discourses in favor of social change. In this sense, the results reported in this research suggest a progressive view of teacher education programs for teachers who teach statistics.

In-service statistics teachers should be professionals with in-depth knowledge of the discipline, but should also know the political role of statistics to make inequalities evident and to address them intentionally in search of social transformation. To comply with this premise, teacher education programs for statistics teachers should consider the social dimension of beings as critical, ethical and political subjects. Technical preparation of statistics teachers is essential, there is no doubt about it, but integrating critical education is a way of recognizing the relationship between being and knowing, as a central issue in the teacher education process. The sense of agency is a transformative capacity that does not develop spontaneously. It is necessary that teacher education programs confront prospective teachers with social crises so that they reflect and question the current conditions of the world and undertake transformative actions towards a more just, democratic and humane society.

The findings indicated in this study are striking and promising for teacher education; however, it is necessary to take them with caution given the cross-sectional nature of the methodological design of this research, which may constitute a limitation.

For future studies, it could be explored, for example, how these indications of statistical thinking and the sense of agency that appear in the teacher education manage to accompany in service statistics teachers in their teaching practice.

Acknowledgements This research was conducted with financial support from Colciencias research grant 438-2017.

References

Bargagliotti, A., Franklin, C., Arnold, P., Gould, R., Johnson, S., Perez, L., & Spangler, D. (2020). *Pre-K-12 guidelines for assessment and instruction in statistics education (GAISE) report II*. American Statistical Association and National Council of Teachers of Mathematics.

Ben-Zvi, D., & Garfield, J. (2004). Statistical literacy, reasoning, and thinking: Goals, definitions, and challenges. In D. Ben-Zvi & J. Garfield (Eds.), *The challenge of developing statistical literacy, reasoning and thinking* (pp. 3–16). Kluwer Academic Publishers.

Campos, C. R. (2016a). *La educación estadística y la educación crítica* [Statistical education and critical education]. Segundo Encuentro Colombiano de Educación Estocástica (2 ECEE).

Campos, C. R. (2016b). *Towards critical statistics education. Theory and practice*. Lambert Academic Publishing.

Campos, C.R., Wodewotzki, M.L., & Jacobini, O. R. (2013). *Educação estatística: Teoria e prática em ambientes de modelagem matemática* [Statistical education: Theory and practice in mathematical modeling environments] (2 edn). Autêntica Editora.

Denzin, N. K., & Lincoln, Y. (2012). *El campo de la investigación cualitativa.* (The field of qualitative research). *Manual de investigación cualitativa, 1.* Gedisa.

Ernest, P. (2015). The social outcomes of learning mathematics: Standard, unintended or visionary? *International Journal of Education in Mathematics, Science and Technology, 3*(3), 187–192.

Estrella, S. (2018). Data representations in early statistics: Data sense, meta–representational competence and transnumeration. In A. Leavy, M. Meletiou-Mavrotheris, & E. Paparistodemou (Eds.), *Statistics in early childhood and primary education. Supporting early statistical and probabilistic thinking* (pp. 239–256). Springer.

Giroux, H. (2006). *La escuela y la lucha por la ciudadanía. Pedagogía crítica de la época moderna* [The school and the struggle for citizenship. Critical pedagogy of the modern age] (4th ed.). Siglo XXI.

Guerrero, O. (2008). *Educación matemática crítica: Influencias teóricas y aportes*. [Critical mathematics education: Theoretical influences and contributions]. *Evaluación e Investigación, 3*(1), 63–78.

Gutiérrez, R. (2009). Embracing the inherent tensions in teaching mathematics from an equity stance. *Democracy & Education, 18*(3), 9–16.

Gutiérrez, R. (2013). The sociopolitical turn in mathematics education. *Journal for Research in Mathematics Education, 44*(1), 37–68.

Gutstein, E. (2003). Teaching and learning mathematics for social justice in an urban, Latino school. *Journal for Research in Mathematics Education, 34*(1), 37–73.

Gutstein, E. (2006). *Reading and writing the world with mathematics: Toward a pedagogy for social justice*. Routledge.

Gutstein, E. (2007). Possibilities and challenges in teaching mathematics for social justice. In *Third annual National Research Symposium of the Maryland Institute for Minority Achievement and Urban Education*. University of Illinois-Chicago.

Jaramillo, D. (2003). *(Re)constituição do ideário de futuros professores de Matemática num contexto de investigação sobre a prática pedagógica*. Tesis Doctoral. Universidade Estadual de Campinas. Brasil.

Lesser, L. M. (2007). Critical values and transforming data: Teaching statistics with social justice. *Journal of Statistics Education, 15*(1), 1–21.
Lesser, L. M. (2014). Teaching statistics for engagement beyond classroom walls. In K. Makar, B. de Sousa, & R. Gould (Eds.), *Sustainability in statistics education. Proceedings of the ninth international conference on teaching statistics (ICOTS 9)*. International Statistical Institute.
Loya, H. (2008). Los modelos pedagógicos en la formación de profesores [Pedagogical models in teacher education]. *Revista Iberoamericana de Educación, 3*(46), 1–8.
Makar, K., & Rubin, A. (2009). A framework for thinking about informal statistical inference. *Statistics Education Research Journal, 8*(1), 82–105.
Pfannkuch, M., & Ben-Zvi, D. (2011). Developing teachers' statistical thinking. In C. Batanero, G. Burrill, & C. Reading (Eds.), *Teaching statistics in school mathematics-challenges for teaching and teacher education. A joint ICMI/IASE study: The 18th ICMI study* (pp. 323–333). Springer.
Pfannkuch, M., & Wild, C. (2004). Towards an understanding of statistical thinking. In D. BenZvi & J. Garfield (Eds.), *The challenge of developing statistical literacy, reasoning and thinking* (pp. 17–46). Kluwer Academic Publishers.
Porciúncula, M., Schreiber, K. P., & Almeida, R. L. (2019). Statistical literacy: A strategy to promote social justice. *Revista Internacional de Pesquisa em Educação Matemática, RIPEM, 9*(1), 25–44.
Ramos, M. A., & Gonçalves, R. E. (1996). As narrativas autobiográficas do professor Como estratégia de desenvolvimento e a prática da supervisao [the teacher's autobiographical narratives as a development strategy and the practice of supervision]. In I. Alarcão (Ed.), *Formação reflexiva de profesores. Estratégias de Supervisão* (pp. 123–150).
Sánchez, S. G. (1998). Fundamentos Para la investigación educativa. Presupuestos epistemológicos que orientan al investigador [Foundations for educational research. Epistemological assumptions that guide the researcher]. *Cooperativa editorial Magisterio*.
Skovsmose, O. (1999). *Hacia una filosofía de la educación matemática crítica.* [Towards a philosophy of critical mathematics education.] (P. Valero, Trad.). Una Empresa Docente (Trabajo original publicado en 1994).
Souza, L., Lopes, C. E., & Pfannkuch, M. (2015). Collaborative professional development for statistics teaching: A case study of two middle-school mathematics teachers. *Statistics Education Research Journal, 14*(1), 112–134.
Weiland, T. (2017). Problematizing statistical literacy: An intersection of critical and statistical literacies. *Educational Studies in Mathematics, 96*, 33–47.
Weiland, T. (2019). Critical mathematics education and statistics education: Possibilities for transforming the school mathematics curriculum. In G. Burrill & D. Ben-Zvi (Eds.), *Topics and trends in current statistics education research. International perspectives* (pp. 391–411). Springer.
Wild, C. J., & Pfannkuch, M. (1999). Statistical thinking in empirical enquiry. *International Statistical Review, 67*(3), 223–248.
Zapata-Cardona, L. (2016a). ¿Estamos promoviendo el pensamiento estadístico en la enseñanza? [Are we promoting statistical thinking in teaching?] *Segundo Encuentro Colombiano de Educación Estocástica (2 ECEE)*. Bogotá, Colombia.
Zapata-Cardona, L. (2016b). Enseñanza de la estadística desde una perspectiva crítica [Teaching statistics from a critical perspective]. *Yupana, 10*, 30–41.
Zapata-Cardona, L. (2018). Students' construction and use of statistical models: A socio–critical perspective. *ZDM, 50*(7), 1213–1222. https://doi.org/10.1007/s11858-018-0967-8
Zapata-Cardona, L., & González-Gómez, D. (2017). Imágenes de los profesores sobre la estadística y su enseñanza [Images of teachers on statistics and its teaching]. *Educación Matemática, 29*(1), 61–89.
Zapata-Cardona, L., & Marrugo, L. M. (2019). Critical citizenship in Colombian statistics textbooks. In G. Burrill & D. Ben-Zvi (Eds.), *Topics and trends in current statistics education research. International perspectives* (pp. 373–389). Springer.

Toward Statistical Literacy to Critically Approach Big Data in Mathematics Education

Carlos Eduardo Ferreira Monteiro and Rafael Nicolau Carvalho

Abstract Big data has commonly been linked to volume, speed, and variety and the way in which data are produced and stored. This chapter aims to explore the trends, classify the research themes, and discuss the limitations of studies that approach big data associated with mathematical and statistical education contexts. The discussion intends to contribute to possible future directions toward a statistics literacy that critically approaches big data in mathematics and statistics education. The text focuses on reflections of an integrative review of articles published in the Web of Science database with the following descriptors: big data, literacy, and teacher education. The results indicate a polysemy referring to the term big data. The review also suggests that studies linking big data to mathematics and statistics teacher education are scarce. In the thematic analysis, some surprising questions emerge, such as the narrative—or even folklore—around big data as a form of knowledge capable of solving many economic, social, and, particularly, educational problems. Finally, the results revealed that most studies have a predominantly non-critical perspective of big data in the interface with mathematics and statistics education.

Keywords Big data · Statistical literacy · Statistics education · Mathematics education · Teacher education

---

C. E. F. Monteiro (✉)
Postgraduate Program in Mathematics and Technology Education (Edumatec), Department of Psychology, Inclusion and Education (Dpsie) Center of Education (CE), Universidade Federal de Pernambuco (UFPE). [The Federal University of Pernambuco (UFPE), Recife, Brazil
e-mail: carlos.fmonteiro@ufpe.br

R. N. Carvalho
Postgraduate Program in Social Work (PPGSS) Department of Social Work (DSS) Center of Human Sciences and Arts (CCHLA), Universidade Federal da Paraíba (UFPB). [The Federal University of Paraíba (UFPB), João Pessoa, Brazil
e-mail: rafael.carvalho@academico.ufpb.br

G. F. Burrill et al. (eds.), *Research on Reasoning with Data and Statistical Thinking: International Perspectives*, Advances in Mathematics Education,
https://doi.org/10.1007/978-3-031-29459-4_18

1 Introduction

In contemporary social contexts, digital technologies are increasingly accessible and support different forms of interaction that, in turn, generate data through systems that register and store information about people's actions. In these contexts, it is worrying that adolescents and young students are unaware of the likely impact of such data tracking, especially regarding commercial practices and opinion formation. Thus, ways of approaching and understanding the massive information being collected become valuable for corporations, governments, as well as for ordinary citizens.

Zeelenberg and Braaksma (2017) state that big data can be defined as large datasets related to social activities not covered by official statistics. For instance, social media users generate enormous datasets by sharing information, writing opinions, contacting friends, and using mobile phones for online shopping or banking transactions. Big data is a dataset continuously fed by various means, thus becoming a wide range of jumbled data that can be analysed and processed through data mining procedures. Such procedures search for consistent patterns and/or systematic relationships between variables and validate them by applying the detected patterns to new subsets of data (Caldas & Silva, 2016). Thus, using statistical techniques and applying specific expertise, it is possible to explore a dataset through data mining to identify patterns, favouring data-based decision-making.

The new social contexts of data production are a challenge for citizens to analyse and make decisions. Therefore, statistical literacy is considered vital because it enables the critical interpretation of information based on arguments related to the data and its context (Gal, 2002). Schools and teachers can play an important role in reflecting on how big data impact people's daily lives. Mathematics and statistics teaching that approaches how to critically interpret and deal with big data can contribute to students' citizen participation at different educational levels.

An increasing number of researchers and educational professionals are approaching big data and arguing that it can be a revolutionary element to help solve many educational problems (e.g., Baig et al., 2020). However, the approaches to big data should not only focus on how to manage a massive amount of data but on finding ways to prepare students to deal with those data that differ in structure and source from the traditional ones (Ainley et al., 2015).

Gal (2002) proposes a statistical literacy model based on people's abilities to interpret and critically evaluate statistical information, arguments related to data, or phenomena they may encounter in different contexts. Furthermore, this statistical literacy presupposes the ability to communicate and discuss an understanding of meanings, opinions about the implications, or concerns about the acceptability of conclusions taken from the data. The teaching and learning of mathematics and statistics that involve big data demand a perspective of statistical literacy that considers the specificities of the big data (François et al., 2020). Other critical skills might also be necessary due to the complex context from which the data are generated and the implications of their utilisation.

Many literature review studies about the relationship between big data and several areas have been conducted, but only a few approached the interface between big data and education (Baig et al., 2020). Monteiro (2021) developed an integrative literature review using the Scopus database aiming to answer research questions about the implications of big data for the education of teachers who teach mathematics and statistics. The analyses of the articles confirmed the assumption that little has been investigated about big data in statistics teacher education, although some of them address relevant elements that future pedagogical approaches should contain. On the one hand, some articles emphasise critical dimensions of statistical literacy and its repercussions for citizen protagonism. In this sense, we see that big data is limited if compared, for example, to open data from statistical research institutions (Ridgway, 2016).

On the other hand, some articles seemed to highlight only the possible advantages of big data without further criticism and based on a somewhat naive view that big data analyses linked to the students' and educational systems' realities could lead to the personalisation of teaching processes (e.g., Ruiz-Palmero et al., 2020). The reflections from that literature review confirm that it is necessary to involve students in teaching and learning about big data so they will be able to analyse social situations in which they are involved. However, most articles analysed did not consider teachers' need to have educational experiences that lead them to reflect on the processes involving big data.

This chapter presents and discusses results from an integrative literature review on big data in the Web of Science database. This review aims to explore the trends, classify the research themes, discuss the limitations, and provide possible future directions in the relationship between big data in mathematics and statistics education. This study differs from Monteiro's (Monteiro, 2021) with respect to several elements. First, the database used is different. Second, it includes journal articles and conference proceedings papers. Third, this investigation includes a more comprehensive range of research studies associated with mathematics and statistics teacher education; it also has the potential to discuss educational processes involved in teaching and learning big data related to mathematics and statistics education. Such elements increased the number of publications in this study.

This chapter is a first step in reflecting on big data and mathematics and statistics education. It contributes to future research that critically analyses the potentialities, limitations, and ethical implications of using big data in mathematics and statistics education.

2 Methodology

The basic search descriptors were: big data, literacy, and teacher education. We combined them through the Boolean connector "AND" of the base search engine Web of Science in all areas. We established the search period from 2015 to 2021 (April) because the object of study is recent, and investigations on educational

Table 1 Publications found from searches with descriptors

Descriptors	Total	Selected
“big data” and “literacy”	323	39
“big data” and “teacher education”	872	74
Total	1195	113

effects are scarce. Furthermore, we defined that the search would be in the English language only and delimited that the publications would be articles from journals and papers published in proceedings.

Table 1 presents the total results of the searches using the combinations of descriptors and the results as per the following criteria: they should be articles from journals or papers published in conference proceedings and have some connection with the search terms in the title, abstract, or keywords of all thematic areas of the Web of Science platform. Using these criteria allowed us to identify 1195 publications.

The next step was to read the abstracts of the 1195 articles identified, which resulted in the exclusion of many publications. One explanation is that big data is a generic term widely used in computer science, informatics, and data science. For example, many of these publications described technical processes on big data involving hardware and software aspects. However, while reading the abstracts, we identified that several papers contained the words “big data” but did not actually discuss big data in education. Since our objective was to explore the connection between big data and statistics and mathematics education, considering the literacy and teacher education context, many publications were excluded.

Although “statistics education” and “mathematics education” were not searched, these areas were taken into account in the more general context of literacy and teacher education and especially related to statistical literacy. In this selection phase, we excluded publications that did not emphasise education or literacy. This procedure also resulted in a large number of exclusions. After reading the primary elements of the publications, we selected 113 publications to read in full, as shown in Table 1.

For the inclusion stage in the review, the publications had to provide elements to answer our research questions, which, in this stage, served as inclusion questions.

- What are the implications of the different meanings of big data for the education of teachers who teach mathematics and statistics?
- What elements constitute literacy in big data as a dimension of statistical literacy?
- What are the experiences of teaching and learning statistics using big data?

After reading the 113 articles and applying the criteria, we excluded 73. Forty articles remained (Table 2).

We read the 40 publications in three stages: (1) exploratory reading, (2) comprehensive reading, and (3) interpretive reading. The exploratory readings aimed to identify some elements of the articles, such as objectives, types of studies,

Table 2 Publications included after the reading process

Descriptors	Selected	Included
"big data" and "literacy"	39	15
"big data" and "teacher education"	74	25
Total	113	40

Table 3 Classification of articles included according to the type of study

Type of study	Frequency
Theoretical essay	19
Research with empirical data collection	05
Literature review	05
Analysis of digital tools	05
Case report	05
Documentary research	01
Total	40

methodology, country, and thematic area. The comprehensive readings sought to raise themes, categories, and connections among the selected publications and the issues raised in the studies. Finally, the interpretative readings required us to integrate the contents of the selected publications with other knowledge to produce a summing up that relates to the research questions.

3 Results

Most of the selected papers were related to the combination of the descriptors: "big data" and "teacher education" (Table 1), which may be explained by their many different meanings. However, it may also indicate the development of research relevant to teacher education linked to big data. This result differs from the one in the integrative literature review performed with the same research questions (Monteiro, 2021) in the Scopus database, which found a higher frequency of journal articles linked to the combination of the descriptors: "big data" and "literacy". However, in that literature review, the search was limited to journal articles. This study included papers published in conference proceedings. Another comparative aspect between Monteiro's review and this research is that the two reviews only shared five articles of interest.

Table 2 shows a total of 40 articles included for the construction of the interpretive summing-up, with 15 articles linked to the Boolean descriptor "big data" and "literacy" and 25 articles with "big data" and "teacher education". Table 3 presents the results related to the classification of the type of study discussed in the article. The most frequent category, theoretical essay, refers to papers that explore the discussion about big data in general, arguing about the benefits or concerns when using them in the educational context.

Articles in the category: "research with empirical data collection" included quantitative studies that sought to analyse the participants' motivation in activities using big data and the understanding and perception of big data, data literacy and the incorporation of big data in education.

The papers that presented a "literature review" explored the different meanings of big data and the challenges for educational fields. The texts started from different theoretical perspectives and explored various situations using big data. Publications classified as "analysis of digital tools" addressed big data in educational situations that included digital tools, such as computer programs and information processing models. In the category "case report", we included papers that brought not just proposed models to approach big data but,, the discussions of projects developed with the participation of students or education professionals. The only article categorised as "documentary research" analysed the education projects and programs of several countries and, by mining educational data, sought to find data that pointed to the development of the "twenty-first-century skills" for school-age children.

An important aspect was knowing the countries of the first authors of the articles included in the summing-up because this information could lead us to infer the international trends related to the discussion of big data. Table 4 presents the list of the first authors' countries and the corresponding frequency of articles.

Table 4 indicates that 50% of the publications had an Asian researcher as the first author (20). This suggested a fast appropriation of big data as an element for the development of the Asian economies. For example, the articles included in the study provided evidence that China has implemented big data strategies to promote the

Table 4 Countries of origin of the authors of the articles included

First author's country	Frequency
China	15
USA	06
UK	04
Spain	03
Russia	02
India	02
Brazil	01
Brunei	01
Cyprus	01
Malaysia	01
The Netherlands	01
Norway	01
Romania	01
Saudi Arabia	01
Total	40

country's economic and social development, based on improving digital infrastructure, broad access to high-speed internet, use of artificial intelligence, and the integration of big data research with industrial production (e.g., Liu & Zhang, 2021). These strategies also seem to support the planning and control of educational processes, such as developing teaching approaches associated with big data. Perhaps, this investment in technologies to expand China's economic and social development can explain, for example, the number of Chinese authors in publications on big data.

Many papers (11) had a European researcher as their first author. Six were from the United States, two from Russia, and one from Brazil. However, as discussed later, only a few papers from Europe and the USA explored more critical aspects and repercussions in approaching big data in education or statistics teaching.

In the thematic analysis of the articles, we identified the frequency of topics related to each article, according to the results presented in Table 5. In this analysis, the same article may have been included in several thematic categories.

Table 5 shows that a large number of papers (22) emphasised the importance of developing education built on big data-based predictions. Usually, these articles only highlighted positive aspects without critically analysing how those data are generated and applied (e.g., Dai et al., 2017). Another frequent thematic category (18 papers) refers to papers that emphasised the technology dimension in the educational discussion of big data.

Although many articles pointed out that teachers need to develop new professional skills that include knowing how to approach big data to improve their performance and that of their students, only nine papers focused on aspects of teacher

Table 5 Themes presented in the articles included

Themes	Frequency
Big data in education	22
Technology	18
Teacher education	9
Innovation and entrepreneurship	7
Educational reform	4
Big data literacy	3
Digital literacy	3
Curriculum	3
Statistical literacy	3
Statistics education	3
Teaching model	3
Information literacy	2
Literacy	2
Ethics and human rights	1
Teaching project	1
Digital inclusion	1
Misinformation	1

professional development. Many articles emphasised the novelty of approaching big data in education and the need for changes in educational institutions. However, only seven highlighted the dimensions of innovation and entrepreneurship, and only four underscored the need for educational reforms. The identification of the other categories suggests the scope of the interfaces that big data can maintain in the context of education. However, as explained, the proposal to include publications that do not specifically discuss big data in statistics education and mathematics education was an attempt to have more articles that could evidence the limitations and provide future directions for research studies.

The detailed appreciation of the emerging themes in the articles allowed us to develop the interpretative analyses of the contents of the articles that is presented below.

4 Discussion

In this section, we present an analysis of both searches, the search with descriptor "Big Data and Literacy" (4.1) and the search with "Big data and teacher education" (4.2), considering their selection, inclusion, and categorisation processes to build an integrative summing-up of the combined descriptors, which we call axes: (1) big data and literacy and (2) big data and teacher education.

The axes generally reflect the big data aspects, such as definitions, relationships with different scientific areas and education, professional education, and skills for the new millennium. However, each axis points to different analytical paths. For example, the axis "big data and literacy" refers to publications that emphasise the formation of a "new field of knowledge" (e.g., big data literacy, digital literacy, data literacy, algorithmic literacy) in the way of doing and carrying out scientific research. Besides addressing conceptual aspects of big data, the publications in the axis "big data and teacher education" presented possibilities and challenges so that teachers could have an introduction and development in big data to expand and improve their performance. It also has objective and subjective effects on teacher education.

4.1 Big Data and Literacy: A Semantics under Construction

In this subsection, we consider the analyses of 15 articles included through the combined descriptor "big data" and "literacy". When reading in full, we also sought to answer the research questions, which was our criterion for including those articles. However, the emerging themes were recorded in our protocol to capture the reflections on what the articles discussed. Therefore, we do not carry out a description of the papers; instead, we will foster a discussion based on the topics covered.

Several papers described big data as having a crucial role in a "new era" or "data era" (e.g., Dai et al., 2017; Liu & Zhang, 2021). The argumentation in some of those papers was based on experiments or projects approaching big data, emphasising only the positive consequences. For example, Ying (2019) explored big data concepts and how they have modified language codes, relationships, and learning. That author argued that data literacy for university students in China is essential to the better development of their country. Generally, publications that overvalue big data have an uncritical perspective of teaching and learning processes; they do not consider the risks of using data related to their ethical, subjective, and value-forming implications for humans (Cope & Kalantzis, 2016; Fiofanova, 2020; Liu & Zhang, 2021; McGowan, 2020; Ying, 2019).

Some of the selected papers reflect the context of introducing big data in schools, considering how personal digital data is perceived in different cultural contexts. For example, Hoel et al. (2020) compared social contexts in which big data were used in schools in China and Norway. The study suggested that Chinese teachers are open-minded and interested in exploring the data for all information they may provide. In contrast, Norwegian teachers are focused on school curriculum scope and concerned about privacy and data surveillance questions. However, although the authors considered social and cultural factors, the conclusions do not seem to go deeper into the discussion of sociocultural issues or educational policies.

Some other papers developed critical analyses on the relationships between big data and education (Carmi et al., 2020; Papacharissi, 2015; Pawluczuk, 2020;Sander, 2020 ; Souza, 2018). They argued that all citizens must recognise the risks and benefits of the increasing implementation of data collection, automation, and systems that can predict behaviours, choices, and tastes (Sander, 2020; Souza, 2018). Therefore, data literacy must overcome the mere learning of digital media and the internet or dealing with large volumes of data. Critical learning related to big data practices would be necessary, mainly to know their risks. For those papers that develop a critical approach, big data must be understood as a phenomenon of our time; therefore, it cannot be deprived of sociocultural, political, and economic assumptions (Papacharissi, 2015).

From a critical perspective on using big data in educational contexts, we identified an argumentation that explores the idea that big data can be an exclusion process (Pawluczuk, 2020). Often, when someone collects data and others have their data collected, they are excluded from the consent process and even from using their own information. This process is called the "splitting of big data", which is an asymmetrical power dynamic between those who collect, analyse, and benefit from the data (e.g., social media companies) and those who are the targets of the data collection process (e.g., social media users). Data literacy can encourage understanding of how data are collected and allow citizens to identify and critically analyse dis–/mis–/mal-information via digital media (Carmi et al., 2020). Education programmes can use examples from real-life situations, taking into account information distortions to promote proactive citizens at the centre.

On the relationship between big data and statistical literacy, we highlight three articles that address this discussion: Ridgway (2016), Ainley et al. (2015), and Tractenberg (2017). Ridgway (2016) emphasises the irrelevance of a static curriculum that offers little help in understanding the data and how the data are collected, used, and manipulated. He explains the difference between big data and open data. The first refers to the large volume of data collected from our personal contact with different technologies, often without our knowledge. Open data, in turn, consists of data collected through official studies and surveys and bring important information about various contexts of people's lives. Both expose the need for even more critical statistical thinking and provide contexts for discussing basic statistical ideas. Therefore, it is necessary to create curricula that devote more attention to the interpretation of open data in the classroom, which may differ from big data with respect to features such as subjecting the entire process of generating and presenting the data to intense scrutiny.

Statistical literacy at school should explore the different meanings of big data and present how it relates to and impacts all aspects of life (Ainley et al., 2015). Big data can therefore be "student data" collected from different devices, smartphones, applications, and websites. The challenge will be how to teach students to understand those data, explore situations in which teachers develop task plans with them, and highlight the notion of sampling and representativeness. Statistical literacy reflects the complexity of logical reasoning —the mental and cognitive skills also necessary for citizen social participation (Tractenberg, 2017). Statistical literacy would also be of great value to scientific practice. Moreover, statistical literacy is a vital human skill that helps people to avoid the risks of falling into the automated analyses that organise and explore large volumes of data.

When reflecting on the articles analysed, we decided to put the expression "semantics under construction" as a subtitle in this subsection. We start from the idea that big data is a complex social phenomenon that involves more than the manipulation of large volumes of data or the handling of technological artefacts to analyse, interpret, and understand social processes based on these data. While this is also an important debate, we highlight how this process creates narratives and defines human life dimensions. The articles present several adjectives and concepts that suggest directions for those narratives, such as "big data era", "data era", "algorithms era", "digital society", "network society", and "information economy". They propose redefining our ways of life and relationships in a perspective of control and predicting our attitudes and behaviours. Some claim that big data is a new way of producing knowledge, redefining our very idea of knowledge (e.g., Fiofanova, 2020). The idea behind this narrative is that the sheer volume of data provides hitherto inaccessible ways of understanding society and its members. It is the positivist dream come true that devalues, for example, world views and sociocultural and subjective aspects. The data capture subjectivity itself. Big data acts invisibly, silently, and insensitively in social situations (e.g., Wassan, 2015). The "big data era" narrative seeks to overcome this gap, claiming that all human behaviour can be rationally apprehended. Thus, we are subjected to a "colonisation of the lifeworld", i.e., of what identifies us as social beings, through the economy and scientific practice of the big data era (Beuving & de Vries, 2020). We can also mention the ethical issues around big data, including the extent to which the information that makes up our lives can be accessed and used to select, delete, and punish.

4.2 Big Data and Teacher Education: Limits and Possibilities

This subsection develops the integrative summing-up of 25 papers, which had been included from the searches with the combined descriptor "big data" and "teacher education." Our discussion of those papers incorporates reflections that have the potential for problematisation in mathematics and statistics teacher education. Some of those reflections are similar to those presented in the previous subsection.

We identified 14 papers that address teacher education related to big data more generally (e.g., Deng, 2017; López-Belmonte et al., 2019; Schildkamp, 2019). Only one article specifically discusses probability teacher education (Mezhennaya & Pugachev, 2019). Some selected papers do not particularly reflect on teacher education (10 papers) but refer to teachers as important subjects for using big data in specific courses, institutions, or educational systems (e.g., Anshari et al., 2015; Li et al., 2019; Logica & Magdalena, 2015; Xidong & Xiaoye, 2016; Yang et al., 2019; Yu & Wu, 2015).

When examining whether the articles critically analyse the use of big data in students' teaching and learning, we identified that almost all emphasise only the supposedly positive elements of said use. In most papers, there is an exacerbated appreciation of big data as paraphernalia that can offer many possibilities to improve the educational performance of educational institutions and their participants. For example, Yawei and Shiming (2019) argued that, in promoting innovative education reform under big data background, teachers play an important role in reaching innovation and entrepreneurship.

The articles highlight the teachers' poor preparation to approach new curricular aspects in a data society. They state that the solution would be the analysis of big data (e.g., Yawei & Shiming, 2019). Still, they do not offer proposals on how teachers could learn to do this analysis on how those actions could be incorporated into the teaching career. However, they show elements of the impact of big data on education and the content of teacher professional development (e.g., Cui & Zhang, 2018; Xu et al., 2018). They exemplify specific strategies that include cultivating awareness of big data, enhancing the wisdom of big data, and developing data analysis capabilities. Nevertheless, the authors did not explain how teachers can develop such abilities.

Most of the included publications are reflection articles or essays, which make comments based on the assumption that big data is fundamental for improving education. Some of those studies are based on a literature review carried out by technological mechanisms of semantic analysis (e.g., Park, 2020) or a narrative review of the literature restricted to authors from a specific country or region (e.g., Huda et al., 2017; Wang & Cai, 2016). Only a few studies are based on empirical data that explore the possibilities of applying big data to improve education, such as the one by López-Belmonte et al. (2019), which aimed to identify the level of digital competence in information and communications technology of 832 teachers and to infer the possibilities for developing their knowledge about big data.

Several studies propose analysis models that have not been effectively tested or developed with teachers (e.g., Dai et al., 2018; Huang et al., 2015; Mohammed et al., 2018; Wassan, 2015).

Only two articles explore the ethical, political, and social challenges of using big data. Their debate involved different issues and interfaces with other fields of knowledge (Schouten, 2017; Zeide, 2017). Zeide (2017) reflects on the concrete effects of specific technologies and examines how big data-driven tools alter the structure of schools' pedagogical decision-making and change fundamental aspects of education enterprise in the USA context. Those data analysis models proposed by the author imply using big data generated by educational institutions and systems, which can have significant ethical implications. According to the author, while pervasive data collection and mining feed learning analytics, this creates ubiquitous surveillance, impacting more on everyday life outside of formal education and on the assessment and representation of student achievement, academic credit, and intellectual mastery.

Schouten (2017) explores big data use in higher education, considering the ethics of aggregating data as a tool for the debate about using statistics in the legal context. The author argues that in the educational context, the moral risks do not generate an absolute prohibition against using big data as evidence. Therefore, according to \ Schouten, big data can help us to know students better, and, if well used, big data may enable us to lessen the moral risk.

Among those papers included with the combined descriptors "big data" and "teacher education", there is a predominance of discussion on the use and analysis of big data in higher education (e.g., Mbombo & Cavus, 2021; Peñaloza-Figueroa & Vargas-Perez, 2017).

Our analyses of these articles provide evidence of the scarce critical analysis of big data use in teacher education contexts. Linked to the narrative of economic development and the "age of data", the use of big data appears more as an artefact aimed to, supposedly, improve teacher education, thus being another skill to be developed as a synonym for an education connected with the new era. Most papers present a perspective that incorporating big data use in teaching contexts depends on the teachers' skills (e.g., Schildkamp, 2019), although the papers do not give alternatives for teachers to learn. The use of big data seems to appear as a magic solution for improving education, positively influencing the social and economic development necessary for the "data era". This uncritical narrative of reality appears to be hegemonic. In this sense, it is urgent to develop reflections and critical educational approaches to big data in schools and in teacher education to counter the non-critical narratives of overvaluation of big data as a problem-solver for education, including mathematics education and statistics.

5 Final remarks

Our study sought to analyse trends, classify research topics, discuss limitations, and provide possible future directions in the relationships between big data and mathematics and statistics education. The discussions are intended to contribute to possible actions toward statistics literacy that critically approach the use and possible role of big data in statistics and mathematics teaching and learning.

Our analysis of the included papers revealed that only a few articles explore the relationship between big data and statistical literacy. Some emphasise a critical notion about big data (Carmi et al., 2020; Papacharissi, 2015; Pawluczuk, 2020; Sander, 2020; Souza, 2018), while most underscore big data as a facilitator for fast economic progress and related technical aspects, such as the use of analysis techniques and big data to improve economic, educational, and social actions (Cope & Kalantzis, 2016; Fiofanova, 2020; Liu & Zhang, 2021; McGowan, 2020; Ying, 2019). Linked to this process of economic development are the uses of big data to control the population through data manipulation and the application of algorithms by governments and large social media corporations. From the analysis of the included publications, we could identify that the ethical aspects of big data use are still not problematised enough and need to be developed in this field.

In the thematic analysis developed, some issues surprised us, such as the narrative —or even folklore— around big data as a form of knowledge capable of solving many economic, social, and, in particular, educational problems. Therefore, we hope our analyses gives evidence that a discourse emphasising a view of "big data" as a problem-solver for many issues can become more powerful if not confronted with critical, ethical, and human rights elements. We realise that an entire system is formatted so that big data can spread more and more through the social field. New areas are constituted, and old ones are developed from the narrative of the algorithmic era as a new digital pedagogy based on data and standardising new skills and abilities for this new era. These big data meanings may impact mathematics and statistics teacher education, as teachers can be trained to approach big data mechanically by reproducing ideas and procedures without critical conscience of the implications for citizens and for students at different educational levels.

In contrast to the trend of overvaluation of big data without critical analyses, we argue that statistical literacy is important for teachers and students to learn how to interpret big data and understand the epistemic and social implications of the much-acclaimed data era. In this sense, we advocate that statistical literacy can collaborate with creating big data literacy (e.g., François et al., 2020). Sociocultural and ethical issues must be considered, especially when there is an unequal distribution of power over big data, for example, if we consider who collects the data and who has their data collected. Those who collect can dispose of the information for their own purposes, dispossessing others from participating and deciding on their own paths. We understand that the use of big data in education can favour some processes. Yet, if all actors, including students and teachers, are unaware of how their data will be used, that collection becomes authoritarian and a violence against human rights.

Literacy should not just be about knowing how to use big data or mining information but also about having a critical understanding of its use. Teachers need to experience educational processes that lead them to reflect on the digital procedures that involve big data.

We draw attention to something deeper: the claim of "data science" to become broad social engineering, thus annihilating other understandings and worldviews. It is our stance that the algorithm not replace our diversity, forms of artistic and cultural expression nor replace the human in us. Mathematics and statistics education can contribute to a more objective definition of big data to make it more understandable for the population. It can strengthen the notion of protecting our data gathered from contact with digital technologies and favour a critical reading of how large corporations use our data. Thus, statistical education could promote greater critical awareness of data protection from a human rights perspective.

The review discussed in this chapter suggests complex relationships between big data and statistical education seems to point to a field or area still being defined. This means that new studies and research are needed to consolidate this relationship, giving more emphasis to this new thematic field.

Acknowledgements This study was partly financed by the Coordenação de Aperfeiçoamento de Pessoal de Nível Superior – Brasil (CAPES) – Finance code 88881.337562/2019-01. We also acknowledge Maria Isabel de Castro Lima for her professional editing.

References

Ainley, J., Gould, R., & Pratt, D. (2015). Learning to reason from samples: Commentary from the perspectives of task design and the emergence of "big data.". *Educational Studies in Mathematics, 88*(3), 405–412. https://doi.org/10.1007/s10649-015-9592-4

Anshari, M., Alas, Y., & Guan, L. S. (2015). Developing online learning resources: Big data, social networks, and cloud computing to support pervasive knowledge. *Education and Information Technologies, 21*(6), 1663–1677. https://doi.org/10.1007/s10639-015-9407-3

Baig, M. I., Shuib, L., & Yadegaridehkordi, E. (2020). Big data in education: A state of the art, limitations, and future research directions. *International Journal of Educational Technology in Higher Education, 17*(1). https://doi.org/10.1186/s41239-020-00223-0

Beuving, J., & de Vries, G. (2020). Teaching qualitative research in adverse times. *Learning and Teaching, 13*(1), 42–66. https://doi.org/10.3167/latiss.2020.130104

Caldas, M. S., & Silva, E. C. (2016). Fundamentos e aplicação do Big Data: como tratar informações em uma sociedade de yottabytes [Fundamentals and application of Big Data: How to handle information in a yottabyte society]. *University Libraries: Research, Experiences, and Perspectives, 3*(1), 65–85. https://periodicos.ufmg.br/index.php/revistarbu/article/view/3086

Carmi, E., Yates, S. J., Lockley, E., & Pawluczuk, A. (2020). Data citizenship: Rethinking data literacy in the age of disinformation, misinformation, and malinformation. *Internet Policy Review, 9*(2). https://doi.org/10.14763/2020.2.1481

Cope, B., & Kalantzis, M. (2016). Big data comes to school: Implications for learning, assessment, and research. *AERA Open, 2*(2), 2332858416641907. https://doi.org/10.1177/2332858416641907

Cui, H., & Zhang, D. (2018). Strategies on teacher professional development in big data era. In *2018 4th International Conference on Education Technology, Management and Humanities Science (ETMHS 2018)*. Atlantis Press.

Dai, Q., Chen, Y., & Hua, G. (2017). Relationship between big data and teaching evaluation. In *Proceedings of the 3rd International Conference on Education and Social Development – ICESD 2017* (pp. 8–9).

Dai, H., Tao, Y., & Shi, T. W. (2018). Research on mobile learning and micro course in the big data environment. In *Proceedings of the 2nd international conference on e-education, e-business and e-technology* (pp. 48–51).

Deng, M. (2017). Analysis on value evolution of higher vocational teachers and its development paths under the big data background. In *3rd International Conference on Arts, Design and Contemporary Education, ICADCE 2017* (pp. 738–740). Atlantis Press.

Fiofanova, O. A. (2020). New literacy and data-future in education: Advanced technology smart big-data. *Revista Inclusiones, 7*, 174–180. http://revistainclusiones.org/index.php/inclu/article/view/1276

François, K., Monteiro, C., & Allo, P. (2020). Big data literacy as new vocation for statistical literacy. *Statistics Education Research Journal, 19*(1), 194–205. https://doi.org/10.52041/serj.v19i1.130

Gal, I. (2002). Adults' Statistical Literacy: Meanings, components, responsibilities. *International Statistical Review / Revue Internationale de Statistique, 70*(1), 1. https://doi.org/10.2307/1403713

Hoel, T., Chen, W., & Lu, Y. (2020). Teachers' perceptions of data management as educational resource: A comparative case study from China and Norway. *Nordic Journal of Digital Literacy, 15*(3), 178–189. https://doi.org/10.18261/issn.1891-943x-2020-03-04

Huang, L., Wei, Y., Zamboni, A., Zhang, J., & Xu, H. (2015). Big data analysis in a social learning platform. In *4th International conference on computer, mechatronics, control and electronic engineering* (pp. 1467–1470). Atlantis Press.

Huda, M., Maseleno, A., Shahrill, M., Jasmi, K. A., Mustari, I., & Basiron, B. (2017). Exploring adaptive teaching competencies in big data era. *International Journal of Emerging Technologies in Learning (IJET), 12*(03), 68. https://doi.org/10.3991/ijet.v12i03.6434

Li, J., Yang, Q., & Zou, X. (2019). Big data and higher vocational and technical education: Green food and its industry orientation. In *Proceedings of the 2019 International Conference on Big Data and Education* (pp. 118–123). https://doi.org/10.1145/3322134.3322150

Liu, F., & Zhang, Q. (2021). A new reciprocal teaching approach for information literacy education under the background of big data. *International Journal of Emerging Technologies in Learning (IJET), 16*(03), 246. https://doi.org/10.3991/ijet.v16i03.20459

Logica, B., & Magdalena, R. (2015). Using big data in the academic environment. *Procedia Economics and Finance, 33*, 277–286. https://doi.org/10.1016/S2212-5671(15)01712-8

López-Belmonte, J., Pozo-Sánchez, S., Fuentes-Cabrera, A., & Trujillo-Torres, J.-M. (2019). Analytical competences of teachers in big data in the era of digitalized learning. *Education Sciences, 9*(3), 177. https://doi.org/10.3390/educsci9030177

Mbombo, A. B., & Cavus, N. (2021). Smart university: A university in the technological age. *TEM Journal, 10*(1), 13–17.

McGowan, B. S. (2020). OpenStreetMap mapathons support critical data and visual literacy instruction. *Journal of the Medical Library Association: JMLA, 108*(4), 649. https://doi.org/10.5195/jmla.2020.1070

Mezhennaya, N. M., & Pugachev, O. V. (2019). Advantages of using the CAS mathematica in a study of supplementary chapters of probability theory. *European Journal of Contemporary Education, 8*(1). https://doi.org/10.13187/ejced.2019.1.4

Mohammed, A., Kumar, S., Singh, S. P., & Sharma, R. P. (2018). Enhancing teaching and learning in educational institutes using the concept of big data technology. In *2018 International conference on computing, power and communication technologies (gucon)* (pp. 1038–1041). IEEE.

Monteiro, C. E. F. (2021). Letramento estatístico e big data: Uma revisão integrativa da literature [Statistical literacy and big data: An integrative literature review]. In C. E. F. Monteiro & L. M. T. L. Carvalho (Eds.), *Temas emergentes em letramento estatístico [Emerging themes in statistical literacy]* (pp. 158–181). UFPE. https://editora.ufpe.br/books/catalog/view/666/677/2080

Papacharissi, Z. (2015). The unbearable lightness of information and the impossible gravitas of knowledge: Big data and the makings of a digital orality. *Media, Culture & Society, 37*(7), 1095–1100. https://doi.org/10.1177/0163443715594103
Park, Y. E. (2020). Uncovering trend-based research insights on teaching and learning in big data. *Journal of Big Data, 7*(1). https://doi.org/10.1186/s40537-020-00368-9
Pawluczuk, A. (2020). Digital youth inclusion and the big data divide: Examining the Scottish perspective. *Internet Policy Review, 9*(2). https://doi.org/10.14763/2020.2.1480
Peñaloza-Figueroa, J. L., & Vargas-Perez, C. (2017). Big-data and the challenges for statistical inference and economics teaching and learning. *Multidisciplinary Journal for Education, Social and Technological Sciences, 4*(1), 64. https://doi.org/10.4995/muse.2017.6350
Ridgway, J. (2016). Implications of the data revolution for statistics education. *International Statistical Review, 84*(3), 528–549. https://doi.org/10.1111/insr.12110
Ruiz-Palmero, J., Colomo-Magaña, E., Ríos-Ariza, J. M., & Gómez-García, M. (2020). Big data in education: Perception of training advisors on its use in the educational system. *Social Sciences, 9*(4), 53. https://doi.org/10.3390/socsci9040053
Sander, I. (2020). What is critical big data literacy and how can it be implemented? *Internet Policy Review, 9*(2). https://doi.org/10.14763/2020.2.1479
Schildkamp, K. (2019). Data-based decision-making for school improvement: Research insights and gaps. *Educational Research, 61*(3), 257–273. https://doi.org/10.1080/00131881.2019.1625716
Schouten, G. (2017). On meeting students where they are: Teacher judgment and the use of data in higher education. *Theory and Research in Education, 15*(3), 321–338. https://doi.org/10.1177/1477878517734452
Souza, R. R. (2018). Algorithms, future and digital rights: Some reflections. *Education for Information, 34*(3), 179–183. https://doi.org/10.3233/efi-180200
Tractenberg, R. (2017). How the mastery rubric for statistical literacy can generate actionable evidence about statistical and quantitative learning outcomes. *Education Sciences, 7*(1), 3. https://doi.org/10.3390/educsci7010003
Wang, L., & Cai, R. (2016). Classroom questioning tendencies from the perspective of big data. *Frontiers of Education in China, 11*(2), 125–164. https://doi.org/10.1007/bf03397112
Wassan, J. T. (2015). Discovering big data modelling for educational world. *Procedia-Social and Behavioral Sciences, 176*, 642–649.
Xidong, W., & Xiaoye, L. (2016). Study of higher education reform under the background of big data. *Innovation in Regional Public Service for Sustainability, 505*. https://doi.org/10.2991/icpm-16.2016.136
Xu, X., Wang, Y., & Yu, S. (2018). Teaching performance evaluation in smart campus. *IEEE Access, 6*, 77754–77766.
Yang, Q., Li, J., & Zou, X. (2019). Big data and higher vocational and technical education: Green tourism curriculum. In *Proceedings of the 2019 international conference on big data and education* (pp. 108–112). https://doi.org/10.1145/3322134.3322149
Yawei, L., & Shiming, Z. (2019). The role and task of innovation and entrepreneurship teachers under the background of big data. In *Proceedings of the 2019 international conference on big data and education* (pp. 98–102). https://doi.org/10.1145/3322134.3322146
Ying, Y. (2019). Research on college students' information literacy based on big data. *Cluster Computing, 22*(S2), 3463–3470. https://doi.org/10.1007/s10586-018-2193-0
Yu, X., & Wu, S. (2015). Typical applications of big data in education. In *2015 International conference of Educational Innovation Through Technology (EITT)* (pp. 103–106). IEEE. https://doi.org/10.1109/EITT.2015.29
Zeelenberg, K., & Braaksma, B. (2017). Big data in official statistics. In T. Prodromou (Ed.), *Data visualisation and statistical literacy for open and big data* (pp. 274–296). IGI Global.
Zeide, E. (2017). The structural consequences of big data-driven education. *Big Data, 5*(2), 164–172. https://doi.org/10.1089/big.2016.0061

Interdisciplinary Data Workshops: Combining Statistical Consultancy Training with Practitioner Data Literacy

Danny Parsons, David Stern, Balázs Szendrői, and Elizabeth Dávid-Barrett

Abstract An approach to interdisciplinary data workshops has been developed that brings together mathematical science students and practitioners to work on problems with real data in the practitioners' area of expertise. It combines ideas from training statisticians for statistical consultancy and developing data literacy skills in non-specialists. The approach was conceived through collaborative research projects aimed at tackling development challenges in Africa using data-based approaches to corruption in public procurement and farmer experimentation in agriculture. Implementation of the approach in four workshops in three African countries is presented. The workshops provided important learning outcomes both for students, experiencing a problem-solving approach to working with real data in a genuine context, and for practitioners to gain data awareness, literacy, and skills.

Keywords Data skills · Statistical consultancy training · Statistics training for non-specialists · Data problem solving · Cooperative learning

1 Introduction

This paper presents an approach to interdisciplinary data workshops that was developed through two projects funded by the Global Challenges Research Fund (GCRF), established by the UK government to support challenge-led interdisciplinary

D. Parsons (✉) · D. Stern
IDEMS International, Reading, UK
e-mail: danny@idems.international

B. Szendrői
Faculty of Mathematics, University of Vienna, Vienna, Austria

E. Dávid-Barrett
School of Law, Politics and Sociology, University of Sussex, Brighton, UK

G. F. Burrill et al. (eds.), *Research on Reasoning with Data and Statistical Thinking: International Perspectives*, Advances in Mathematics Education,
https://doi.org/10.1007/978-3-031-29459-4_19

research that addresses problems faced in developing countries (Department for International Development, 2015).

The two projects, led by the Mathematical Institute, University of Oxford, focused on novel ways to use the mathematical sciences to address development challenges through research. The first was on data-based approaches to measuring corruption risks in public procurement and brought together mathematical scientists and social scientists. A follow-on project, focusing on data-driven agriculture, explored novel approaches in the use of statistics to support the analysis of large data sets from on-farm trials with large numbers of farmers, involving farmer organisations in the process.

The collaborative nature of these projects informed the design of interdisciplinary data workshops that bring together mathematics and statistics students and practitioners to gain experience working together on problems with real data in the practitioners' area of expertise.

The paper presents this approach and implementation in and beyond these projects in four workshops in three African countries. The learning outcomes for students and practitioners are discussed.

2 Background

2.1 Contextual Background

The two projects leading up to the workshops were about bringing people from the mathematical sciences and other disciplines together around development challenges using tailored tools and relevant data. The first project on corruption risks in public procurement took advantage of the increase in government procurement data being digitised and made open combined with new methods of analysis to detect risk of corruption (Fazekas & Kocsis, 2020). A large (250,000 contracts and 50 variables), open data set of World Bank-funded contracts in 170 countries was used as a basis for developing a tailored menu for procurement analyses within R-Instat, a front-end menu system to R (Stern, 2017). The menu system of R-Instat combined with the tailored procurement menu was designed to enable a wider range of practitioners to easily work with procurement data to produce results and visualisations.

Two workshops took place in Tanzania during this project in 2017. One introduced the analysis of public procurement data and the World Bank data set to mathematical science students. The other was a meeting with practitioners to discuss the potential and limitations of procurement data. One conclusion reached was that the experience of participants at both these workshops would have been enhanced by an interdisciplinary approach, where students and practitioners interacted, discussing both technical and contextual issues with each other. This informed the design of the interdisciplinary data workshops presented here.

The second project focused on supporting FUMA Gaskiya, a farmer federation in Niger with 17,000 members, to use novel approaches for analysing the data from 1700 farmers' on-farm experiments of three low-cost fertilisers. FUMA Gaskiya implemented digital data collection with farmers to construct a database with data at the plot level and the farmer level, which led to high quality data structures. A tailored menu in R-Instat was developed to support FUMA Gaskiya, and others, to easily produce appropriate analyses for such data. To conclude the project, meetings and interdisciplinary data workshops were held in Niger to share the results of this work with a wider range of agricultural organisations.

2.2 Pedagogical Background

Data has become ubiquitous in society and people are increasingly bombarded with data and information in their work and everyday life (Wolff et al., 2016). Understanding the role of data and the ability to interpret and interrogate data and use it for decision making are no longer specialist skills needed by a few (Boyd & Crawford, 2012). Data literacy is increasingly recognised as a set of skills needed by practitioners in many sectors where their primary training is not statistics or data science. Data literacy is generally defined in terms of the ability to understand or interpret data and to use it for decision making (Frank et al., 2016; Wolff et al., 2016).

There is a growing body of research related to the teaching and learning of data literacy and statistics for non-statisticians. Many studies suggest that activities should be based on a context which the learners know. Tygel and Kirsch (2015) suggest that data literacy should be presented and taught in contexts with which learners are familiar. This approach supports people to see the relevance of data to their own lives, even before they use and interpret data themselves using technology (Bhargava et al., 2016). Da Valle and Osti (2016) used games in magazines to engage non-specialists in official statistics. The games were chosen to be similar to typical games found in general puzzle magazines in order to ensure familiarity for the readers.

D'Ignazio (2017) defines five tactics to support "creative data literacy" to empower non-technical learners. This includes use of community-centered data, use of messy data to discuss the concept of uncertainty, use of learner centered tools, and use of non-standard visualisations.

Bromage et al. (2022) identified lack of motivation and "statistics anxiety" as key challenges in teaching statistics and statistical literacy to non-specialists. Again, they suggest beginning with learners' personal experiences and interests to overcome this. They also suggest that computational software and interactive tools can support more active learning approaches for non-specialists.

There is also extensive literature on teaching statistical consulting skills for statisticians. Good statistical consultants require a wide range of skills, including an interest and ability to problem solve, listening and communication skills, the ability to speak a client's language, computer skills and an ability to teach others these skills, as well as good statistical knowledge and understanding (Kotz et al., 2005). Kenett and Thyregod (2006) note that university education generally focuses on data analysis, which is just one of several steps in the statistical consulting cycle. Other steps include problem elicitation, data collection and formulation, and presentation of findings. They also state that consulting is collaborative, and good communication is the key factor in determining success. There have also been efforts to give students statistical consulting experience within the classroom. For example, Bilgin and Petocz (2013) developed a capstone unit that gave students the opportunity to act as statistical consultants. As well as the skills mentioned previously, they identified other important skills that students mentioned gaining from this experience. These include ethics and sustainability, the ability to learn how to learn when faced with unfamiliar statistical methods, and developing their confidence as a professional. Vance et al. (2022) emphasise the importance of creating shared understanding through common knowledge for successful collaboration between statisticians/data scientists and domain experts. They provide a five-step process for teaching shared understanding to statistics students. Again, many of the steps focus on good communication, including asking good questions and listening, paraphrasing information to develop common knowledge, summarising information, and then applying shared understanding. Examples of applying these steps in courses for undergraduate and graduate students are presented that include in-class exercises and collaboration meetings. Although our concept of interdisciplinary workshops pre-dates this work, some of the ideas are similar to that of Vance et al. (2022), who expressed them in a more clear and succinct language. We borrow some of this language to improve the explanation of our workshop design in the following section.

Our workshop design brings together aspects of both teaching data literacy to non-specialists and statistical consultancy training for statisticians/data scientists. While the design mainly draws on the approaches described above, our innovation is to combine aspects of both approaches in order to achieve parallel learning outcomes for both groups working jointly together.

3 Workshops

3.1 Workshop Design

Following the experience of earlier workshops in the first project where participants of different disciplines were not mixed, an idea for interdisciplinary data workshops was conceived to bring together statistical students and practitioners to work on problems involving real data in the practitioners' area of expertise. Interdisciplinary data workshops involve three main activities:

1. Activities to develop a shared understanding of the practitioners' area of expertise
2. Discussions on the role and use of data in the practitioners' area of expertise, and
3. Hands-on activities working with data, including communication of findings

The initial activities begin with the practitioners' area of expertise and do not involve data. They could include plenary or small group discussions/brainstorming sessions with mixed groups of statisticians and practitioners. These are designed to develop common knowledge among the two groups leading to a shared understanding of the practitioners' domain, in line with the more recent suggestions from Vance et al. (2022).

Before doing hands-on data activities, the potential role and importance of data to the practitioners' domain is explored through discussions. According to Bhargava et al. (2016), non-specialists need to first understand the potential role of data and embrace its value to them. This is an opportunity to discuss the value of data and data specialists and also the challenges and limitations of any data, and to understand that data are not "silver bullets" that can neatly answer any question posed. Hence, this step is crucial as it builds trust through open discussion and questioning and develops motivation and interest for then working directly with data.

A large part of the workshop is devoted to hands-on activities working with data from the practitioners' domain in interdisciplinary groups of statisticians and practitioners. The groups formulate a question of interest, which must be grounded in an understanding of the domain and be possible to investigate statistically with the data available in the workshop. This requires shared understanding within the interdisciplinary team of both the context of the domain and potential analyses with the available data. Groups can use the tools of their choice to analyse the data, but they are introduced to R-Instat as a simple point-and-click based general statistical software. R-Instat's tailored menus for public procurement data and on-farm agriculture research data are shown, and they aim to make it easy to do specific analyses of interest with data from these domains. These analyses are often more complex to produce in general statistical software. In particular, these tailored menus aim to make non-standard visualisations, that are relevant to the type of data being used, easy to produce. Such visualisations are uncommon in traditional statistics teaching but easier for non-specialists to understand and may be more commonly encountered in everyday work or life, such as in reports or news articles. Non-standard visualisations available in R-Instat's tailored menus include mosaic plots, tree maps, lollipop charts, and connected paired points (on two boxplots) (Fig. 1).

Workshops are led jointly by a specialist in statistics and a specialist in the practitioners' domain. This is done to show that both groups have learning opportunities and contributions to make, and therefore neither group should feel like the "junior partner" within an interdisciplinary group.

The workshop design serves multiple purposes, providing different learning opportunities and outcomes for different types of participants. For students, the workshops are intended to provide opportunities to gain practical experience with real world problems involving real data. The inclusion of real data on its own is not sufficient to improve the teaching of statistics, but by facilitating interaction with practitioners, students can be exposed to a genuine context and purpose for the data, leading to exposure to a broader set of statistical skills. This includes translating problems and questions from a practitioner into statistical problems, interpreting results in a real-world context, and communication within a team and to non-technical audiences.

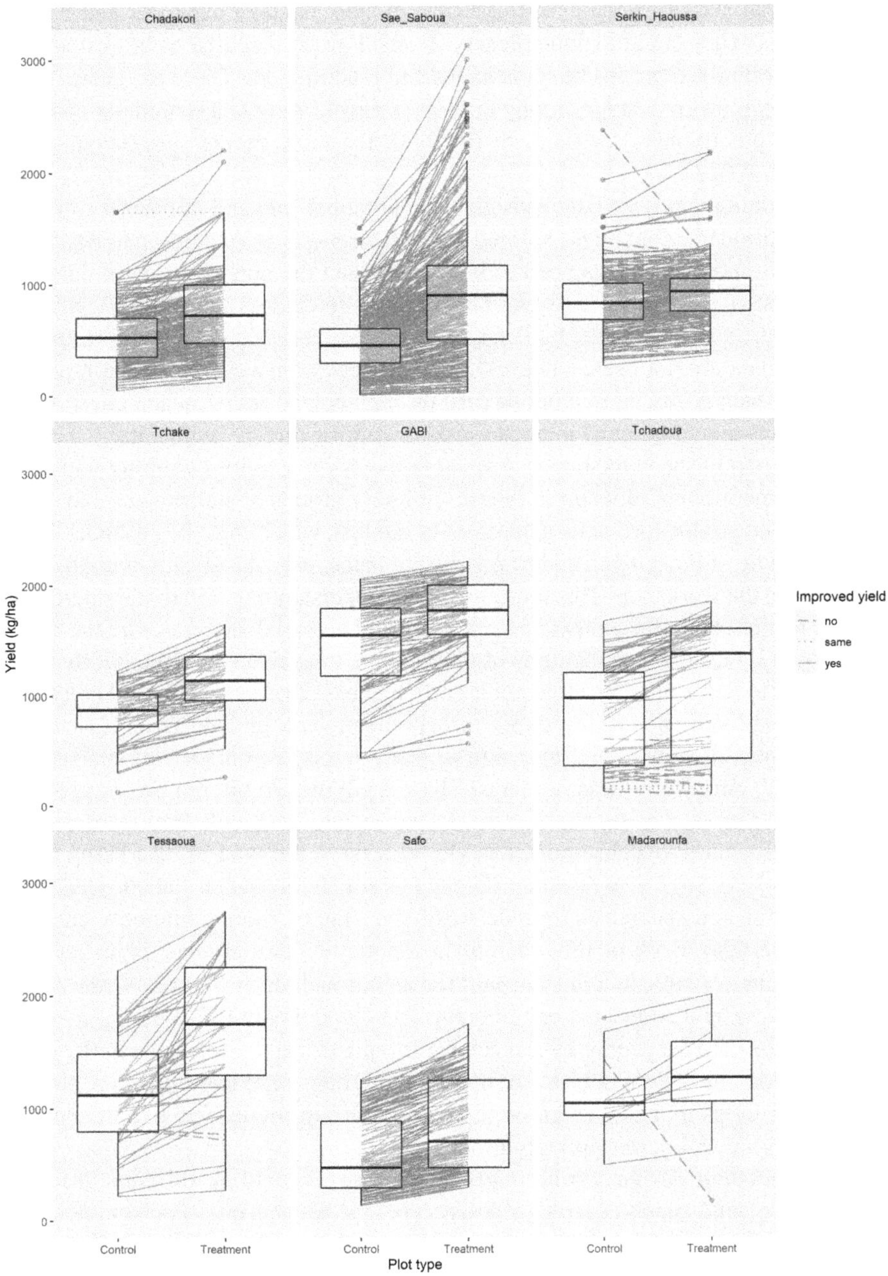

Fig. 1 A connected paired points graph with boxplots at 11 sites in Niger. Each line represents a farmer's yield from a treatment and control plot. The slope of the line indicates whether the yield increased with the treatment (positive, solid), decreased (negative, dashed) or was the same (flat, dotted)

The main outcome for practitioners is data literacy. This may be to varying levels as the practitioner group is often more varied in their background and experiences. However, a core outcome is to understand the potential and limitations of data and the role of data experts in their work, which we define as "data awareness", a component of data literacy. Some practitioners will gain further data literacy skills including interpreting, critiquing, and reasoning with data, as well as technical skills working with data, producing visualisations, and using statistical methods. Practitioners who are able to express an appreciation and an understanding of the role of data and data experts in their field will have demonstrated a level of data awareness. Those who further show an ability to meaningfully interact with data demonstrate data literacy. If they can also use software and statistical methods to work with data they demonstrate data skills. We note that demonstration of these skills can also relate to prior knowledge of participants and not necessarily be solely attributed to the workshop. However, formal and informal evaluation of participants can help to understand the effect of the workshop.

Four workshops that followed this design are presented in this paper. Each workshop was held over two consecutive days with 4–6 hours of sessions each day. In general, day one focused on activities 1 and 2 as given above, ending with initial practical work with data. Day two focused on activity 3 involving group practical work and presentation and discussion.

The Results section is based on responses to an evaluation questionnaire completed by the participants in the two procurement workshops, as well as observations from the workshop trainers across the four workshops. The evaluation questionnaire included 12 questions (eight rating scale and four open text questions) grouped into 3 parts: "Workshop objectives and content", "Logistics", and "Going Forward". The Results section mainly draws from the "Workshop objectives and content" part, which included questions on the added value of the workshop to understanding procurement data analysis, the workshop length and workload, the R-Instat tool, the clarity of the objectives, the most useful aspects of the workshop, and what, if anything, the workshop made participants think differently about.

3.2 Procurement Workshops

The first two-day interdisciplinary data workshop took place in May 2018 at the African Institute for Mathematical Sciences (AIMS) Ghana. The 50 participants comprised of 37 AIMS Ghana MSc Mathematical Science students from several African countries and 13 practitioners from anti-corruption civil society organisations (CSOs) in Ghana. The first day involved discussions and group activities on defining public procurement and corruption and on the ways that the procurement process can be manipulated for corrupt gain. This began with a presentation introducing corruption risks in public procurement and red flag indicators. This gave the mathematical science students the necessary background and initiated interaction with the practitioners to develop common knowledge about the context to build shared understanding. A World Bank dataset containing details of 250,000 public procurements in 170 countries was presented to instigate discussions on what the role of data and data specialists could be in public procurement. The afternoon session was dedicated to practical data activities. Participants were shown R-Instat as a tool to explore and understand the structure of the World Bank data, and they used it to carry out exploratory analyses, such as single variable summaries, two-way frequency and summary tables, boxplots, bar charts, mosaic plots, and tree diagrams. They also learned data cleaning and manipulation including observation filtering to reduce the dataset or remove discovered errors, converting variables to appropriate data types (numeric/categorical) and summarising data, for example, annual or country-level statistics.

On day two, interdisciplinary groups of three to five participants were formed to work together on a chosen problem with groups given the morning session to work on this while trainers were available to discuss with groups individually. Groups began by formulating a question to investigate that was both of interest to the practitioners and students in the group and practical to investigate with the available data. Part of this exercise also involved defining the data needed to investigate the question, that is, which variables and which subset of the data were needed. This also helped identify proposed questions that could not be tackled with the available data. Then, by using R-Instat and the tailored Procurement menu, participants created visualisations related to their question including mosaic plots to display multiple categorical variables, multi-way tables, and summaries at yearly and country levels. They were required to interpret their results and prepare a short presentation with their analysis and findings, which was presented to the other participants in the afternoon session before a final plenary discussion. An example of a mosaic plot showing the proportion of contracts let through different procurement procedure types (e.g. from open competition to sole-source) in Ghana over the years 1998 to 2014 produced by one of the groups using R-Instat is shown in Fig. 2. Groups were able to produce standard and non-standard visualisations as part of their work.

This workshop was replicated at Makerere University, Uganda in October 2018 and followed the same structure. The 30 workshop participants were final-year undergraduate mathematics and statistics students together with practitioners from CSOs in Uganda and a representative of the Ugandan Public Procurement and Disposal of Public Assets Authority (PPDA), with practitioners making up one third of the group. An additional dataset of Ugandan national public procurement contracts that had been web-scraped from an open data portal was also included in the workshop. Participants were free to choose to use the World Bank dataset and/or the Uganda dataset for their group projects. This allowed participants to conduct both in-depth studies of Ugandan contracts and cross-country comparisons with the World Bank data.

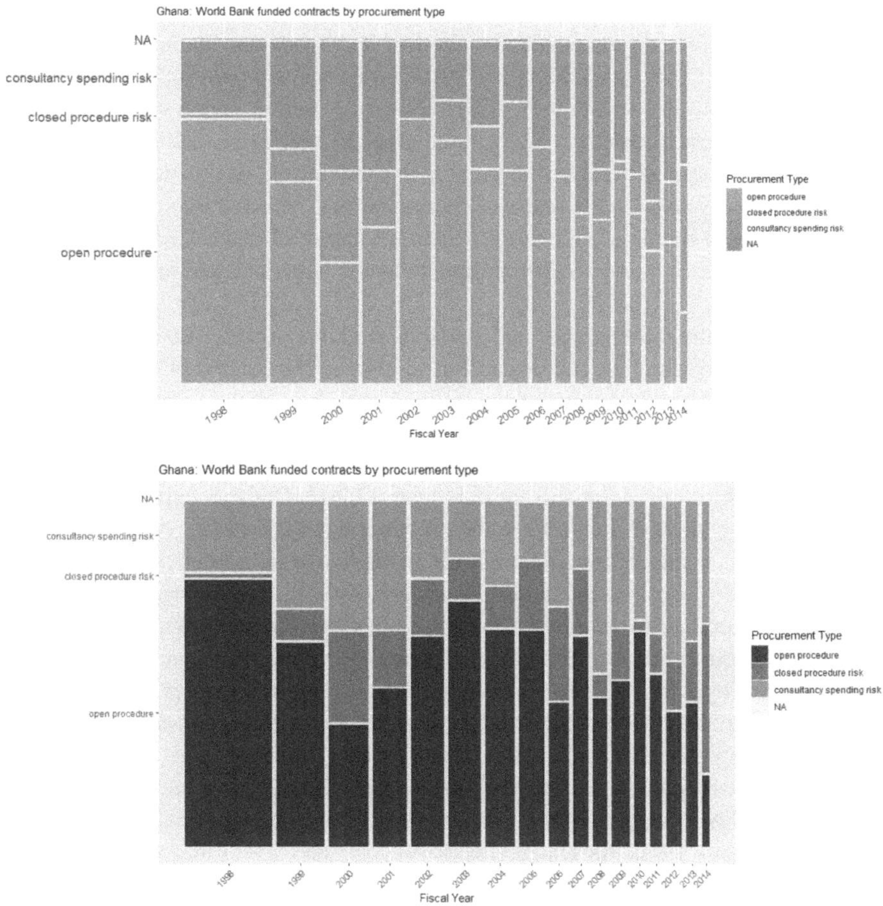

Fig. 2 A mosaic plot of World Bank funded contracts in Ghana by procedure type: open procedure, closed procedure risk, consultancy spending and NA/unknown for each year from 1998 to 2014. The width of the bars is proportional to the number of contracts in that year

3.3 Agriculture Workshops

Two interdisciplinary data workshops focusing on agriculture were held in Niger in June 2018. The structure of these workshops was similar to the procurement workshops but adapted to suit the context through collaborative discussions with the participant organisations. For example, it was agreed that some workshop time with the farmer practitioners separately would be beneficial.

The first workshop was a two-day workshop for 24 farmers from a farmers' organisation, with 10 mathematics and statistics students from the University of Maradi joining for the second day only. The farmers were members of the farmer federation FUMA Gaskiya who had been collecting data digitally as part of on-farm experiments in their communities. The aim of the first session on day one was to show the value of the digital data collection that had been done by the farmers in the days leading up to the workshop. The participants were shown how the data collected via their devices was automatically sent and stored in a central online repository and could therefore be accessed immediately. In a plenary session the data were investigated in a spreadsheet by the trainers to show initial results interspersed with participant discussions on the real-world implications. This included detecting possible mistakes in the data collection, which led to useful discussions between the participants, such as trying to differentiate between a mistake and an outlier that is a true value.

On day two, the mathematics and statistics students joined. The sessions began with a plenary presentation and discussion session on the context of the farmer federation, the experiments being carried out and the data collection being done. The aim of this was to create common knowledge and shared understanding between the two groups of participants. Interdisciplinary teams of three to five participants were formed to analyse the (anonymised) data from FUMA Gaskiya experiments using R-Instat and the tailored menu. As in the procurement workshops, participants were first introduced to the dataset and the tools available by the trainers and were given the opportunity to explore the data in small groups with support from the trainers. Each group then had to generate a question of interest to investigate. The groups were challenged to produce the graph of connected pairs of points on top of two boxplots shown in Fig. 1. The connecting lines represent a farmer carrying out a fertilizer treatment and a control on adjacent plots. In the afternoon session, plenary discussions took place on interpreting visualisations, including Fig. 1, and insights into what these meant for the farmers' own contexts were drawn out.

The second workshop was a two-day workshop attended by 15 mathematics and statistics students from the University of Niamey and 25 farmer representatives from farmer organisations and practitioners from agriculture non-governmental organizations. The workshop began with presentations and discussions on the potential role of the mathematical sciences and data experts in agriculture and development and how research can and should interact with development in Niger. Discussions began and concluded the day in plenary, with small groups in between on specific topics. On the second day, interdisciplinary groups of students, farmers and practitioners were formed to investigate the same data as the first workshop. In most groups, students manipulated the software to produce results and visualisations, and the farmers and practitioners discussed the interpretations of the results by relating it to their own context.

4 Results

Twenty-five out of 50 participants from the first procurement workshop completed the evaluation questionnaire, and 25 out of 30 participants from the second procurement workshop completed the same questionnaire.

The workshops provided students with experience of analysing real data with a genuine context in a problem-solving setting. As one student noted in the workshop evaluation, "that was really excellent and added value to me as a mathematician…it's needed in all institution to see the applicability of statistics and mathematics in their own life" with another saying "Participating to this workshop made me understand how relevant data analysis is to inform policy makers." All participants who completed the workshop evaluation from the two procurement workshops agreed or strongly agreed that the workshop added value to their understanding of procurement data analysis.

In both sets of workshops, the interactions with practitioners appeared to be valuable in exposing students to statistics in a broader setting than is usually taught. The workshop trainers observed that practitioners challenged students when they drew conclusions from their analyses that overstated what the data showed or made unsubstantiated assumptions about the context. This was valuable learning for the students on the importance of accurate communication, in understanding the limits of what can be concluded from data and the value of engagement with practitioners and importance of developing a shared understanding. The opportunity to practice communication in a real world setting and the ability to receive feedback directly from expert practitioners may be important in supporting students to improve their communication skills. As one student noted "I have been able to learn how to give reports to authorities without over confirming… what am not inclusive in and also made me see that really gaps can exist but reporting makes a difference." This may be important for economic development given the growing role of evidence in public policy-making (Cairney, 2016).

The practitioners were more varied than the students in terms of their skills, interests, roles and experiences. The variety of learning outcomes of the practitioners can be broadly grouped into three aspects of data literacy. Firstly, data awareness, where practitioners gained the experience of seeing the role of data in their subject area and an understanding and appreciation for how data, and data experts, could enhance their work. In the workshops in Niger, representatives from the agriculture organisations said to the workshop trainers that they learned the value a data manager could bring to their organisation from hearing about the value FUMA Gaskiya gained from using digital data collection. In Uganda, a participant noted "The workshop enabled me to appreciate how using existing data can promote the fight against corruption. There is great potential for collaboration between academia and CSO in dealing with social challenges such as corruption".

A second outcome was data interpretation skills. Practitioners gained experience interpreting and questioning results, using their subject knowledge to engage in discussions around data. In Uganda, practitioners were able to question students' broad conclusions about corruption using their knowledge of procurement, exemplified through comments such as "this workshop made me understand how relevant data analysis is to inform policy makers." In Niger, the farmers used their knowledge of the local landscape and weather to explain to researchers present why the graph in Fig. 1 showed poor results in some areas. It was interesting for the workshop trainers to note that non-standard visualisations such as paired connected points were easier for farmers to understand than standard visualisations such as boxplots. We conjecture that this relates to the former graph displaying individuals whereas the later displays summaries, and therefore requires a further layer of abstraction. This may have important implications for the teaching of basic statistics and what visualisations should be included in training of non-specialists.

Some practitioners also gained hands-on data skills through working with R-Instat and its tailored menus. In the short time frame of a workshop, a menu-driven statistics software was key to enabling participants to quickly produce results. The tailored menus meant that participants could produce complex and non-standard visualisations relating to questions of interest. This appeared to be a successful approach to exposing practitioners to statistical concepts. At the procurement workshop in Ghana, some practitioners had experience using statistics software, and they were able to learn how to use R-Instat quickly to investigate the data themselves. One participant commented "R-Instat is a good statistical software which is easy to use for non-statisticians…it is more specific to analysing procurement data".

Across the workshops, the trainers identified the importance of valuing the prior knowledge and skills of both sets of participant groups equally, as a crucial element of a successful workshop. Efforts were made to establish from the beginning through the initial presentations and discussions that participants in each group had value to contribute to the other with neither groups' skills seen as more valuable than the other, rather that both were needed for successful collaboration. This was particularly important for the practitioner groups, where a fear of, or anxiety towards mathematics is often observed.

Also, while the element of having time for practical group work on the majority of the second day was successful in providing hands on experience, in future workshops, the trainers would recommend interspersing these with a number of short plenary discussions based on observations of the groups. When this was done in some workshops, participants appeared to benefit from hearing about a success or challenge encountered by another group, and this provided extra learning opportunities during practical sessions. Ensuring that trainers look out for such opportunities and include them in the practical sessions would benefit future workshops.

5 Conclusions

The interdisciplinary data workshop design provides an approach to simultaneously engaging statistics students to develop statistical consultancy skills and practitioners/non-specialists to develop data literacy. This builds on existing methods and knowledge of teaching and learning related to these two sets of skills but combines them into a single workshop approach.

The approach was implemented in four workshop events in three African countries. In each workshop, there was evidence that students gained valuable experiences working with data in a rich, real-world context. Students were also exposed to the value of communication with practitioners and the importance of developing shared understanding, which are critical skills for statistical consulting (Vance et al., 2022).

The practitioner groups were more varied, but in each workshop practitioners were able to gain some components of data literacy. For some, this was limited to data awareness, including understanding the potential of data specialists for their work. This may be sufficient for the needs of practitioners in senior or managerial positions, particularly if others in their team gain further data literacy skills. Some practitioners were also able to develop data investigation and hands-on data skills, and they saw that these skills are in reach, and not restricted to those with mathematical and statistical backgrounds. We suggest that other studies following this approach could be useful to further understand the development of data literacy skills by participants.

Other interdisciplinary initiatives such as Virginia Tech's laboratories for interdisciplinary statistical analysis (Vance & Pruitt, 2016), provide opportunities to institutionalise such workshops in universities to further support improved tertiary statistics education. In many areas, collaboration is often limited to practitioners handing over data to a data expert for analysis with little input from the practitioner. The approach presented here shows how genuine interdisciplinary collaboration can be beneficial to all, drawing on expertise from both groups of participants. In this study, two distinct content areas were used as the focus of the workshops, and we believe the approach would be relevant to any other area there is a need for data expertise.

In many countries, including those in which our workshops took place, there is a perception that studying mathematics is only useful to become a teacher or lecturer and that it has little real-world application (Mutodi & Ngirande, 2014). In such contexts, our experience suggests that these workshops could be particularly valuable in changing this perception, allowing for students to see new career possibilities, and practitioners to see a role for mathematical sciences and its graduates in their field. Increasing the engagement of mathematical scientists in interdisciplinary work and development, as was done in these projects, could lead to wider changes in perception and benefits to disciplines that embrace it.

References

Bhargava, R., Kadouaki, R., Bhargava, E., Castro, G., & D'Ignazio, C. (2016). Data murals: Using the arts to build data literacy. *The Journal of Community Informatics, 12*(3). https://doi.org/10.15353/joci.v12i3.3285

Bilgin, A., & Petocz, P. (2013). Students' experience of becoming a statistical consultant. In *Proceedings of the Joint IASE/IAOS Satellite Conference Statistics Education for Progress* (pp. 1–6).

Boyd, D., & Crawford, K. (2012). Critical questions for big data: Provocations for a cultural, technological, and scholarly phenomenon. *Information, Communication & Society, 15*(5), 662–679. https://doi.org/10.1080/1369118X.2012.678878

Bromage, A., Pierce, S., Reader, T., & Compton, L. (2022). Teaching statistics to non-specialists: Challenges and strategies for success. *Journal of Further and Higher Education, 46*(1), 46–61. https://doi.org/10.1080/0309877X.2021.1879744

Cairney, P. (2016). *The politics of evidence-based policy making*. Palgrave Pivot. https://doi.org/10.1057/978-1-137-51781-4

Da Valle, S., & Osti, S. (2016). Statistica enigmista: An ISTAT puzzle magazine to introduce. In J. Engel (Ed.), *Proceedings of the roundtable conference of the International Association of Statistics Education (IASE)*.

Department for International Development. (2015). *UK aid: Tackling global challenges in the national interest*.

D'Ignazio, C. (2017). Creative data literacy: Bridging the gap between the data-haves and data-have nots. *Information Design Journal, 23*(1), 6–18. https://doi.org/10.1075/idj.23.1.03dig

Fazekas, M., & Kocsis, G. (2020). Uncovering high-level corruption: Cross-national objective corruption risk indicators using public procurement data. *British Journal of Political Science, 50*(1), 155–164. https://doi.org/10.1017/S0007123417000461

Frank, M., Walker, J., Attard, J., & Tygel, A. (2016). Data literacy-what is it and how can we make it happen? *The Journal of Community Informatics, 12*(3), 4–8. https://doi.org/10.15353/joci.v12i3.3274

Kenett, R., & Thyregod, P. (2006). Aspects of statistical consulting not taught by academia. *Statistica Neerlandica*, 396–411. https://doi.org/10.1111/j.1467-9574.2006.00327.x

Kotz, S., Balakrishnan, N., Read, C. B., & Vidakovic, B. (2005). *Encyclopedia of statistical sciences* (Vol. 1, 2nd ed.). John Wiley and Sons.

Mutodi, P., & Ngirande, H. (2014). The influence of students' perceptions on mathematics performance. A case of a selected high school in South Africa. *Mediterranean Journal of Social Sciences, 5*(3), 431. https://doi.org/10.5901/mjss.2014.v5n3p431

Stern, D. (2017). Seeding the African data initiative. In *Proceedings of the IASE satellite conference "Teaching statistics in a data rich world"*.

Tygel, A., & Kirsch, R. (2015). Contributions of Paulo Freire for a critical data literacy. In *Proceedings of web science 2015 workshop on data literacy* (pp. 318–334).

Vance, E. A., & Pruitt, T. (2016). *Virginia Tech's laboratory for interdisciplinary statistical analysis annual report 2015–16*. Virginia Tech. Laboratory for Interdisciplinary Statistical Analysis. http://hdl.handle.net/10919/72099. Accessed 19 Mar 2022.

Vance, E. A., Alzen, J. L., & Smith, H. S. (2022). Creating shared understanding in statistics and data science collaborations. *Journal of Statistics and Data Science Education, 30*, 1–17.

Wolff, A., Gooch, D., Montaner, J. J., Rashid, U., & Kortuem, G. (2016). Creating an understanding of data literacy for a data-driven society. *The Journal of Community Informatics, 12*(3). https://doi.org/10.15353/joci.v12i3.3275

Part V
Statistical Learning, Reasoning, and Attitudes

Distinctive Aspects of Reasoning in Statistics and Mathematics: Implications for Classroom Arguments

AnnaMarie Conner and Susan A. Peters

Abstract Reasoning plays an important role in mathematics and statistics, but the kinds of reasoning used to establish results differ between mathematics and statistics. In general, we see more probabilistic and contextual reasoning in statistics, whereas in algebra and other areas of mathematics, results rely on deductive reasoning, perhaps after inductive or abductive reasoning is used to examine patterns. Secondary school teachers are expected to teach topics from both mathematics and statistics, and they are asked to use collective argumentation in their teaching. It is important for teachers to support their students in making arguments that use appropriate reasoning for the subject in which they are engaged. In this paper, we discuss distinctions between reasoning in statistics and mathematics, use diagrams of argumentation to illustrate these differences in practice, and propose that warrants, rebuttals, and qualifiers are important aspects of distinguishing arguments that involve statistical reasoning.

Keywords Statistical reasoning · Mathematical reasoning · Argumentation

1 Background

All students should learn to reason appropriately in all subjects, but reasoning plays a particularly central role in mathematics. Reasoning in K-12 mathematics has been discussed in discipline-specific ways, such as algebraic reasoning (e.g., Kaput, 2008; Nathan & Koedinger, 2000), statistical reasoning (e.g., Ben-Zvi & Garfield,

---

A. Conner (✉)
University of Georgia, Athens, GA, USA
e-mail: aconner@uga.edu

S. A. Peters
Porter Education Building, Louisville, KY, USA
e-mail: s.peters@louisville.edu

G. F. Burrill et al. (eds.), *Research on Reasoning with Data and Statistical Thinking: International Perspectives*, Advances in Mathematics Education,
https://doi.org/10.1007/978-3-031-29459-4_20

2004; Garfield, 2002), and geometric reasoning (e.g., Barrett et al., 2006; Battista, 1999). Recent frameworks and policy documents call for a continued focus on reasoning in schools (e.g., Dreyfus et al., 2021; Organization for Economic Co-operation and Development, 2018). For example, the Common Core State Standards for Mathematics (National Governors Association Center for Best Practice and Council of Chief State School Officers, 2010), which includes attention to practices such as making sense of problem situations, reasoning in multiple ways, and creating and critiquing arguments, brings attention to the importance of students reasoning, communicating their reasoning, and critiquing the reasoning of others. Similarly, the National Council of Teachers of Mathematics' (2009) *Focus in High School Mathematics* describes reasoning and sense making as cornerstones of mathematical understanding for all students. To engage students in reasoning, teachers need both advanced knowledge of the broad range of content they are teaching and knowledge of ways to reason about that content and how to support students' reasoning through argumentation.

Historically, algebra has been a high priority in school mathematics (e.g., Stanic & Kilpatrick, 2003). Mathematics teacher education often has reflected this priority. Algebra and calculus topics have been prioritized (Mathematics Association of America and National Council of Teachers of Mathematics, 2017), and many teachers have had limited opportunities to develop deep understandings of statistics (see, e.g., Lovett & Lee, 2017). As a result, teachers are likely to use the pedagogical strategies and content knowledge developed in their teaching of other mathematical topics when they teach statistics content. Because there are fundamental differences between mathematics and statistics, however, the reasoning necessary to authentically engage in the teaching and learning of statistics is very different from what is used in the teaching and learning of algebra, geometry, and precalculus. Scholars contrast the importance of deduction and abstraction in mathematics with the dependence on inference, probability, and context in statistics (e.g., del Mas, 2004). However, these distinctions are rarely explicated in teacher education. Teacher education would benefit from clear images of mathematical and statistical reasoning and their similarities and differences along with examples of discourse that illustrate mathematical and statistical reasoning.

In concert with an emphasis on reasoning, engaging students in collective argumentation is fundamental to learning because it promotes participation in collaboratively constructing mathematical and statistical ideas. We define collective argumentation as a group arriving at a conclusion through consensus. A growing body of literature in mathematics education focuses on collective argumentation, and calls have been made for a greater focus on collective argumentation in statistics education. The promotion of classroom discourse that includes statistical arguments and establishes norms for statistical argumentation was identified as a key design principle for supporting the development of K-8 (Cobb & McClain, 2004), secondary, and tertiary (Garfield & Ben-Zvi, 2009) students' statistical reasoning even though "the skills of argumentation are typically not part of statistics taught in school" (Pfannkuch & Ben-Zvi, 2011, p. 329). The growing prominence of data science has prompted calls for additional focus on argumentation in statistics.

For example, Pfannkuch (2018) suggested that "fostering statistical argumentation in statistics and other disciplines is a high priority," with a particular need for "more insight into fostering statistical argumentation" (p. 407). We draw on research in mathematics education (e.g., Conner et al., 2014b) and statistics education (e.g., Fielding-Wells & Makar, 2015) using Toulmin's (1958/2003) characterization of arguments to examine argumentation and reasoning. Specifically, we use Conner et al.'s (2014b) Teacher Support for Collective Argumentation (TSCA) framework to explore both the structure of arguments and the ways teachers support students' reasoning in contributing to argumentation. By focusing on collective argumentation, we can foreground the kinds of reasoning that are occurring in classrooms; in this paper we specifically respond to calls for more investigation of statistical argumentation by demonstrating the importance of qualifiers, rebuttals, and warrants in statistical arguments (e.g., Otani, 2019) and the role of the teacher in supporting students to make arguments using statistical reasoning. In this paper, we describe collective argumentation and a framework for supporting collective argumentation, outline multiple kinds of reasoning and highlight differences between mathematical and statistical reasoning, and then illustrate how the framework and conceptualization of argumentation can be used to support the development of statistical reasoning within collective statistical argumentation. Due to the theoretical nature of this paper, we provide both hypothetical examples and examples grounded in data from several different research projects to illustrate the nature of statistical reasoning and the role of the teacher in facilitating *collective* argumentation.

2 Collective Argumentation

Toulmin (1958/2003) introduced a conceptualization of arguments across disciplines as comprised of a set of common components. We use the following set of definitions of components drawn from Toulmin's diagrams of arguments to situate our work. *Claims* are statements one is attempting to establish. *Data* are information on which a claim is based. *Warrants* are justifications linking data to claims. *Qualifiers* (what Toulmin called *modal qualifiers*) indicate the strength with which claims are made. *Rebuttals* address circumstances under which warrants may not hold. While many researchers focus on claims, data, and warrants as core components of arguments, we view *qualifiers* and *rebuttals* as relevant to statistical arguments (see also Inglis et al., 2007). Krummheuer's (1995) adaptation of Toulmin's model has been used in mathematics education research as an analytical tool for examining mathematical arguments (e.g., Conner et al., 2014b; Inglis et al., 2007; Rasmussen & Stephan, 2008). Argumentation in statistics education has been explored using Toulmin diagrams or adaptations of Toulmin's model to consider the argumentation needed for formal and informal inference (LeMire, 2010; Otani, 2019; Zieffler et al., 2008); the argumentation that students use when exploring statistics and probability (Gómez-Blancarte & Tobías-Lara, 2018; Krummenauer & Kuntze, 2018; Schnell, 2014; Tunstall, 2018); how a pedagogical focus on argumentation can strengthen students' statistical and probabilistic reasoning (Ben-Zvi, 2006; Fielding-Wells, 2014; Fielding-Wells & Makar, 2015; Osana et al., 2004) or provide opportunities for learning (Cobb, 1999; Weber et al., 2008); and to characterize teachers' discussions about teaching (Groth, 2009). Research on argumentation in statistics education has focused primarily on the arguments constructed in statistical contexts; this paper foregrounds the role of the teacher in facilitating collective argumentation as well as the kinds of reasoning exhibited in that argumentation.

Conner et al. (2014a) used Toulmin's model to examine the structure of arguments and thus distinguish among several kinds of reasoning (deductive, inductive, abductive, and analogical). Extended Toulmin diagrams (Conner, 2008), which add color and line style to denote participants in argumentation, allow for exploration of characteristics and structure of arguments and for investigating how teachers and students participate in constructing arguments (see Key in Fig. 1). Figure 1 provides an example of an extended Toulmin diagram, illustrating how claims, data, warrants, qualifiers, and rebuttals can appear in relation to one another. Figure 1 also illustrates another important aspect of the analysis of argumentation using extended Toulmin diagrams: Arguments often contain chains of reasoning, in which one part of an argument can function in multiple ways. For instance, often a student's utterance can function first as a claim, needing evidence and reasoning to be provided by data and warrant, and then it can function as data as the argument continues,

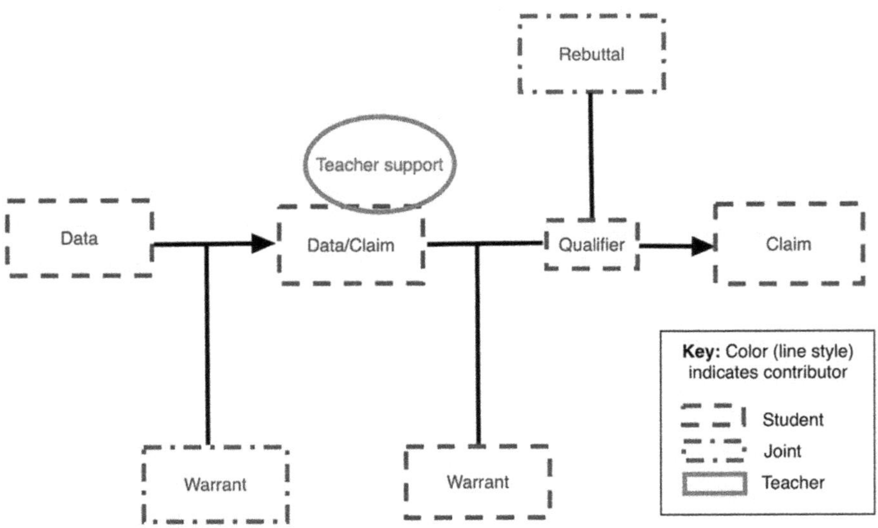

Fig. 1 Generic extended Toulmin diagram. *Note.* In Extended Toulmin Diagrams, color (line style) indicates who contributed the component (teacher, student, or co-constructed jointly). The position of a component with respect to line segments and arrows indicates the function of the component (as claim, data, etc.). For more information about extended Toulmin diagrams, see Conner (2008), Conner et al. (2014b), or Conner et al. (2022)

providing evidence for additional claims. We refer to such components of arguments as data/claims to indicate the multiple roles they play within an argument.

The Teacher Support for Collective Argumentation (TSCA) framework (Conner et al., 2014b) characterizes three ways teachers support collective argumentation: (1) directly contributing components of the argument (e.g., claim, data, warrant), (2) using questions to elicit components of the argument, or (3) using other supportive actions to encourage students in developing components of the argument (e.g., drawing representations, revoicing). A strength of the TSCA framework is providing the opportunity to see across types of teacher support and how they help students in the construction of arguments. Like others (e.g., Fielding-Wells & Makar, 2015), we hypothesize that teachers' actions are important as students learn to reason statistically.

3 Reasoning

Philosophers have identified many kinds of reasoning, and some of these kinds of reasoning have been studied in mathematics and statistics education. Most sources describe reasoning in general as the process of drawing conclusions from information such as assumptions or evidence (e.g., Galotti, 1989; NCTM, 2009). We adopt the following definitions of different kinds of reasoning from Peirce (1956) and others to consider the multiple ways in which conclusions might be drawn in mathematics and statistics. *Deductive reasoning* is drawing conclusions as a logical consequence of established assumptions or conclusions and is the only way to arrive at a conclusion with certainty. *Inductive reasoning* involves abstracting or generalizing from individual observations. *Abductive reasoning* can be seen as the reverse of deductive reasoning and involves hypothesizing what rule would lead to a particular case. *Analogic reasoning* (reasoning by analogy) involves recognizing similarities or correspondences between (in mathematics) two mathematical entities, relationships, or solution methods. Another description of kinds of reasoning arises from explicitly considering certainty and variability in specific situations. *Deterministic reasoning* is focused on getting answers with certainty, seeking explainable causes while potentially neglecting variability; whereas *nondeterministic reasoning* takes variability and unexplainable causes into consideration for answers to be obtained with some degree of uncertainty. *Probabilistic reasoning* is reasoning about uncertainty through consideration of both randomness and variability (Savard, 2014) and thus is encompassed by but more constrained than nondeterministic reasoning. Finally, *contextual reasoning* involves using elements of the context of a problem or elements of the data collection context, acknowledging the variability inherent in a situation, to engage in problem solving. These sets of distinctions are not mutually exclusive. That is, inductive reasoning can be (and often is) involved in probabilistic reasoning; deductive reasoning is often part of deterministic reasoning. Table 1 provides simplistic examples of each kind of reasoning.

Table 1 Simplistic example of each kind of reasoning

Kind	Example
Deductive reasoning	*ABCD* is a square; All squares are rectangles; ∴ *ABCD* is a rectangle
Inductive reasoning	Multiple examples of triangles with interior angles sums of 180; ∴ (Probably) all triangles have interior angle sum of 180
Abductive reasoning	All the items in this bag are white; I have a white item; ∴ The item must be from the bag
Analogical reasoning	The angles of an equilateral triangle are equal; A regular polygon is like an equilateral triangle; ∴ The angles of a regular polygon are equal
Deterministic reasoning	This phenomenon can be modeled by the equation $y = 3x + 4$; I know x is 3; ∴ y is 13
Probabilistic reasoning	I get fouled while playing basketball and have two free throws; I rarely make free throws; ∴ I probably will not make both shots
Contextual reasoning	Our model for the relationship between the sales price for a car and the car's mileage predicts that a car of this type will have a negative value at 130,000 miles; A car cannot have a negative value; ∴ We should find a better model for the relationship
Nondeterministic reasoning	A sample of data is randomly selected from a population; The sample will not match the population exactly; ∴ I cannot describe the population exactly

Many definitions of statistical reasoning focus on combinations of ideas about data and chance to reason beyond data and draw inferences from data. For example, Scheaffer (2006) suggests that statistical reasoning includes thinking about variability in data and design, resulting in probabilistic solutions to contextual problems. Other descriptions focus on the probabilistic and contextual nature of statistical reasoning by identifying additional factors such as considering how to collect data or how data have been collected and combining statistical and probabilistic ideas to make inferences (e.g., Bargagliotti et al., 2020; Franklin et al., 2007; Garfield & Ben-Zvi, 2008). Unlike the deductive reasoning used to reach certain conclusions, statistics involves using inductive reasoning to reach uncertain (but probable) conclusions in general, although students might employ multiple types of reasoning when studying statistics. Students who calculate summary statistics or create representations for one or multiple sets of data may do so mechanically with minimal reasoning or may reason deterministically by strictly using these measures and inscriptions to describe data distributions. Other students might recognize relationships in the data that are similar to those previously studied (e.g., concluding the value of the mean is greater than the value of the median for a right-skewed distribution; describing center and spread of distributions with outliers or skew by using median and interquartile range; comparing overlap in the middle 50% of data to describe differences in distributions) and employ analogical reasoning to describe or compare distributions. Still others might employ contextual reasoning to interpret data within the context in which the data are situated. Students who begin to reason beyond the data at hand may be beginning to engage in informal inference, which includes consideration of variability and uncertainty and might result from abductive reasoning (to provide a causal explanation for the data) or inductive reasoning (to make generalizations about a population).

Mathematical reasoning "is often understood to encompass formal reasoning, or proof, in which conclusions are logically deduced from assumptions and definition" (NCTM, 2009, p. 4). Algebraic reasoning, for example, involves reasoning about and with structures (e.g., Harel, 2013); generalizing arithmetic operations, properties, and relationships; and generalizing, representing, justifying, and acting on representations of relationships between quantities (Blanton & Kaput, 2011). The end goal of reasoning in algebra, as in most mathematics, often is a deductive argument ranging from informal explanation and justification to formal proof, although several kinds of reasoning might occur in service of this goal. Although mathematical reasoning can take many forms, what we often see in schools is mathematical reasoning in the form of context-free deterministic reasoning, arriving at answers with certainty. Inglis et al. (2007) argued that mathematical argumentation can involve different kinds of reasoning, but when it is not deductive, the warrant (the part of the argument that contains the reasoning) should be accompanied by an appropriate qualifier, indicating that the argument is not certain. We extend this argument to statistics, and like Otani (2019), argue that statistical arguments should contain qualifiers that align with the kind of reasoning contained in the warrant. We observe that the rebuttal might also contain important aspects of statistical reasoning. Additionally, one role of the teacher is to support students in constructing arguments that acknowledge the uncertainty and variability inherent in statistical reasoning. Thus, perhaps, in statistical reasoning, a teacher may need to emphasize the qualified nature of claims by prompting or contributing qualifiers.

4 Classroom Arguments Illustrating Different Kinds of Reasoning

The TSCA framework (Conner et al., 2014b) offers a powerful analytical tool to construct differentiated descriptions of a teacher's support for collective argumentation in mathematical and statistical contexts. In order to illustrate the importance of qualifiers and warrants in distinguishing argumentation that includes statistical reasoning and to highlight the importance of the teacher's support for argumentation, we present a series of arguments, based in data from a variety of projects, but some of which have been altered to more succinctly illustrate our points. We invite the reader to engage with us in a series of thought experiments to examine arguments characterized by a range of reasoning and the teachers' actions that might support such arguments.

4.1 Argumentation Examples

The first argument occurred in a ninth-grade integrated mathematics class. Ms. Bell, the teacher, and Mike, a student, participated in this argument verbally. Two members of Mike's group were present. On one student's paper were a list of sums of measures of interior angles of polygons, generated by various members of the class as they measured the angles and wrote their results on the board. We diagrammed this episode as shown in Fig. 2 and categorized the reasoning as abductive. Note that the student contributed the claim, data, warrant, and qualifier of the argument, and the teacher asked two questions that supported the argumentation. She asked for a conjecture (What did you figure out?) that prompted the claim, and she requested justification (Why do you think that?) that prompted the warrant. Our analysis of the episode suggests that the teacher was looking for a report of students' inductive reasoning, that is, how they reached the conclusion that the interior angle sums are multiples of 180 degrees. Instead, she got the unanticipated argument, which, when she asked for justification, illustrated abductive reasoning. The role of the teacher is important here because she pressed for justification. If she had not asked, "Why do you think that?" we would not have been able understand the structure of the argument and the abductive nature of the reasoning. When we look at the episode using a mathematical and statistical reasoning frame, we conclude the algebraic reasoning exhibited in this episode involves a search for structure and patterns.

The inclusion of the qualifier in this episode of mathematical reasoning aligns with Inglis et al.'s (2007) suggestions because the reasoning expressed is not deductive. The qualifier is "I think," which indicates the student was less than certain but does not reflect the kinds of qualifiers most crucial to statistical arguments.

Our second set of arguments was derived from observations of a beginning algebra teacher's instruction focused on linear regression (line of best fit). To introduce her students to regression, she began with a task that asked students to visually determine a line that would best fit a set of data. Students then were asked to estimate the "goodness" of fit of the line by measuring the distance from the line to each point and summing the measured distances. After this introductory activity, students collected data by contributing their heights and shoe sizes. In the original

Ms. Bell: What did you figure out?
Mike: They [the sum of the measures of the interior angles] all end up being a multiple of 180.
Ms. Bell read aloud the list of interior angle sums recorded on the board: 360, 540, 720, and 897.
Mike: Although, I think it's supposed to be 900.
Ms. Bell: Why do you think that?
Mike: Because 900 is a multiple of 180.

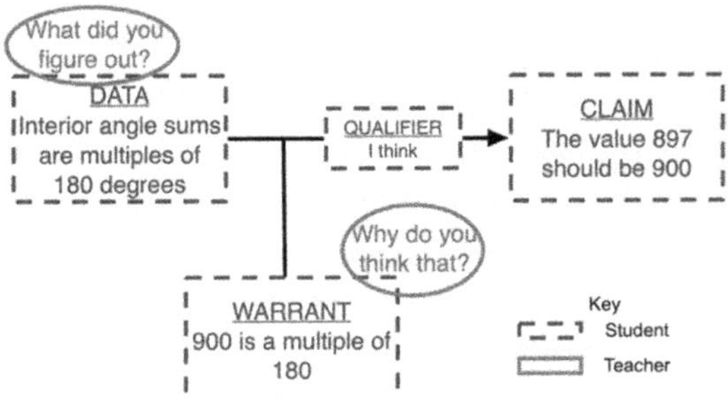

Fig. 2 Argument in an integrated mathematics class illustrating abductive reasoning

classroom, students and teacher worked by hand and engaged in an argument with some of the following characteristics. For the purposes of this paper, we propose the following scenarios, based loosely on what happened in class, but using current technology. In our first scenario, students entered the data into CODAP (2014) and proceeded to use a moveable line to approximate a line of best fit and predict a height for a person who had size 12 shoes. One student entered the data, which was displayed for the class. See Fig. 3 for transcript and diagram of the ensuing argument.

The conversation ended with agreement that the predicted height was 73.28 inches. This argument exhibited deterministic reasoning with little awareness of the context evidenced. There is no qualifier, suggesting that students stated their conclusion with certainty rather than acknowledging the uncertainty associated with making a prediction from sample data. The two warrants are kinds that are commonly observed in mathematics classes. "That's where the line is closest to the dots" (labeled 3 in Fig. 3) is an observation; "Plug 12 into the equation" (labeled 5 in Fig. 3) suggests a calculation for a single value. While this exchange is, we argue, typical for a first introduction to a line of best fit in many classes, it does not contain the kinds of reasoning in warrants or the qualifiers that we would expect in an argument demonstrating statistical reasoning. The teacher supports the argument with questions that prompt parts of the argument; however, in this situation, the teacher does not foreground the context or variability in the data that would lead to more statistical reasoning.

Teacher: Here's our data. How are we going to make this prediction?
Carl: We need to graph it.
Teacher: What kind of a graph? Where do the variables go?
Carl inserts a graph and drags the shoe size to the horizontal axis and height to vertical axis.
Jessica: Oh, we put in a line now!
The students negotiate the position of the line until they come to consensus based on how close the line is to the data points. [CODAP provides the equation for the line: Height in inches = 2.09 Shoe size + 48.2]
Malik: Plug in 12.
Jonathan: We get 73.28.
Teacher: Do we all agree?

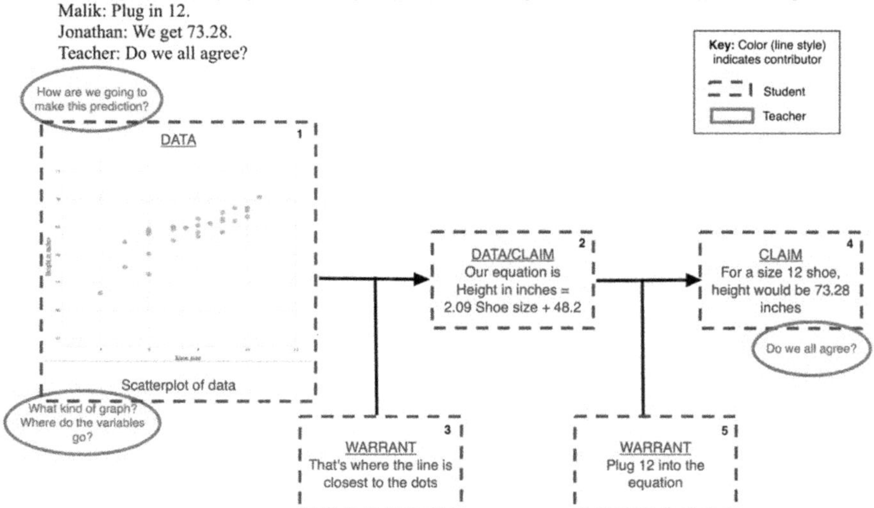

Fig. 3 Argument in an algebra class illustrating deterministic reasoning

Jonathan: We get 73.28.
Teacher: How sure are you?
Mia: I don't know if that's right. I think it could be a little lower or a little higher.
Lateefah: Yeah, 73 is possible, but maybe it could be 71 or 74. I have friends who have the same shoe size but are different heights.
Teacher: What in this data suggests that you might not be sure about the prediction?
Carl: Oh, so at size 8, we have three different heights.
Mia: There are lots of sizes that have different heights.
Teacher: So, what is the difference in heights in those cases?
Students investigate the range of heights for shoe sizes with multiple heights and find the typical range is 3 inches.
Lateefah: Our line predicts 73.28, so it probably would be between 71 and a half and 74 and a half inches.
Mia: But, wait, it seems like the intervals in the heights are getting smaller as the shoe size gets larger.
Malik: Yeah, we noticed that, too, so wouldn't the possible height be closer to 73 since the intervals at 9.5 and 10 are less than 2?
Teacher: What do you think they could be, then, if the interval is getting smaller?
Mia: I think a size 12 shoe would probably be between 72 and 74 inches.
Teacher: Who does this apply to?
Mia: People like us, in ninth grade.

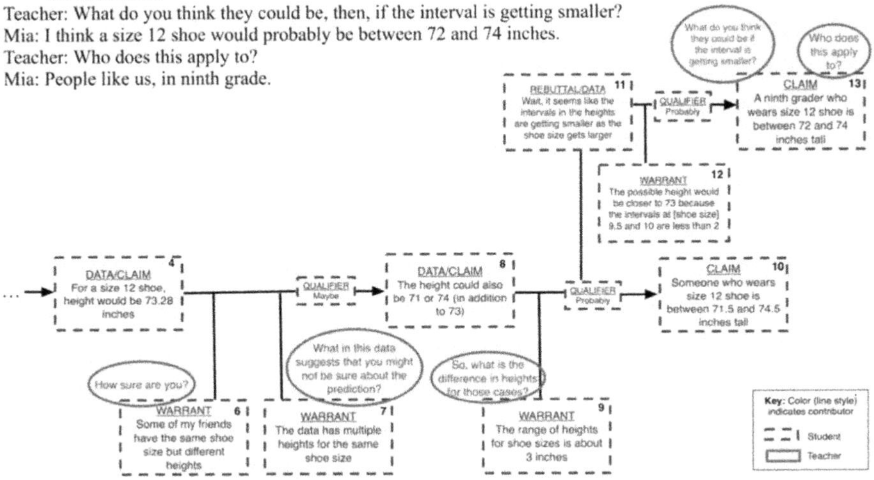

Fig. 4 Argument in an algebra class illustrating statistical reasoning. *Note:* See Fig. 3 for beginning of this argument diagram; Claim labeled 4 in Fig. 3 is labeled Data/Claim 4 in this diagram. See https://tinyurl.com/DiagramConnerPeters for diagram of the complete argument

In order to illustrate a class discussion that contained statistical reasoning, we propose the alternative ending to the previous discussion shown in Fig. 4. In our alternative ending, the teacher asks a question, "How sure are you?" that draws attention to the uncertainty with which a prediction can be made. This prompts a warrant, "Some of my friends have the same shoe size but different heights" (labeled 6 in Fig. 4), that uses student knowledge of the context to reason about the situation. The teacher then explicitly draws student attention to the data by asking, "What in this data suggests that you might not be sure about the prediction?" This prompts the warrant labeled 7 in Fig. 4, which supports the claim that the height could be 71 or 74 (labeled 8 in Fig. 4). This claim is accompanied by a qualifier, "maybe," which indicates the claim is made with some uncertainty. This qualifier is similar to the one in Fig. 2, in that it illustrates some uncertainty in the contribution of this claim, but not necessarily in a statistical sense. However, the next qualifier, "probably," seems to relate to the student's reasoning in a way that does reflect the probabilistic nature of statistical reasoning when combined with the warrant (labeled 9 in Fig. 4) that describes the average range of heights for shoe sizes. We hypothesize that arguments that contain statistical reasoning will often contain qualifiers that reflect a level of uncertainty for the claim, such as *probably*, *possibly*, or even *likely*.

The next part of this hypothetical argument contains a rebuttal that demonstrates even closer attention to the graphical display of data. A student, Mia, questions the warrant by observing that the intervals of data at the same shoe size seem smaller as the shoe size increases (rebuttal labeled 11 that also serves as data for the following claim). This leads the students to claim that the predicted height for a shoe size of 12 should be between 72 and 74. The teacher asks a question that prompts the specific values of this claim (labeled 13). The teacher also redirects students' attention to the context by her last question, prompting students to modify their final claim to specify for whom this claim might be valid. This question could lead to additional statistical reasoning if she fostered a discussion about uncertainty in relation to sampling variability, extrapolation, and sample representativeness. Notice that this sub-argument originating from the rebuttal also contains a qualifier, "probably," indicating uncertainty about the claim in a statistical sense. In addition to characterizing the reasoning in this episode as contextual and nondeterministic, we describe this argument as exhibiting inductive reasoning in that the students are using the data from the class to make a generalization about the relationship between height and shoe size for the population.

Our fourth argument occurred during the opening activity of a weeklong professional development program for middle and high school mathematics teachers in statistics. Facilitators used the activity to stimulate thought and to formatively assess teachers' knowledge of statistics. Teachers had previously examined a claim about the average length of a teachers' workday—a claim that was based on data from the American Time Use Survey (ATUS; https://www.bls.gov/tus/). They considered how the data were collected and speculated that elementary and secondary teachers might have different workday lengths on average. To further consider differences in workday lengths between elementary and secondary teachers, the teachers were presented with histograms displaying the number of minutes a sample of elementary and secondary teachers who participated in ATUS recorded for one workday. After examining the histograms and concluding they did not think they could support a claim about a meaningful difference between the average amount of time worked by elementary and secondary teachers, the participants were asked to construct posters containing displays of data that would cause them to conclude there is a meaningful difference in average work time between elementary and secondary teachers. These posters were displayed around the room and teachers examined them prior to the discussion in Fig. 5.

This argument contains two claims supported by deterministic and deductive reasoning, accompanied by rebuttals that contain probabilistic and contextual reasoning. The participants are reasoning about whether their hypothetical distributions contain the characteristic of "meaningful difference" requested by the facilitator (who takes the role of the teacher). The first claim-warrant pair (labeled 2 and 3 in Fig. 5), by Mary, focuses attention on whether mean or standard deviation is relevant to a claim of meaningful difference. Her warrant is deterministic, based

Facilitator: Alright, so we are looking around, seeing what we see that is similar about different groups' interpretations and what would be a meaningful difference…

Mary: I am not sure I agree with the one over at the purple and orange group [referencing the chart paper labeled Over-achievers] being specifically a significant difference. [*pause for 6 seconds*] 'Cause they, to me, those would have a different mean, but their standard deviation is the same.

Nico: The standard deviation could be exactly same, but the number of hours can still be significantly different.

Sam: Yeah, I mean, if you look at how many hours on average a secondary teacher works compared to the average that an elementary teacher works, there is, to me, that's a meaningful difference in work time mean. That, maybe it's not a meaningful difference in anything else, but the interpretation of work time means, that would be a meaningful difference.

Facilitator: So, in terms of what's similar, what do all these graphs have in common in terms of saying what makes a meaningful difference? Anything?

Adri: The centers…

Facilitator: What's that? The cen.. the measures of centers? What about them?

Adri: They make them very hugely different.

Facilitator: So they are distanced, right?

Adri: Yeah.

Facilitator: So we are focusing on centers. Does center have to be the only difference? [*pause*]

Mary: Well I think if you are talking of, what his point was is that we are looking at the average time, so… you'd want to be looking at the measures of center…

Taylor: If you looked at measures of dispersion, such as range, that would show the lower the range, the more consistent work day. It would go back to our conversation that we had earlier about, that how, you know, some teachers work, you know, a longer day, you know a newer teacher might work, and if you ask a teacher in the fall, versus the winter. It might—you would get to the consistency aspect of it.

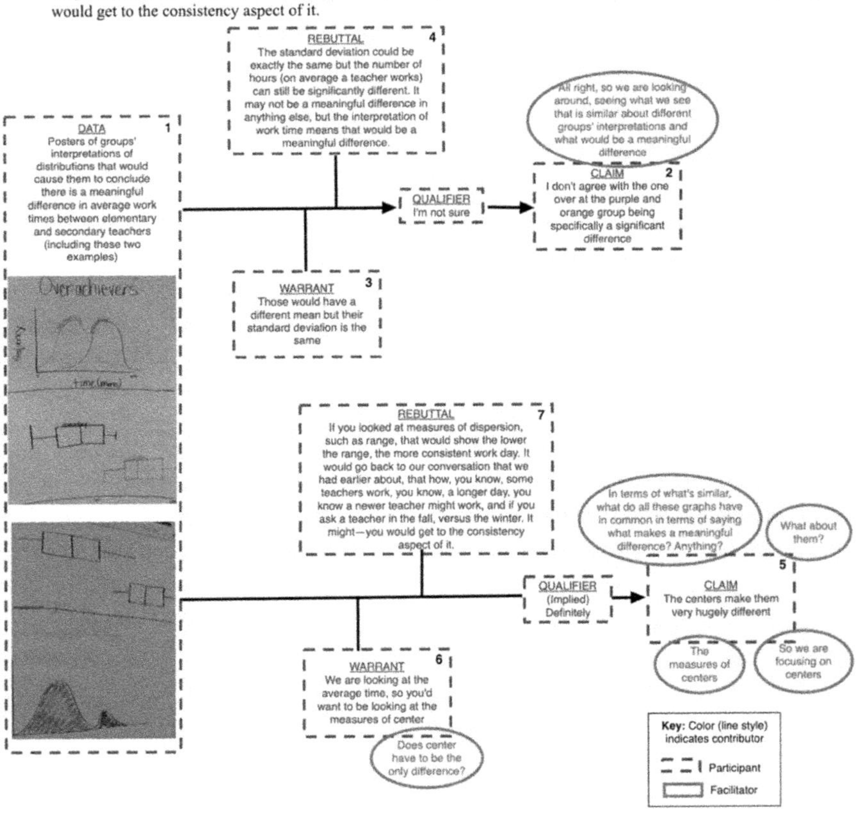

Fig. 5 Argument in a professional development session illustrating multiple kinds of reasoning

on observations from the poster. Nico and Sam, who provide the rebuttal labeled 4 in the argument, bring in probabilistic elements by suggesting that the average number of hours a teacher works can still be significantly different even if the standard deviation is the same. Mary's qualifier indicates that she is not sure about her claim, although this appears to be in a non-statistical sense. It is similar to the "I think" qualifier in the argument in Fig. 2. The second claim (labeled 5 in Fig. 5) responds to a question by the facilitator (teacher), focusing participants' attention on the relevant aspects of the displays in the posters. Adri makes a claim about the centers being "very hugely different" as the common characteristic of these displays. Adri's contribution does not include an explicit qualifier, although her statement supports an implied "definitely" or "certainly" as a qualifier in this case. Mary provides the warrant labeled 6, which supports this as being reasonable because they were asked to look at the "average time," which relates to a measure of center. The reasoning in this warrant points to a definition of average as a measure of center, illustrating again deterministic reasoning. The rebuttal labeled 7 in Fig. 5 brings in a contextual element as Taylor clarifies what would happen if one examined a measure of dispersion rather than a measure of center (providing information about consistency rather than information about average lengths being different).

The teacher's role in this episode of argumentation included asking questions to focus the participants' attention on relevant characteristics of the data displays and ensuring participants heard and understood other participants' contributions by repeating and summarizing their contributions. Because this was the opening activity of a week-long professional development session, the facilitator (teacher) did not push for more discussion of the various statistical ideas at this time, although many of these ideas reappeared throughout the week.

4.2 Looking Across the Examples

According to Toulmin (1958/2003), arguments in any field may contain qualifiers and rebuttals. Qualifiers explicitly express uncertainty, and rebuttals indicate circumstances under which a warrant may not hold (or, as we see in practice, potential aspects of a warrant that may not yet have been explored). Our experience with arguments containing statistical reasoning suggests that qualifiers (such as "probably") and rebuttals that address various aspects of uncertainty within a warrant are potentially more prevalent in statistical arguments than mathematical ones. Additionally, as the argument in Fig. 5 illustrates, some arguments in statistical situations contain qualifiers that are similar to those in mathematical arguments. The rebuttals in Fig. 5 contain aspects of probabilistic and contextual reasoning not apparent in the rest of the arguments. These rebuttals could be examined with a finer grain size and diagrammed as arguments themselves (this could not be done in this chapter due to space constraints). Examining the rebuttals present within statistical arguments is a potentially promising area of further research.

The teachers' support for collective argumentation in these four example arguments that we have chosen to highlight can mainly be classified as questions that prompt argument components. In the first example argument, illustrating mathematical (and abductive) reasoning, the teacher prompted the student to articulate his data and warrant (see Fig. 2). In the second example argument, illustrating deterministic reasoning in a statistical context, the teacher again prompts the generation of data for the argument in the form of the graph (see Fig. 3). The teacher in this example concludes the argument by asking if everyone agrees with the conclusion. We contend that there was an opportunity to highlight the uncertainty of the conclusion that was missed in this example (Fig. 3), which prompted us to generate the third hypothetical example argument (Fig. 4), in which the teacher asked questions that highlighted variability and uncertainty as well as potentially other important statistical ideas such as extrapolation, sampling variability, and representativeness. The questioning in the example illustrated by Fig. 4 draws attention to the strategic role of the teacher in facilitating argumentation that includes statistical reasoning. The questioning and other supportive actions in Fig. 5 demonstrate the important role of the teacher in focusing the direction of the argumentation. Figure 5 illustrates a productive argument in which some aspects of the statistical ideas are left unresolved for future discussion.

5 Discussion

When indicating the certainty with which a claim is made, Toulmin's (1958/2003) qualifier becomes very important. In both mathematical and statistical arguments, claims can be made with varying levels of certainty, depending on the kind of reasoning being employed as well as the extent to which the claimant trusts the reasoning. Thus, in a classroom, qualifiers can reflect both disciplinary and social aspects of the discourse.

When an argument contains deductive reasoning, such as an argument that is a proof, the claim can be made with certainty, and one can argue that it is not necessary to state a qualifier (or if one is stated, it would be *certainly*, or *necessarily*). However, inductive reasoning, such as when a pattern is inferred from a set of examples, might result in a qualifier such as 'it seems' or 'it is plausible that.' This might be accompanied by a rebuttal, either spoken or implied (see Inglis et al., 2007 for examples of qualifiers and rebuttals in mathematical reasoning and LeMire, 2010 and Otani, 2019 for qualifiers and rebuttals appropriate for formal inference). For example, in Fig. 2, we illustrated a student's abductive reasoning and inclusion of the qualifier 'I think,' and in Fig. 5 we illustrated the qualifier 'I'm not sure.' These qualifiers indicate that the speakers were uncertain about their conclusions but were willing to state them. This uncertainty could result from a student's acknowledgement that the reasoning was not deductive. It also could have a social rationale.

The student in Fig. 2 may have said "I think" because he was not sure how his conclusion would be accepted by his peers or his teacher, so he was not yet willing to commit completely to this claim. Likewise, the participant in Fig. 5 may have said "I'm not sure" because she did not want to definitively state that another group's representation was problematic (to her). From a disciplinary standpoint, it may be important to disentangle such reasons.

Inglis et al. (2007) described three different warrant-types in mathematical reasoning, two of which require appropriate qualifiers (structural-intuitive and inductive), and one of which does not require a qualifier (deductive). Their conclusions are based on a study of talented mathematicians' proving activity, and they suggest that more warrant-types are possible within mathematical reasoning. Additionally, Inglis et al. suggest attention should be paid to how warrant-types are paired with qualifiers. Using a non-absolute qualifier within an argument containing complete deductive reasoning is as problematic as using an absolute qualifier (or no qualifier) with inductive reasoning.

Although Inglis et al. (2007) studied mathematicians' qualifiers in proving situations, their conclusions about warrant-qualifier pairings can be extended to arguments containing statistical reasoning. Statistical reasoning is defined to be probabilistic and contextual. As such, it is not deductive and thus should include qualifiers to indicate the level of certainty with which claims are made. Qualifiers are one way to acknowledge variability in statistical arguments; as illustrated in Fig. 4, an argument containing statistical reasoning may contain multiple qualifiers. Sometimes these qualifiers may indicate simply that a student is less certain about their claim, in a social sense, such as the 'maybe' in Fig. 4. Other times, students may accurately use qualifiers such as 'probably' along with warrants that acknowledge variability to indicate that their statistical conclusions are not certain. The content of the rebuttal may also contribute to the statistical reasoning present in an argument, particularly when multiple people contribute to the argument. Facilitating collective argumentation includes taking contributions from multiple speakers and assisting other stakeholders in understanding how that reasoning is relevant in the given situation.

The teacher's role in statistical arguments includes asking questions to prompt argument components, as is illustrated in Figs. 2, 3, 4, and 5. Many of the teacher's questions prompt students to provide warrants, which aligns with findings of previous studies (e.g., Conner et al., 2014b; Gomez Marchant et al., 2021). An open question, and one worthy of exploration, is the extent to which teachers prompt students to provide qualifiers and the potential attention teachers pay to warrant-qualifier and warrant-rebuttal pairings. Amidst the complexities of supporting collective argumentation in lessons requiring statistical reasoning, attending to these pairings may seem daunting. However, attending to qualifiers, warrant-qualifier pairings, and warrant-rebuttal pairings may assist in supporting arguments that contain well-articulated statistical reasoning. The TSCA framework (Conner et al., 2014b) describes a variety of questions and other supportive actions that teachers use when supporting mathematical arguments. Many of these are also useful in supporting statistical argumentation, specifically in prompting the contributions of warrants. However, the framework may need to be expanded to examine, for instance, the kinds of questions needed to prompt contributions of qualifiers (something not observed in the teachers' practice from which the framework was developed).

More research is necessary to examine how teachers support collective statistical argumentation. Many studies have examined statistical argumentation from the perspective of statistical reasoning and constructing statistical arguments (e.g., Ben-Zvi, 2006; Osana et al., 2004), especially with respect to specific interventions (e.g., Fielding-Wells & Makar, 2015); however, there is a need to expand our knowledge of how teachers can support *collective* statistical argumentation. Expanding frameworks from other fields, such as the TSCA framework, is one way to begin this work.

Introducing the TSCA framework (Conner et al., 2014b) and Toulmin's (1958/2003) conceptualization of argumentation to teachers and prospective teachers allows them to reflect on their teaching and support for collective mathematical argumentation in productive ways (see, e.g., Foster et al., 2020; Gomez Marchant et al., 2021; Park et al., 2020). It is reasonable to suggest the same would be true for supporting collective statistical argumentation. Introducing these ideas, specifically related to argumentation in statistical contexts, has potential to address Pfannkuch and Ben-Zvi's (2011) and Pfannkuch's (2018) concerns about prioritizing the teaching of argumentation in conjunction with statistics in school. Such attention to argumentation in teacher education and research on the same is more critical than ever given today's proliferation of data and the growing urgency for students to become data literate (and the need for teachers who can facilitate students' progress toward literacy).

6 Conclusion and Implications

Distinguishing between algebraic (and other mathematical) and statistical reasoning is crucial for teachers who teach concepts from both disciplines. However, although the differences can be enumerated, they are not often understood by classroom teachers. More research needs to be done to clarify the differences between these kinds of reasoning and to prepare teachers to engage their students in appropriate argumentation for statistical and mathematical topics. The TSCA framework offers a powerful analytical tool for researchers to investigate teacher education initiatives to facilitate teachers' construction of arguments that promote students' development of appropriate reasoning in mathematics and statistics. Investigating how teachers can support students in contributing powerful statistical reasoning and appropriate qualifiers and rebuttals is a needed contribution to research in statistics education.

References

Bargagliotti, A., Franklin, C., Arnold, P., Gould, R., Johnson, S., Perez, L., & Spangler, D. (2020). *Pre-K–12 guidelines for assessment and instruction in statistics education II (GAISE II): A framework for statistics and data science education.* American Statistical Association. https://www.amstat.org/asa/files/pdfs/GAISE/GAISEIIPreK-12_Full.pdf

Barrett, J. E., Clements, D. H., Klanderman, D., Pennisi, S.-J., & Plaki, M. V. (2006). Students' coordination of geometric reasoning and measuring strategies on a fixed perimeter task: Developing mathematical understanding of linear measurement. *Journal for Research in Mathematics Education, 37*(3), 187–221. https://doi.org/10.2307/30035058

Battista, M. T. (1999). The importance of spatial structure in geometric reasoning. *Teaching Children Mathematics, 6*(3), 170–177. https://doi.org/10.5951/TCM.6.3.0170

Ben-Zvi, D. (2006). Scaffolding students' informal inference and argumentation. In A. Rossman & B. Chance (Eds.) *Proceedings of the 7th international conference on teaching statistics.* International Statistical Institute. Retrieved from http://www.stat.auckland.ac.nz/~iase/publications/17/2D1_BENZ.pdf

Ben-Zvi, D., & Garfield, J. (2004). Statistical literacy, reasoning, and thinking: Goals, definitions, and challenges. In D. Ben-Zvi & J. Garfield (Eds.), *The challenges of developing statistical literacy, reasoning, and thinking* (pp. 3–15). Kluwer. https://doi.org/10.1007/1-4020-2278-6_1

Blanton, M. L., & Kaput, J. J. (2011). Functional thinking as a route into algebra in the elementary grades. In J. Cai & E. Knuth (Eds.), *Early algebraization: A global dialogue from multiple perspectives* (pp. 5–23). Springer. https://doi.org/10.1007/978-3-642-17735-4_2

Cobb, P. (1999). Individual and collective mathematical development: The case of statistical data analysis. *Mathematical Thinking and Learning, 1*(1), 5–43. https://doi.org/10.1207/s15327833mtl0101_1

Cobb, P., & McClain, K. (2004). Principles of instructional design for supporting the development of students' statistical reasoning. In D. Ben-Zvi & J. Garfield (Eds.), *The challenge of developing statistical literacy, reasoning, and thinking* (pp. 375–395). Kluwer. https://doi.org/10.1007/1-4020-2278-6_16

Common Online Data Analysis Platform [Computer software]. (2014). *The Concord Consortium.* https://codap.concord.org/app/static/dg/en/cert/index.html

Conner, A. (2008). Expanded Toulmin diagrams: A tool for investigating complex activity in classrooms. In O. Figueras, J. L. Cortina, S. Alatorre, T. Rojano, & A. Sepulveda (Eds.), *Proceedings of the joint meeting of the international group for the psychology of mathematics education 32 and the North American chapter of the international group for the psychology of mathematics education XXX* (Vol. 2, pp. 361–368). Cinvestav-UMSNH.

Conner, A., Singletary, L. M., Smith, R. C., Wagner, P. A., & Francisco, R. T. (2014a). Identifying kinds of reasoning in collective argumentation. *Mathematical Thinking and Learning, 16*(3), 181–200. https://doi.org/10.1080/10986065.2014.921131

Conner, A., Singletary, L. M., Smith, R. C., Wagner, P. A., & Francisco, R. T. (2014b). Teacher support for collective argumentation: A framework for examining how teachers support students' engagement in mathematical activities. *Educational Studies in Mathematics, 86*(3), 401–429. https://doi.org/10.1007/s10649-014-9532-8

Conner, A., Tabach, M., & Rasmussen, C. (2022). Collectively engaging with others' reasoning: Building intuition through argumentation in a paradoxical situation. In *International Journal of Research in Undergraduate Mathematics Education.* Advance online publication. https://doi.org/10.1007/s40753-022-00168-x

del Mas, R. C. (2004). A comparison of mathematical and statistical reasoning. In D. Ben-Zvi & J. Garfield (Eds.), *The challenge of developing statistical literacy, reasoning, and thinking* (pp. 79–95). Kluwer. https://doi.org/10.1007/1-4020-2278-6_4

Dreyfus, T., Kouropatov, A., & Ron, K. (2021). Research as a resource in a high-school calculus curriculum. *ZDM – Mathematics Education, 53*, 679–693. https://doi.org/10.1007/s11858-021-01236-3

Fielding-Wells, J. (2014). Where's your evidence? Challenging young students' equiprobability bias through argumentation. In K. Makar, B. DeSousa, & R. Gould (Eds.), *Sustainability in statistics education. Proceedings of the ninth international conference on teaching statistics (ICOTS9, July, 2014), Flagstaff*. International Statistical Institute. https://iase-web.org/icots/9/proceedings/pdfs/ICOTS9_2B2_FIELDINGWELLS.pdf

Fielding-Wells, J., & Makar, K. (2015). Inferring to a model: Using inquiry-based argumentation to challenge young children's expectations of equally likely outcomes. In A. Zieffler & E. Fry (Eds.), *Reasoning about uncertainty: Learning and teaching informal inferential reasoning* (pp. 1–27). Catalyst Press.

Foster, J. K., Zhuang, Y., Conner, A., Park, H., & Singletary, L. (2020). One Teacher's Analysis of Her Questioning in Support of Collective Argumentation. In *Mathematics education across cultures: Proceedings of the 42nd annual meeting of the North American chapter of the international group for the psychology of mathematics education* (pp. 2067–2071). Cinvestav/AMIUTEM/PME-NA. https://doi.org/10.51272/pmena.42.2020

Franklin, C., Kader, G., Mewborn, D., Moreno, J., Peck, R., Perry, M., & Scheaffer, R. (2007). *Guidelines for assessment and instruction in statistics education (GAISE) report*. American Statistical Association.

Galotti, K. (1989). Approaches to studying formal and everyday reasoning. *Psychological Bulletin, 105*(3), 331–351. https://doi.org/10.1037/0033-2909.105.3.331

Garfield, J. (2002). The challenge of developing statistical reasoning. *Journal of Statistics Education, 10*(3). https://doi.org/10.1080/10691898.2002.11910676

Garfield, J., & Ben-Zvi, D. (2008). *Developing students' statistical reasoning: Connecting research and teaching practice*. Springer. https://doi.org/10.1007/978-1-4020-8383-9

Garfield, J., & Ben-Zvi, D. (2009). Helping students develop statistical reasoning: Implementing a statistical reasoning learning environment. *Teaching Statistics, 31*(3), 72–77. https://doi.org/10.1111/j.1467-9639.2009.00363.x

Gomez Marchant, C. N., Park, H., Zhuang, Y., Foster, J., & Conner, A. (2021). Theory to practice: Prospective mathematics teachers' recontextualizing discourses surrounding collective argumentation. *Journal of Mathematics Teacher Education, 24*, 1–29. https://doi.org/10.1007/s10857-021-09500-9

Gómez-Blancarte, A., & Tobías-Lara, M. G. (2018). Using the Toulmin model of argumentation to validate students' inferential reasoning. In M. A. Sorto, A. White, & L. Guyot (Eds.), *Looking back, looking forward. Proceedings of the tenth international conference on teaching statistics (ICOTS20, July, 2018), Kyoto*. International Statistical Institute. https://iase-web.org/icots/10/proceedings/pdfs/ICOTS10_8C1.pdf

Groth, R. E. (2009). Characteristics of teachers' conversations about mean, median, and mode. *Teaching and Teacher Education, 25*, 707–716. https://doi.org/10.1016/jtate.2008.11.005

Harel, G. (2013). DNR-based curricula: The case of complex numbers. *Journal of Humanistic Mathematics, 3*(2), 2–61. https://doi.org/10.5642/jhummath.201302.03

Inglis, M., Mejia-Ramos, J. P., & Simpson, A. (2007). Modelling mathematical argumentation: The importance of qualification. *Educational Studies in Mathematics, 66*(1), 3–21. https://doi.org/10.1007/s10649-006-9059-8

Kaput, J. J. (2008). What is algebra? What is algebraic reasoning? In J. J. Kaput, D. W. Carraher, & M. L. Blanton (Eds.), *Algebra in the early grades* (pp. 5–18). Lawrence Erlbaum Associates.

Krummenauer, J., & Kuntze, S. (2018). Primary students' data-based argumentation—An empirical reanalysis. In E. Bergqvist, M. Osterholm, C. Granberg, & L. Sumpter (Eds.), *Proceedings of the 42nd international group for the psychology of mathematics education* (Vol. 3, pp. 251–258). PME.

Krummheuer, G. (1995). The ethnography of argumentation. In P. Cobb & H. Bauersfeld (Eds.), *The emergence of mathematical meaning: Interaction in classroom cultures* (pp. 229–269). Erlbaum.

LeMire, S. D. (2010). An argument framework for the application of null hypothesis statistical testing in support of research. *Journal of Statistics Education, 18*(2). https://doi.org/10.1080/10691898.2010.11889492

Lovett, J. N., & Lee, H. S. (2017). New standards require teaching more statistics: Are preservice secondary mathematics teachers ready? *Journal of Teacher Education, 68*(3), 299–311. https://doi.org/10.1177/0022487117697918

Mathematical Association of America & National Council of Teachers of Mathematics. (2017). *The role of calculus in the transition from high school to college mathematics: Report of the workshop held at the MAA Carriage House, Washington, DC, march 17–19, 2016*. Mathematical Association of America. https://www.maa.org/sites/default/files/RoleOfCalc_rev.pdf

Nathan, M. J., & Koedinger, K. R. (2000). Teachers' and researchers' beliefs about the development of algebraic reasoning. *Journal for Research in Mathematics Education, 31*(2), 168–190. https://doi.org/10.2307/749750

National Council of Teachers of Mathematics. (2009). *Focus in high school mathematics: Reasoning and sense making*. Author.

National Governors Association Center for Best Practice & Council of Chief State School Officers. (2010). *Common core state standards for mathematics*. Author.

Organization for Economic Co-operation and Development (OECD). (2018). *PISA 2022 mathematics framework (draft)*. Author. https://pisa2022-maths.oecd.org/files/PISA%202022%20Mathematics%20Framework%20Draft.pdf

Osana, H. P., Leath, E. P., & Thompson, S. E. (2004). Improving evidential argumentation through statistical sampling: Evaluating the effects of a classroom intervention for at-risk 7th-graders. *Journal of Mathematical Behavior, 23*, 351–370. https://doi.org/10.1016/jmathb.2004.06.005

Otani, H. (2019). Comparing structures of statistical hypothesis testing with proof by contradiction: In terms of argument. *Hiroshima Journal of Mathematics Education, 12*, 1–12.

Park, H., Conner, A., Foster, J. K., Singletary, L., & Zhuang, Y. (2020). One Teacher's Learning to Facilitate Argumentation: Focus on the Use of Repeating. In *Mathematics education across cultures: Proceedings of the 42nd annual meeting of the North American chapter of the international group for the psychology of mathematics education* (pp. 1961–1962). Cinvestav/AMIUTEM/PME-NA. https://doi.org/10.51272/pmena.42.2020

Peirce, C. S. (1956). Sixth paper: Deduction, induction, and hypothesis. In M. R. Cohen (Ed.), *Chance, love, and logic: Philosophical essays* (pp. 131–153). G. Braziller. (Original work published 1878).

Pfannkuch, M. (2018). Reimagining curriculum approaches. In D. Ben-Zvi, K. Makar, & J. Garfield (Eds.), *International handbook of research in statistics education* (pp. 387–413). Springer. https://doi.org/10.1007/978-3-319-66195-7_12

Pfannkuch, M., & Ben-Zvi, D. (2011). Developing teachers' statistical thinking. In C. Batanero, G. Burrill, & C. Reading (Eds.), *Teaching statistics in school mathematics—Challenges for teaching and teacher education: A joint ICMI/IASE study: The 18th ICMI study* (pp. 323–333). Springer. https://doi.org/10.1007/978-94-007-1131-0

Rasmussen, C. L., & Stephan, M. (2008). A methodology for documenting collective activity. In A. Kelly, R. Lesh, & J. Baek (Eds.), *Handbook of design research methods in education: Innovations in science, technology, engineering, and mathematics teaching and learning*. Routledge. https://doi.org/10.4324/9781315759593.ch10

Savard, A. (2014). Developing probabilistic thinking: What about people's conceptions? In E. J. Chernoff & B. Sriraman (Eds.), *Probabilistic thinking: Presenting plural perspectives* (pp. 283–298). Springer. https://doi.org/10.1007/978-94-007-7155-0

Scheaffer, R. L. (2006). Statistics and mathematics: On making a happy marriage. In G. F. Burrill (Ed.), *Thinking and reasoning with data and chance: Sixty-eighth yearbook* (pp. 309–321). National Council of Teachers of Mathematics.

Schnell, S. (2014). Types of arguments when dealing with chance experiments. In C. Nicol, S. Oesterle, P. Liljedahl, & D. Allan (Eds.), *Proceedings of the joint meeting of PME 38 and PME-NA 36* (Vol. 5, pp. 113–120). PME.

Stanic, G. M. A., & Kilpatrick, J. (2003). *A history of school mathematics* (Vol. 1 & 2). National Council of Teachers of Mathematics.

Toulmin, S. E. (2003). *The uses of argument* (Updated ed.). Cambridge University Press. (Original work published 1958).

Tunstall, S. L. (2018). Investigating college students' reasoning with messages of risk and causation. *Journal of Statistics Education, 26*(2), 76–86. https://doi.org/10.1080/10691898.2018.1456989

Weber, K., Maher, C., Powell, A., & Lee, H. S. (2008). Learning opportunities from group discussions: Warrants become the objects of debate. *Educational Studies in Mathematics, 68*, 247–261. https://doi.org/10.1007/s10649-008-9114-8

Zieffler, A., Garfield, J., DelMas, R., & Reading, C. (2008). A framework to support research on informal inferential reasoning. *Statistics Education Research Journal, 7*(2), 40–58. https://doi.org/10.52041/serj.v7i2.469

Sustainable Learning of Statistics

Hanan Innabi, Ference Marton, and Jonas Emanuelsson

Abstract Sustainable learning is learning that persists and grows over time. It continues beyond formal instruction, and what has been learned in one situation can be expanded and also used in other situations. We address one of the fundamental questions in statistics education research: What makes learning of statistics sustainable? Our conjecture is that enabling learners to focus on variability in statistics will most likely facilitate more sustainable learning of statistics. This conjecture has two sources of support. Learning and teaching of statistics involves two lines of reasoning: (1) deterministic, oriented towards exact numbers and causal explanations, and (2) stochastic, oriented towards uncertainty and variability, where both are important. However, there seem to be a general tendency among people to reason deterministically rather than stochastically. Research shows that a focus on variability in statistics can help in learning to reason stochastically. The second source concerns variation in learners' experience. Research drawing upon phenomenography and Variation Theory argues that experiencing differences and similarities (in this order) is a necessary condition of learning. Further, we suggest that using Variation Theory as a pedagogical tool to emphasize variability in statistics to support students' development of stochastic reasoning will potentially increase sustainable learning of statistics.

Keywords Sustainable learning · Variation theory · Learning statistics · Teaching statistics · Stochastic reasoning

---

H. Innabi (✉) · F. Marton · J. Emanuelsson
Department of Pedagogical, Curricular, and Professional Studies,
The University of Gothenburg, Gothenburg, Sweden
e-mail: hanan.innabi@gu.se; ference.marton@ped.gu.se; jonas.emanuelsson@ped.gu.se

G. F. Burrill et al. (eds.), *Research on Reasoning with Data and Statistical Thinking: International Perspectives*, Advances in Mathematics Education,
https://doi.org/10.1007/978-3-031-29459-4_21

1 Introduction

With the accelerated development of science, technology, and communications, the importance of the discipline of statistics is growing. In today's data-rich world, statistics has become a central topic of study. There have been repeated calls from the statistics education community during the last three decades for high-quality learning of statistics that includes statistical reasoning and making sense of data rather than simply focusing on rote skills, computations, and procedures. (See for example: GAISE II (Bargagliotti et al., 2020), the American Statistical Association (ASA)'s Statistical Education of Teachers (SET) (Franklin et al., 2015)). Here we are talking about high-quality learning of statistics that helps learners to reason statistically to interpret and critically evaluate statistical information and data-based arguments that appear in learners' lives both now and in the future.

When considering the importance of high-quality learning of statistics that can be used in real-life contexts, which are increasingly emphasized by the statistics education community, the question arises as to how to contribute to such high-quality learning of statistics. According to our line of reasoning, when discussing the issue of high-quality learning, two intertwined challenges arise. First, people frequently forget what they have learned. Second, people learn but fail to apply their knowledge, except in the original context they are familiar with. Thus, they have trouble using the knowledge in new situations. Being able to successfully overcome these two challenges is characteristic of high-quality learning—that is, learning that enables students to handle novel situations in powerful ways both now and in the future. Such learning we call "sustainable learning", which refers to the general meaning of sustainability: the propensity of something to continue and grow over time. Sustainable learning continues beyond formal instruction; what has been learned can be expanded and used in other situations, that is, learning to solve problems that you have not yet learnt to solve (Fülöp, 2019). The two characteristics of high-quality learning mentioned above and subsequently used to define sustainable learning are closely related. The first characteristic is that learning is less prone to being forgotten, and the second is that learning can be used in novel situations to address novel problems. The latter may be seen as a function of the learners' ability to make sense of novel problems. The more the students learn from novel situations, the more their original learning will become enriched instead of fading away. Here, learning becomes a preparation for future learning in which learners can begin to handle various situations.

In this chapter, we discuss what might make learning of statistics sustainable. In addressing this issue, we draw upon results of previous research to support our conjecture that enabling learners to focus on variability in statistics to a greater extent will most likely facilitate more sustainable learning of statistics. Our argument rests on two sources of support. First, we see the learning and teaching of statistics in terms of the juxtaposition of two lines of reasoning: deterministic reasoning, which is oriented towards exact numbers and causal explanations, and stochastic reasoning, oriented towards distributed alternatives and probabilistic reasoning. To reason statistically, both are important (Biehler, 1989). However, people are in general more inclined to reason in deterministic ways (Nilsson, 2013; Sánchez et al., 2018; Shaughnessy, 1977). Hence, the stochastic line of reasoning has to be strengthened, because working for a better balance in the two lines of reasoning is working towards better learning and teaching of statistics (Pfannkuch & Brown, 1996). This leads us to the problem that the stochastic line of reasoning is less common when people handle issues around them, as well as the question of what we can do about this, and in particular how we can facilitate learning of a stochastic line of reasoning. Arguments are frequently voiced in the statistics education community that a key is to focus more on the variation in data in statistics (Garfield & Ben-Zvi, 2008; Moore, 1990; Snee, 1999). By paying greater attention to variability, we will contribute to better stochastic reasoning. Here we are closing in on the question of what can be done regarding teaching and school practice to help students to see and acknowledge variability in statistics.

The second source of support concerns a learning theory: Variation Theory (VT). Before clarifying this source, it is important to mention that in this chapter, we are using the term *variation* in two different contexts. The first one is the context of statistics: the omnipresence of variability in data. Variation in statistics is defined as describing or measuring observable characteristics of variability (Reading & Shaughnessy, 2004). The second context is variation in learners' experiences within a learning theory, that is, VT. To distinguish between the two contexts of variation, for the first context, we use the term variability in data, and for the second, the term variation in learning. As we shall see later in this chapter, a substantial number of studies have been carried out indicating that variation (in learning) is a necessary condition of learning, and it has proven to be a powerful pedagogical tool with potential for making learning sustainable. Below, we briefly characterize the two research specializations that are being developed within this framework, namely: phenomenography and Variation Theory. To summarize, in this article, we discuss what might make statistical learning sustainable. In addressing this issue, we are drawing upon two main sources: (1) learning and teaching of statistics as a juxtaposition of deterministic and stochastic perspectives, and (2) findings in the research approaches of phenomenography and Variation Theory.

1.1 Sustainable Learning of Statistics – An Example

Let us start by giving an example of what we have in mind when talking about our primary aim, which is facilitating "sustainable learning". The same example is also used to differentiate stochastic and deterministic ways of reasoning. This example is found in Chap. 10 of Nobel laureate Daniel Kahneman's best-seller *Thinking, Fast and Slow* (Kahneman, 2011). One of the cases presented in this chapter concerns results from a study funded by the Gates Foundation, which awarded $1.7 billion for research on what makes schools successful. The main finding was that successful schools are mostly small schools. Four times as many small schools were found among the best schools than expected. With the support of such findings, many small schools were built in the US; and many big schools were divided into smaller units. And, of course, causal conclusions may easily appear logical. We might, for instance, think that the individual students receive more attention and are thus more stimulated in small schools. This is an example of deterministic reasoning. Such conclusions, can however, easily shown to be misleading by applying stochastic reasoning and exploring alternative outcomes: If we try to find not only the best schools but also the worst (i.e., the least successful) schools, these frequently turn out to be small schools too. But how on earth can both the best and the worst schools be small? As pointed out above, the results from the research financed by the Gates Foundation can, from a deterministic perspective, be interpreted in terms of a logical explanation of why small schools are more successful. However, although the explanation may sound logical, it is wrong. The paradox is solved when a stochastic perspective is adopted, and the learner becomes aware of the fact that small schools make smaller samples and smaller samples spread more widely (vary more).

If we now combine the stochastic and the deterministic perspectives in a more careful comparison of small and large schools, this time adopting a deterministic perspective, we might find somewhat better results in the larger schools, thanks to a wider range of subjects from which to choose (Kahneman, 2011).

Most people know that a bigger random sample is more precise than a smaller sample, but "knowing" might mean different things, Kahneman argues. Even if you know the preceding sentence, "*...you must try hard to realize that the following statements tell us exactly the same thing:*

**Bigger samples are more exact than smaller.*
**Smaller samples yield more often extreme outcomes than bigger.*

The first statement sounds entirely fine, but before you understand the second intuitively you have neither understood the first in actual fact" (Kahneman, 2011, p.164–5). Using our line of reasoning, understanding the second statement is an example of sustainable learning. Understanding the first statement, but not the second, is an example of non-sustainable learning (i.e., it is not a deep or powerful understanding). Our view of sustainable learning with its two characteristics (that it persists and allows students to deal with new situations) reflects a deep and powerful understanding. The finding that small schools had better results than large schools is not always immediately seen as an artifact of a well-known principle in statistics, the Central Limit Theorem, and the relationship between sample size and variability. Although all statisticians should understand this, the subjects in this study did not because there was a new context which was related to the statistical principle but in a way they had not thought about. Later in this chapter, we will continue with this example to show how we can support such learning.

2 The Juxtaposition of Statistical Reasoning

Statistics is associated with uncertainty: we are uncertain when we draw a conclusion about a population based on a sample; we are uncertain when we predict the future based on the past or present. Furthermore, we are also uncertain about the precision and accuracy of our measurements when we measure a phenomenon. Statistics — in the definition adopted by the American Statistical Association (ASA) — "is the science of learning from data, and of measuring, controlling and communicating uncertainty" (Wild et al., 2018, p.6).

2.1 Deterministic and Stochastic Reasoning

The term "deterministic" suggests that some future events can be calculated with precision and without randomness. In deterministic reasoning, under certain assumptions, from a given input, a definite output can be produced. For example, the relationship between the circumference (c) and radius (r) of a circle is deterministic because it is an exact formula that will always give you the correct answer (assuming that the calculation is performed correctly in Euclidean space, and on measuring 'distance' L2 norm): $c = 2\pi r$. Conversely, a random event cannot be precisely predicted with a determined formula. While an informed guess can be made, this is done with uncertainty; for example, the chance of meeting your brother (who lives abroad) tomorrow or rolling the number six with a die. In this chapter, we refer to reasoning concerned with randomness and variability as stochastic (or probabilistic) reasoning.

To further clarify the meaning of deterministic and stochastic mindsets, consider an example related to a statistical indicator: the mean. When the "sample mean" is seen as a number representing a specific population, and this number can be calculated by adding all the sample values and dividing by the number of values without seeing the sampling error, a deterministic way of reasoning is present. However, when considering the validity of this calculation procedure, errors relating to measuring accuracy or identifiable differences if another sample is selected can occur; this view is consistent with stochastic reasoning.

In statistics, mathematical models are usually constructed to describe the data and the variations it contains. Using these models, the statistician tries to identify the factors that account for the observed variation. However, these factors usually cannot explain all the observed variations. Thus, two parts form these models: the deterministic part identifies explainable patterns, and the stochastic part identifies the not yet identified sources or the non-systematic sources underlying this variation that cannot be analyzed directly, which is considered random (Pfannkuch & Brown, 1996). An example to clarify this is the linear relationship, whereby if the response (y) and predictor variable (x) have an exact linear relationship, then that relationship is deterministic, and x can fully explain the variation in y. However, most things in real life are a mixture of random and deterministic relationships. For instance, forecasting weather is based on deterministic patterns and some degree of randomness and uncertainty.

The terms *signal* and *noise* (see Konold & Pollatsek, 2002) constitute a metaphor for deterministic and stochastic reasoning, which originates from the process of receiving a signal and tuning into a radio station, where "noise" describes the unwanted signal heard when the radio is not exactly tuned to the correct frequency. Therefore, "signal" refers to the meaningful information that we are actually trying to detect, and "noise" is the random, unwanted variation that disturbs the signal. For example, any value obtained by measurement contains two components: one carries the information of interest (the signal), and the other comprises random errors (noise). Such random errors are unwanted because they reduce the precision and accuracy of the signal.

Stochastic reasoning is related to what is left after explaining what can be explained. For example, to explain why certain differences appear, one may examine the explained variance because this can be understood. However, some differences cannot be explained; therefore, what is left is the stochastic variance (Pfannkuch & Brown, 1996; Wild et al., 2011).

> *The distinction between 'deterministic' and 'random' is, however, not as clear cut as it might appear. In statistics in particular, it is very difficult to make this distinction. The only pure deterministic reasoning may be in mathematics, such as the earlier example of the formula for circumference, or in a geometric proof. Depending on the statistical model we are considering, there can also be "noise in the signal". For example, if we are talking about linear regression modeling, then the above analogy about the "explained part" and the "unexplained part" makes sense, as we can account for "causes" for part of the variation. However, suppose we consider statistical models for estimating a population mean from a sampling distribution. Then the "signal"—namely our sample means—vary from sample to sample, so there is some "noise" in our estimator—and the best we can do is to reason stochastically about our "signal" and make statements using confidence intervals for "catching" that estimated population mean.* (J. M. Shaughnessy, personal communication, February 6, 2022)

2.2 Statistical Reasoning Requires Stochastic and Deterministic Reasoning

Over the last 30 years, the statistics education community has been calling for statistical reasoning to be taught as a balance between stochastic and deterministic reasoning (e.g., Biehler, 1989; Pfannkuch & Brown, 1996; Wild et al., 2011). Pfannkuch and Brown (1996) state that our task is to enable our students to be comfortable thinking probabilistically and deterministically. The idea that both deterministic and stochastic reasoning must be considered in statistical reasoning was clarified by a metaphor used by Wild et al. (2011), that looking at the world using data is like looking through a window with ripples in the glass. What is seen in the data is not quite how things really are in the populations the data comes from.

Biehler (1989) called for students to be allowed to practice dual thinking: deterministic and non-deterministic. Teaching statistics should be directed at helping students to see the probabilistic model, as it usually originates from a mathematical model that accounts for randomness. For example, when teaching the straight-line probabilistic model (linear regression), students have to learn that when representing a relationship between two variables, x and y, in some situations, this relationship can be represented exactly as a mathematical linear function model ($y = \beta_\circ + \beta_1 x$), where y indicates the study variable and x is the explanatory variable, and the terms $\beta_\circ$ and β_1 are the parameters of the model. Simultaneously, students have to experience that in many other situations in real life (for example, the relationship between a person's arm span and their height), the data points do not lie on a straight line for several reasons, such as the inherent randomness in the observations and the effect of all other neglected variables in the model. Thus, a random or another probabilistic component must be factored into the model ($y = \beta_\circ + \beta_1\, x + \varepsilon$). Students have to discern that ε indicates the unobservable error component representing the difference between the expected values and the observed values of y.

2.3 *Tendency toward Deterministic Reasoning*

This section clarifies that when people examine a problem in the real world, they tend to search for a deterministic cause, pattern, or model to justify or judge events, even those events involving uncertainty and randomness. Research in psychology and education has provided evidence that people generally demonstrate less developed ways of reasoning amid uncertainty (Shaughnessy, 1977). During the 1970s, Daniel Kahneman and Amos Tversky demonstrated, through a series of experiments, that people estimate the likelihood of events using judgmental heuristics, which sometimes leads to multiple systematic errors (Tversky & Kahneman, 1974). One of the major heuristics of Kahneman and Tversky is "availability." People use this heuristic when they assess the frequency of a class or the probability of an event based on the ease with which instances or occurrences can be imagined. Therefore, it is assumed that if examples are quickly visualized, there must be many. Another heuristic is "representativeness", whereby people estimate the likelihood of events based on how well an outcome represents some aspect of its parent population (Tversky & Kahneman, 1973).

More recent studies show some evidence of students' tendency to use deterministic reasoning (e.g., Henriques & Oliverira, 2016; Nilsson, 2013; Sánchez et al., 2018; Schulz & Sommerville, 2006). Sánchez et al. (2018) describe determinism in chance as interpreting theoretical propositions about chance as if they were causal statements or as if they predicted results that could only occur in one way. They explain this misconception in terms of students' inability to address the uncertainty, randomness, or variability present in a situation. Some other research has shown two extreme views of the relationship between sample and population: that any sample can either always or never represent the population (e.g., Innabi, 2007).

Both of these two extreme views can be explained by a deterministic mindset, as it reflects an over-reliance on rules and theorems. Meletiou-Mavrotheris and Lee (2002) connect the view that *any sample will never represent the population* to "people who distrust statistics completely because, unlike mathematics, it deals with uncertainty" (p. 23), while the other extreme view, that *any sample represents the population*, is associated with "people who use statistical methods for solving real-world problems in the same way they would use an artificial mathematics problem coming out of a textbook." (p. 23).

From a phenomenographic perspective, we see that the weakness identified in people's stochastic reasoning is due to the conflict between the inbuilt heuristics found in research (e.g., Kahneman and Tversky) and the deterministic forms continuously encountered in the formal education system. These heuristics can be seen as seeds of stochastic reasoning, and not considering these heuristics in education, and a lack of direction and guidance to a deeper level of stochastic reasoning, could be the reason for human weaknesses when reasoning amid uncertainty (see Shaughnessy (1992)). For example, representativeness is a basic and important idea in statistics, where a sample should represent the population in order for it to be possible to draw valid conclusions. However, because deterministic reasoning is often enhanced at the expense of stochastic reasoning, what people have already learned by interacting with the uncertain world directs them to the less effective path, which ultimately results in the errors and fallacies found in research. This legacy contributes to the dominance of the deterministic view of statistics and hinders the teaching of statistics based on uncertainty and stochastic reasoning. Watson (2016), for example, indicates that statistics textbooks contribute to strengthening the deterministic view. She clarifies that until now, the focus has been on handling data without context. Giving sets of numbers to students to identify statistical indicators (such as the mean and standard deviation) will lead students to perceive these statistics merely as calculated numbers instead of the characteristics of the data.

Furthermore, statistics has always been associated with mathematics (see Cobb and Moore (1997)). Sánchez et al. (2018) have shown that deterministic reasoning is promoted by the conception of mathematics as a set of definitions and procedures that are unrelated to reality. They demonstrate this with a misleading learning example where if the probability of rolling "3" with a die is 1/6, then students expect that when rolling a die six times, one "3" will be obtained. Another example is the belief that a random sample is a miniature replica of a population. The traditional view that considers mathematics as fixed (and not dynamic) and the deep-rooted beliefs about the nature of mathematics and the nature of its teaching and learning (Innabi & Emanuelsson, 2021) are carried over into statistics. This is probably what motivated the call to separate statistics from mathematics and consider statistics as a stand-alone science: to free it from the specific answer and the static explanation of variance. Some believe that recognizing statistics as a branch of mathematics is not good for the discipline (Glencross & Binyavanga, 1997). Whether statistics should be part of mathematics or not, both disciplines should develop a wider perspective enriched with more fallibilistic and humanistic views, which is important for a clearer and broader perspective on the real world.

2.4 Stochastic Reasoning and Variability

The importance of the concept of variability in statistics education is increasingly becoming recognized. Dealing with the omnipresence of variability in data is a fundamental base for statistical reasoning. In any dataset, the elements are not usually alike, and the extent to which they vary is of basic importance in statistics. Statistical problem solving and decision making depend on understanding, explaining, and quantifying variability in the data within the given context, such as variability within a group, variability between groups and sample-to-sample variability (Bargagliotti et al., 2020). Konold and Pollatsek (2002) define variation as noise in a dataset. Variation in the statistical sense may come from different sources, as Franklin et al. (2007) point out, such as measurement variation (when repeated measurements performed on the same entity vary), natural variation (where individuals are different), induced variation (when outside forces or factors act to change naturally occurring variability), and sampling variation (when measurements of different samples vary). In the same vein, Shaughnessy (2007) states that students' reasoning about variability in statistics focuses on three areas: variability within data, variability across samples, and variability between distributions.

In the 1990s, many statistics educators recognized the lack of attention paid to variability by curricula and national assessments (e.g., Batanero et al., 1994; Green, 1993; Shaughnessy, 1997), which led to the creation of a movement promoting the importance of variability in statistics education. Statistics education reform began (and continued) by calling for an increased focus on variability, bringing it more to the fore in teaching. Moore (1990) describes statistical thinking as recognizing the omnipresence of variability, while Snee (1999) states, "If there was no variation, there would be no need for statistics and statisticians" (p. 257). Several scholars (e.g., Cobb & Moore, 1997; English & Watson, 2016; Garfield & Ben-Zvi, 2008) have considered variability to be the fundamental phenomenon underlying the entire field of statistics. Accordingly, notable changes have been made to several curriculum documents around the world. Despite awareness among statistics educators of the importance of focusing on variation in statistics education, much effort is still needed in teaching and learning statistical reasoning (Ben-Zvi, Makar, & Garfield, 2018a; Carter et al., 2017; Cobb, 2015).

In this regard, it is worth mentioning the increasing role that simulation now plays in many statistics courses. A simulation-based approach, including resampling and re-randomisation, focuses more on variability than the traditional formula-based methods. Giving students of all ages and levels a suitable dynamic visual tool they can use to answer questions related to change and differences—such as what is happening in the present, what might happen in the future, and what would happen if this or that variable changed—will help them open up new dimensions and expand their vision. A substantial number of studies have shown that dynamic visual methods can be highly powerful analytic tools for teaching statistical reasoning (e.g., Konold & Miller, 2005; Marton & Pang, 2006; Paparistodemou & Meletiou-Mavrotheris, 2008; Ridgway, 2015). For example, Ekol (2015) found that the dynamic sketching and dragging action mediated students' informal understanding of the meaning of standard deviation and variability, and helped students check their conjectures, which was not possible in a static environment.

Variability in statistics and stochastic reasoning are two sides of the same coin. Developing stochastic reasoning means developing students' ability to acknowledge and handle randomness and uncertainty. This is related directly to considering differences. Such differences might be statistics relating to different samples of a specific population or residuals of the observed and expected values of the dependent variable. So, paying attention to variability in statistics is the answer to the question: how can we affect students' deterministic ways of thinking about statistical problems so that students will learn to adopt a certain form of statistical reasoning that reflects stochastic reasoning? Hence, learning opportunities are required to encourage appropriate ways to realize, understand, and model the variability of data (Carter et al., 2017; Dierdorp et al., 2017). Classroom experiences that focus, for example, on identifying values from several samples rather than on expecting a value from just one sample or focusing attention on multiple aspects of the sampling distribution (center, shape, and variability) (Noll & Shaughnessy, 2012) are instructional examples that might help to develop stochastic reasoning.

We might not be adding anything new when talking about the fact that variability in statistics is a fundamental idea that should be considered in statistics education. However, theorizing this fact within a theory of instructed learning should not just explain it but also show how to design teaching in order to consider variability in statistics. This is what we discuss in the following sections.

3 Variation in Learning

The qualitative research approach Phenomenography and its theoretical development Variation Theory both consider variation as a necessary condition of learning. Hence, this chapter handles *variation* in different contexts. In phenomenography, variation is about differences between ways of understanding and in VT, it is about variation in the dynamics of understanding, that is, variation within a way of understanding. In both cases, we are dealing with variation in human experience. On the other hand, variation as the term is used in statistics is about variability related to data. From the perspective of VT, when learners are exposed to variability in data, this automatically includes variation in learners' experiences of what they are being asked about. In the following, we will outline the theoretical framework used to relate these forms of variation to each other.

3.1 Phenomenography

Phenomenography is a qualitative, non-dualistic and explorative research approach that describes how learners experience and perceive a phenomenon (Marton, 1981; Marton, 2000). Phenomenography does not make statements about the world but about people's conceptions of the world (Marton, 1986). Each way of experiencing, seeing, or understanding can be understood in terms of which aspects of the phenomenon are discerned and not discerned through the learner's awareness (Marton & Booth, 1997). Phenomenography aims at describing qualitatively different ways in which people experience, see, understand, conceptualize, or perceive phenomena in the world around them (Marton, 1986). The result of a phenomenographic analysis is called an outcome space. "Such structures (a complex of categories of description) should prove useful in understanding other people's understandings" (p. 34). According to the phenomenographic approach, "a careful account of the different ways people think about phenomena may help uncover conditions that facilitate the transition from one way of thinking to a qualitatively 'better' perception of reality" (p. 33). The knowledge generated by phenomenographic research has direct educational implications that help teachers to help their students learn better. "Encouraging teachers to pay attention to students' ways of thinking and to facilitate students' realization that there are different ways of thinking may be the most important pedagogical implications of a phenomenographic view of learning" (Marton, 1986, p. 47).

Phenomenography makes it possible to describe the learners' conceptions from the learners' viewpoint, whatever these conceptions are. In phenomenography, we do not search for right or wrong answers; we try to see what the learners see. These perceptions might appear erratic from the viewpoint of external observers, but from the viewpoint of the learner, what you see is simply their reality. Thus, we can help learners develop their statistical reasoning by understanding their perspectives better. According to VT, the learning theory that emerged from phenomenography, teaching can best be designed by first finding out what these views are and uncovering individual perspectives on relevant issues (Lo, 2012).

To clarify what was mentioned in Sect. 2.3 regarding the tendency toward deterministic reasoning, from a phenomenographic perspective, we consider that the findings that demonstrate less developed intuition concerning stochastic reasoning should not be considered as revealing a disappointing feature of human reasoning. In contrast, it should be seen positively because people constitute these heuristics as a result of their interaction with an uncertain world. These heuristics can be seen as seeds that have to be directed to a qualitatively more powerful perception of reality. As described below, this can be supported by several studies showing that students' stochastic reasoning can develop much further.

3.2 Making Learning Possible

VT is a learning theory developed from the phenomenographic research approach. VT can provide both a theoretical framework and a tool for designing learning opportunities. It explains *how* learners come to understand or see phenomena in certain ways. The core tenet of VT is that people learn through experiencing differences and similarities (in this order). For instance, in order to experience something as sweet, one must be exposed to variation in taste, such as sourness, bitterness, or saltiness. It would not be possible to notice a sweet taste if all tastes were identical. Furthermore, the very concept of taste would be meaningless if there were not different kinds of tastes. Hence, it also implies that learning can be brought about by affording learners the experience of such differences and similarities. It thus represents a theoretical framework that can direct teachers' attention toward what must be done to provide learners with the necessary learning opportunities (Marton & Booth, 1997). VT is hence both a theory on learning and a theory of teaching.

According to phenomenography and Variation Theory, learning has taken place when the learner's way of understanding changes from one way to another more complex way. VT clarifies and explains how such changes can be made possible. To see something in a specific way, the learner must discern certain aspects or attributes of the object of learning and the only opportunity to discern them is when they vary (Marton, 2015). Critical aspects are aspects that must be discerned by a learner to develop a certain understanding of an object of learning; these aspects are critical for understanding the phenomenon in a more developed way but have not yet been discerned by the learner (Holmqvist & Selin, 2019). For example, if the object of learning is to understand the concept "mean" in statistics, the learner has to discern some aspects such as: the mean is representative of all the other values; the sum of deviations from the mean equals zero; it is influenced by the extreme values thus it is not always a good idea to use it to represent data; to find the mean, add up all the values and then divide by the number of values. If a learner can calculate the mean but does not see it as a representative indicator of the data, the former aspect is not a critical aspect but the latter is a critical aspect that a student has to discern. If a learner discerns certain aspects of something and another learner discerns different aspects, we say that the two learners see the same thing in different ways. So, a way of seeing something can be defined in terms of the aspects that are discerned at a certain point in time.

We argue with Marton and Tsui (2004) that the most effective way to help learners learn is to focus on providing opportunities for them to experience variation in critical aspects of the object of learning. Here, we want to clarify that deliberately using specific patterns of variation and invariance does not guarantee that learning will happen. Instead, we argue that patterns of variation make learning, in terms of seeing an object of learning in a specific way, possible. Hence it is a necessary but not sufficient condition.

In VT, the *object of learning* is defined as a "specific insight, skill, or capability that the learners are expected to develop" (Marton & Pang, 2006, p. 2). The object of learning can be formulated in three different ways with increasing precision: in terms of content, in terms of educational objectives, and in terms of critical aspects (Marton, 2015 p. 22). Within Variation Theory, learning is always learning of something in particular, a content that delimits what is to be learned. In statistics, examples of objects of learning described at the level of content might be standard deviation, sampling, t-test, or linear regression. These terms, however, do not give many clues as to how more specifically the learners should master the content. If the object of learning is also formulated in terms of what the learners can do with the content, the precision increases. This might involve for example, *remembering* the formula, *calculating* standard deviation or *understanding* how its size is related to the distribution. A well-developed understanding of how the size of the standard deviation of a variable is related to the form of the distribution of the same variable requires the learner to see several aspects (which is something different from being able to calculate the standard deviation). A more developed way of handling standard deviation requires the learner to be aware that if the spread is high, many values lie far from the mean of the variable. If the spread is smaller, values tend to group closer to the mean. Further, the spread needs to be related to the size of the calculated value of the standard deviation. If a learner has not previously attended to spread, this aspect is critical. If the same learner has not previously discerned that standard deviation can take on different values, this aspect is also critical. If these two aspects are discerned simultaneously, the learner's understanding is more developed than if one of the aspects is not attended to. Mastering the object of learning formulated as critical aspects means that learners have to be afforded the opportunity to discern its critical aspects (Marton, 2015).

To understand what is made possible to learn and what is not possible to learn in a learning situation, it is necessary to pay close attention to what aspects are varied and what aspects are invariant in the situation (Marton, 2015). This variation and invariance we term a *pattern of variation*. Within VT, there is a set of such identified patterns:

First, *contrast*. For a person to experience something, he or she must experience something else that contrasts with it. In order to understand what "three" is, for instance, a person must experience something that is not three: "two" or "four", for example. This illustrates how a value (such as three) is experienced within a certain dimension of variation, which corresponds to an aspect (numerosity or "manyness"). Thus, comparing different things is a central idea in this learning theory. For instance, to understand what "normal distribution" is, the learner must experience something that is not normally distributed: "uniform distribution" or "skewed distribution". This illustrates how a value (normal distribution) is experienced within a certain dimension of variation that corresponds to an aspect (probability distributions).

Second, *generalization*. In order to fully understand what "three" is, we must also experience varying appearances of "three", for example three apples, three monkeys, three toy cars, three books, and so on. This variation is necessary in order for us to be able to grasp the idea of "threeness" and separate it from irrelevant features (such as the color of the apples or the very fact that they are apples). The function of this pattern of variation and invariance is not to afford a completely new meaning but to generalize a meaning once the meaning has been found through contrast. By discerning aspects that vary independently from the focused aspect, the learners generalize the focused aspects over the non-focused aspects. In generalization, we systematically vary the different possible aspects while keeping the focused aspect invariant. Suppose our focused aspect is "normal distribution," and we want the learners to realize that it is a distribution pattern occurring in many natural phenomena and in several contexts (not just the context mentioned). We open up the dimension of contexts that have a normal distribution (IQ, height, weight, test scores, etc.). Notice here that "normal distribution" is invariant and that which is normally distributed (the context) is varied. Another critical aspect might be that the "normal distribution" can have different "standard deviations." Here, we keep the distribution (normal) invariant and vary the standard deviation.

Third, *fusion*. If there are several critical aspects that the learner has to take into consideration at the same time, they must all be experienced simultaneously. In everyday life, it is seldom that only one aspect of something varies at a time, and so the way in which we respond to a situation, such as hitting a target with a ball, springs from a more general holistic perception of the situation. For an object of learning that comprises several critical aspects, learning is about experiencing a whole rather than the parts of which it is composed. This requires simultaneous discernment of critical aspects, where several varying aspects need to be fused. Lo (2012) provides the example that a teacher can make use of variation to help students to discern that the price will increase if there is greater demand, provided that the supply is kept constant, and that the price will drop if there is greater supply, provided that the demand is kept constant. However, helping students to experience the effect on price when both supply and demand vary simultaneously, bringing about fusion, means that students will become aware that the price of a given commodity is determined by the relative magnitude of changes in both the supply and demand of that commodity. As an example in teaching statistics, if the object of learning is "making predictions about or estimating population proportions from a sampling distribution," students should be able to integrate multiple aspects of expected sampling distributions: center, shape, and variability (see Noll & Shaughnessy, 2012, p. 547). From the perspective of VT, this requires the fusion pattern, which involves simultaneously varying the mean, spread, and shape of the sampling distribution.

In short, in the context of the current research interest, VT is what might be called a *theory of instructed learning*, more precisely learning that takes place with the aid of and in interaction with other people. VT lends itself to being a design tool when designing teaching, instruction, and other learning activities. It indicates what patterns of variation and invariance should be made available for the learners to discern in order to see an object of learning in a certain way. It is a theory that was developed from phenomenographic results through analysis of the categories of description in terms of what dimensions of variation were discerned. It also became a theory of teaching, since teaching or instruction is about arranging activities which afford a certain pattern of variation and thus make learning possible (Lo & Marton, 2012). VT is a pedagogical theory, in the sense that it gives indications of how instruction should be designed to achieve this kind of learning. It is not a theory in the sense that it addresses sociological issues or psychological issues or issues of gender and so on: it addresses questions of how teaching should be organized in terms of handling the content in order to afford the learning of certain and specified understandings.

This chapter is a theoretical discussion within the Variation Theory (VT) framework to show how sustainable learning can be developed by enhancing stochastic reasoning. We argue that if teaching is designed from the point of view of VT, in particular where we introduce variation in critical aspects, this will increase the potential for sustainable learning. Variation Theory "looks at the kind of learning that enables learners to deal with future, novel situations in powerful ways." (Marton, 2015, p.27).

4 Enhancing the Sustainable Learning of Statistics

As mentioned above, sustainable learning is characterized by (1) learning that enables students to deal with novel situations in powerful ways and (2) learning that persists and continues to grow. Many studies have been carried out about the effect of learning on other learning (transfer, learning to learn, generative learning, etc.) Transfer in the classical sense is the effect of learning one thing on the learning of another thing as a function of similarity (Marton, 2006), while sustainable learning is the effect of learning one thing on the learning of another thing as a function of difference and similarity. "Generative learning," as conceptualized by Holmqvist et al. (2007), is similar to "sustainable learning", but we use the latter term to emphasize the close relationship between continuity and change in this kind of learning. "Learning to learn" is about the effect of learning certain things on the learning of other things, while sustainable learning is the effect of learning on the learning of the same kind of things. "Learning to learn" is about becoming good at learning in general, while "sustainable learning" is content-specific; it is about becoming good at learning something particular. Hence, when talking about sustainable learning, we have in mind a particular object of learning, which is a description of what is to be learned. Our concern here is learning within the broad field of statistics.

4.1 Developing Stochastic Reasoning by Applying VT

To support students in reasoning stochastically, teaching has to make it possible for students to experience uncertain situations and handle random variables. Biehler (1989) criticizes the focus on ideal random situations even when teaching probability as a prerequisite for statistics. Our point in this chapter is to open students' eyes to stochastic ways of reasoning. The main tool for bringing about this change is to place greater emphasis on variability in data and hence, according to VT, also variation in the learners' experience. In the following, some examples from the literature are provided to support how considering variation can promote stochastic reasoning. Notice that none of these studies used VT. However, in exploring these studies from the VT perspective, patterns of variation and invariance can be noticed and hence conclusions on what may be learned can be drawn. In this regard, we acknowledge that patterns of variations are used by many teachers and many researchers without being aware of or using the same terms as we do in VT.

The first example is the study of Prodromou and Pratt (2013), which shows how experiencing variation helps in developing stochastic reasoning. A case study of students aged 14–15 years used an animated computer display and graphs generated by a basketball simulation. Students attempted to make sense of distribution by adopting a range of causal meanings for the variation observed. To offer an apparently determined outcome alongside the possibility of random error (but all in a playful context), a basketball game was chosen. The animation of the basketball player was controlled by dragging the handles on the sliders for the release angle, speed, height, or distance. The students had access to a line graph of the success rate of throwing balls into the basket. The simulation also allowed the students to explore various types of graphs, relating the values of parameters to the frequencies of attempts and the frequencies of success. The results indicated that the computer simulation could offer new ways of harnessing causality to facilitate students' understanding of variation in distributions of data. To bridge deterministic and stochastic reasoning, the students transferred agency to active representations of distributional parameters, such as average and spread. The study showed that giving students the chance to experience differences in release angles helped in developing stochastic reasoning. Students were free to vary or fix several variables (angle, speed, height, or distance). Also, they could see different representations of the variables. Through the lens of VT, patterns of variation and invariance can be observed in this study, including the fusion pattern, where students were able to experience the four variables (angle, speed, height, and distance) varying simultaneously.

In the second example, Pfannkuch and Brown (1996) conducted a pilot study on how students of statistics must be allowed to experience the omnipresence of variation, and experience the dual modes of thinking probabilistically and deterministically to explain that variation. In the first phase of this study, five students majoring in psychology who enrolled in a first-year statistics course were interviewed to investigate their understanding of variation. It was found from this interview that these students tended to think deterministically; they had little understanding of variability and its relationship to sample size. Two weeks after the interview, a one-day, five-hour course was held to experiment with what could be done to increase students' understanding of variation. The course focused on deliberately thinking both deterministically and probabilistically about problems, experiencing experiments that reveal that small samples are not representative of the population, experiencing probability based on models and real data, and clarifying variation and chance. First, an overview of deterministic and probabilistic thinking was explicitly given, showing that it is necessary to do both when one is presented with a situation. To enhance students' awareness of variation, several situations were presented in which they were encouraged to think about the problem both deterministically and probabilistically. For instance, students were asked to measure the length of a page, and the results were plotted on a graph. Three weeks after the one-day course, a second interview took place to further investigate the students' statistical thinking, beliefs, and intuitions. The results showed that asking students to consider situations from a non-deterministic perspective might improve their probabilistic thinking. Similarly, it may be possible, through experiments and logical explanations, to raise students' awareness of the tacit intuitive models that lead them astray. Applying the terminology of VT, the pattern of "contrast" is clear here, where stochastic reasoning was explicitly introduced to students as a contrast to deterministic reasoning several times.

A third example from a study by Meletiou-Mavrotheris and Lee (2002) was motivated by the results of a previous study of an introductory statistics course. In that previous study, the results showed that the students had poor stochastic intuition and tended to think deterministically, regardless of whether they came from a lecture-based classroom or a course following the Projects-Activities-Cooperative Learning-Exercises (PACE) model. PACE is an approach that attempts to provide a structured framework for integrating projects and hands-on activities conducted cooperatively in a computer-based classroom environment. It was concluded that the students' difficulties might stem from inadequate emphasis by the instructors on the notion of variation and the connections among statistical ideas. This led to the decision to modify the PACE model. A course was taught that retained a format similar to that used in the previous study; however, it differed considerably in emphasizing variation as its central tenet. A five-week introductory statistics course in higher education was conducted with 32 students. This course placed greater emphasis on helping students improve their intuition about variation and its relevance to statistics. For example, the idea of making conjectures ran throughout the course. Students would state what they believed might or might not be true and then look critically at the data to evaluate their statements. In the previous study, the PACE results showed superficial knowledge of statistical concepts and a tendency to think deterministically and seek causes behind ephemeral patterns in the data. In contrast, students in the modified study were found to better understand the relationship between chance and regularity, and to apply stochastic reasoning much more effectively.

Lastly, we offer an example from Nilsson (2013) that can support the importance of students experiencing the appropriate pattern of variation when learning stochastic reasoning. This experiment-based study examined how students try to make sense of real-world, random dependent situations by analyzing how students balance deterministic and probabilistic ways of reasoning. The study examined an episode of probability teaching located in an outdoor setting. The activity involved 12 students creating their own frequency data by planting 15 sunflower seeds each. The sets of seeds were planted each in their own square and the number of seeds that grew was marked in a diagram. The analysis showed that the students did not use this sample information in the experiment; instead, deterministic features contextualized within the frame of ecology dominated their reasoning. The author concluded that, although the students were asked to decide on whether one seed will grow or not, they were not invited or challenged to reflect on the random variation or degrees of certainty. The students were asked to decide on a particular outcome (whether one seed would grow or not). To clarify the author's justification by using VT, we say that varying the number of the planted seeds (first 15 and then 1 seed) and not focusing on the variation of the successful seeds as it appeared in the diagram, puts the aspect "number of planted seeds" to the fore in the attention of the students, and this is what made them stress ecological factors regarding data frequency and randomness. In Nilsson's study, we see that to help students think probabilistically, the physical and biological factors should have been kept invariant while showing that the number of successful seeds varies.

Using VT as a general theory of learning, we explain the results of the above studies related to students developing their stochastic reasoning as follows: in VT, what is made possible to learn can be described by what aspects are made possible to discern in a lesson. We cannot make it possible for the learners to discern an aspect if it is not brought up as a dimension of variation where a learner can encounter differences. In a lesson, how the content is handled can indicate "what is made possible to learn" and "what is not made possible to learn". If you can make a statement about what is possible to learn, you may even understand why students fail to learn what the teacher intends them to learn (Kullberg et al., 2017). Thus, the importance of elucidating the pedagogical implementation of the VT vision of learning is what we are trying to argue for in this chapter.

4.2 A Conjecture

In this section, we suggest a conjecture that is: developing stochastic reasoning is likely to facilitate sustainable learning of statistics. As mentioned before, sustainable learning is learning that is supposed to enable students to handle novel situations now (immediately after formal instruction) and in the future (beyond formal instruction). It can be assessed by means of delayed measurement and by the use of novel problems. Many studies have indicated that people learn new things through the perception of differences. The role of differences and sameness in learning has been widely considered in the literature. Schwartz and Bransford (1998) indicate the positive effect of using variation, such as contrasting, in preparing students for future learning. Schwartz and Martin (2004) investigated how interventions may prepare students to learn, and showed that by comparing datasets, students learn to discern relevant features. The authors also noticed that contrasting cases of small datasets by highlighting key quantitative distinctions relevant to specific decisions could help students notice important quantitative features that they might otherwise overlook. Such studies using different theoretical positions support work within VT.

Several studies explicitly working with VT have provided evidence that when people learn something, they may use it in situations they have never seen before. Using systematic patterns of variation and invariance has shown positive effects in delayed assessments in many cases. Some of these studies were conducted with primary school children in different subjects, such as reading comprehension, mathematics, Swedish, and English (Holmqvist et al., 2007; Holmqvist & Lindgren, 2009; To and Pang, 2019). Other studies were conducted with upper-secondary students in different disciplines, such as reading comprehension (Marton and EDB Chinese Language Research Team, 2010), financial literacy (Pang, 2010; Pang, 2019), economics (Marton & Pang, 2007), and mathematics (Fülöp, 2019).

The positive results of these studies occurred because when students interacted with patterns of variation and invariance, they were likely to develop more powerful ways of seeing the object of learning and become better at discerning its critical aspects. Furthermore, every time they encountered instances of the object of learning after the end of the learning occasion, the likelihood of discerning those aspects grew higher. The studies suggest that every time the learners discern critical aspects, they become better at discerning new critical aspects (Marton, 2006) and support the idea of VT as a powerful pedagogical tool for improving learning.

We are using two kinds of arguments to support our conjecture; one is the way we describe learning as taking place through changes in the learner's awareness, and second, we refer to studies where this principle has been used successfully. Hence, we are using logical and empirical arguments.

5 A Pedagogical Call

We argue that teaching will be more likely to improve if it is informed by theory, especially theory that targets learners' opportunities to learn. Variation Theory provides such a theoretical grounding that is helpful in understanding necessary conditions for learning (Lo & Marton, 2012). The multiple pedagogical opinions about what should be done in order for students to learn better often assume that what is required is to change the way in which teaching is organized and what tools are used (more project work, peer learning, more problem-based learning, more use of technology, more or less homework, grouping or not grouping the students etc.). Instead, VT suggests that researchers and teachers first need to address what it is that has to be learned in each case, and find the different conditions that are conducive to different kinds of learning (Marton & Tsui, 2004). Ben-Zvi, Gravemeijer, and Ainley (2018b) analyzed the literature on statistics education related to learning environments and designed a way to support the teaching and learning of statistical reasoning. They suggested integrating features that show successful results in classroom learning, such as real datasets, technological tools, and trustworthy assessment methods. We agree that such factors can be useful for learning, especially when they are integrated and considered together. However, we argue that there are more important issues for teaching statistics in a powerful and sustainable way to students of all levels. One such issue relates to how the learning content is structured and managed.

To clarify how VT can be used as a pedagogical tool, let us assign the complex object of learning: *students are to be capable of reasoning stochastically*. To achieve this, teachers have to design learning opportunities that make it possible for students to discern the critical aspects of this object of learning. In order to provide the reader with a rather straightforward illustration let us return to the Kahneman (2011) example (school size) in which one of the critical aspects of the object of learning might be related to *understanding the role of sample size*.

Schrage (1983) shows that the effect of sample size on probability and variation is not a factor for those who are statistically naive. Some people may believe that the chance of getting at least seven white balls in 10 draws from a population of 50% black and 50% white is the same as the chance of getting at least 70 whites in 100 draws. In some cases, they may not realize the difference between the chance of getting at least two heads in three tosses of a coin and at least 200 heads in 300 tosses. It is not necessarily apparent that extreme events (such as all heads in a given number of coin tosses) are more likely to occur within smaller sample sizes than larger ones. A difficulty with this type of reasoning has also been called "the illusion of linearity" (Van Dooren et al., 2003).

Using terminology from VT, we can say that in order to overcome this difficulty, students need to experience different sample sizes (varying sample size) for the same population. On the one hand, we want students to see that if we pick a small sample, we run a greater risk of the small sample being unusual just by chance. This could be achieved through such learning activities as: *If you have 100 mixed gumballs—50 red and 50 white, in which case is there a greater chance of getting all red gumballs: selecting 2 gumballs or selecting 50 gumballs? Explain why*. On the other hand, students need to experience how the spread of sampling distribution increases when the sample size reduces. Hence, learning opportunities could be offered by varying the sample size of the sampling distribution.

We suggest that if teachers want students to develop statistical reasoning, the learning content should be designed in a specific and systematic way that is guided by VT. In this sense, the teacher must become aware of and educated about the critical aspects, how to open them up as dimensions of variation and invariance, and how to determine the values in those dimensions (Kullberg et al., 2017, p. 567). In this case, "[t]he learner herself, a teacher, other students, a task, or a set of examples could make it possible for the learner to experience differences (as well as similarities) in relation to critical aspects." Variation Theory is about what we call *instructed learning*, not in the sense of starting with the instruction and then checking how it works but designing the instruction on the basis of our understanding of learning. VT has not yet been explicitly used within statistics and probability education research, despite some indicators from research in probability and statistics education of the benefits of VT (for example, see Nilsson, 2007, 2020) mentioned that the design of an instructional activity was based on principles of VT and that design offers the learner certain opportunities to discern, explore, and compare features of the experienced phenomena with corresponding elements previously experienced (p. 300).

The statistics education community has already acknowledged the importance of students working with variability in data. This chapter offers an invitation to also consider the possibilities that VT offers regarding designing instruction in an even more systematic way. Opportunities for learners to interact with patterns of variation and invariance could be provided through noticing, acknowledging, measuring, describing, modeling, explaining, and predicting the variability in data. VT can serve as a pedagogical tool for researchers and teachers when planning and enacting teaching to facilitate the sustainable learning of statistical reasoning, which requires both deterministic and stochastic reasoning.

Acknowledgments The work reported here was supported by the Swedish Research Council (Grant no: 2020-03128). We want to express our gratitude for invaluable comments from Michael Shaughnessy, Lyn English, Mona Holmqvist, Per Nilsson, and Joanne Lobato. We also thank Angelika Kullberg and the VT working seminar at the Department of Pedagogical Curricular and Professional studies at the University of Gothenburg. We also acknowledge Catherine MacHale Gunnarsson for her professional editing.

References

Bargagliotti, A., Franklin, C., Arnold, P., Gould, R., Johnson, S., Perez, L., & Spangler, D. (2020). *Pre-K-12 guidelines for assessment and instruction in statistics education (GAISE) II*. American Statistical Association and National Council of Teachers of Mathematics.

Batanero, C., Godino, J. D., Vallecillos, A., Green, D. R., & Holmes, P. (1994). Errors and difficulties in understanding elementary statistical concepts. *International Journal of Mathematics Education in Science and Technology, 25*(4), 527–547. https://doi.org/10.1080/0020739940250406

Ben-Zvi, D., Makar, K., & Garfield, J. (Eds.). (2018a). *International handbook of research in statistics education*. Springer Cham. https://doi.org/10.1007/978-3-319-66195-7

Ben-Zvi, D., Gravemeijer, K., & Ainley, J. (2018b). Design of statistics learning environments. In D. Ben-Zvi, K. Makar, & J. Garfield (Eds.), *International handbook on research in statistics education* (pp. 473–502). Springer Cham. https://doi.org/10.1007/978-3-319-66195-7_16

Biehler, R. (1989). Educational perspectives on exploratory data analysis. In R. Morris (Ed.), *Studies in mathematics education: The teaching of statistics* (Vol. 7, pp. 185–201). UNESCO.

Carter, J., Brown, M., & Simpson, K. (2017). From the classroom to the workplace: How social science students are learning to do data analysis for real. *Statistics Education Research Journal, 16*(1), 80–101. https://doi.org/10.52041/serj.v16i1.218

Cobb, G. W. (2015). Mere renovation is too little, too late: We need to rethink the undergraduate curriculum from the ground up. *The American Statistician, 69*(4), 266–282. https://doi.org/10.1080/00031305.2015.1093029

Cobb, G. W., & Moore, D. S. (1997). Mathematics, statistics, and teaching. *The American Mathematical Monthly, 104*(9), 801–823. https://doi.org/10.2307/2975286

Dierdorp, A., Bakker, A., Ben-Zvi, D., & Makar, K. (2017). Secondary students' considerations of variability in measurement activities based on authentic practices. *Statistics Education Research Journal, 16*(2), 397–418. https://doi.org/10.52041/serj.v16i2.198

Ekol, G. (2015). Exploring foundation concepts in introductory statistics using dynamic data points. *International Journal of Education in Mathematics, Science and Technology, 3*(3), 230–241. https://doi.org/10.18404/ijemst.15371

English, L. D., & Watson, J. M. (2016). Development of probabilistic understanding in fourth grade. *Journal for Research in Mathematics Education, 47*(1), 28–62. https://doi.org/10.5951/jresematheduc.47.1.0028

Franklin, C., Kader, G., Mewborn, D. S., Moreno, J., Peck, R., Perry, M., & Scheaffer, R. (2007). *Guidelines for assessment and instruction in statistics education (GAISE) report: A pre-K-12 curriculum framework*. American Statistical Association.

Franklin, C., Bargagliotti, A. E., Case, C. A., Kader, G. D., Schaeffer, R. L., & Spangler, D. A. (2015). *The statistical education of teachers*. American Statistical Association.

Fülöp, É. (2019). *Learning to solve problems that you have not learned to solve: Strategies in mathematical problem solving*. Doctoral Theses from University of Gothenburg. http://hdl.handle.net/2077/60464

Garfield, J., & Ben-Zvi, D. (2008). *Developing students' statistical seasoning: Connecting research and teaching practice*. Springer. https://doi.org/10.1007/978-1-4020-8383-9

Glencross, M. J., & Binyavanga, K. W. (1997). The role of technology in statistics education: A view from a developing region. In J. Garfield & G. Burrill (Eds.), *Research on the role of technology in teaching and learning statistics* (pp. 301–308). International Statistical Institute.

Green, D. (1993). Data analysis: What research do we need? In L. Pereira-Mendoza. (Ed.), Introducing data analysis in the schools: Who should teach it and how? *Proceedings of the international statistical institute round table conference, August 10–14, 1992.*

Henriques, A., & Oliverira, H. M. (2016). Students' expressions of uncertainty in making informal inference when engaged in a statistical investigation using Tinkerplots. *Statistics Education Research Journal, 15*(2), 62–80. https://doi.org/10.52041/serj.v15i2.24115(2)

Holmqvist, M., & Lindgren, G. (2009). Students learning English as second language: An applied linguistics learning study. *Problems of Education in the 21st Century, 2009*(18), 86–96.

Holmqvist, M., & Selin, P. (2019). What makes the difference? An empirical comparison of critical aspects identified in phenomenographic and variation theory analyses. *Palgrave Communications, 5*(71), 1–8. https://doi.org/10.1057/s41599-019-0284-z

Holmqvist, M., Gustavsson, L., & Wernberg, A. (2007). Generative learning: Learning beyond the learning situation. *Educational Action Research, 15*(2), 181–208. https://doi.org/10.1080/09650790701314684

Innabi, H. (2007). Factors considered by secondary students when judging the validity of a given statistical generalization. *International electronic journal of mathematics education*. Special Issue. Emerging Research in Statistics Education, *3*(2), 168–186.

Innabi, H., & Emanuelsson, J. (2021). Enrichment in school principals' ways of seeing mathematics. *International Journal of Mathematical Education in Science and Technology, 52*(10), 1508–1539. https://doi.org/10.1080/0020739X.2020.1782496

Kahneman, D. (2011). *Thinking, Fast and Slow*. Farrar, Straus & Giroux.

Konold, C., & Miller, C. D. (2005). *TinkerPlots: Dynamic data explorations*. Key Curriculum Press.

Konold, C., & Pollatsek, A. (2002). Data analysis as the search for signals in noisy processes. *Journal for Research in Mathematics Education, 33*(4), 259–289. https://doi.org/10.5951/jresematheduc.33.4.0259

Kullberg, A., Runesson Kempe, U., & Marton, F. (2017). What is made possible to learn when using the variation theory of learning in teaching mathematics? *ZDM Mathematics Education, 49*, 559–569. https://doi.org/10.1007/s11858-017-0858-4

Lo, M. L. (2012). *Variation theory and the improvement of teaching and learning*. Göteborgs Universitet, Acta Universitatis Gothoburgensis. http://hdl.handle.net/2077/29645.

Lo, M. L., & Marton, F. (2012). Towards a science of the art of teaching: Using variation theory as a guiding principle of pedagogical design. *International Journal for Lesson and Learning Studies, 1*(1), 7–22. https://doi.org/10.1108/20468251211179678

Marton, F. (1981). Phenomenography - describing conceptions of the world around us. *Instructional Science, 10*, 177–200. https://doi.org/10.1007/BF00132516

Marton, F. (1986). Phenomenography – A research approach to investigating different understandings of reality. *J. Thought, 21*, 28–49. http://www.jstor.org/stable/42589189

Marton, F. (2000). The structure of awareness. In J. Bowden & E. Walsh (Eds.), *Phenomenography* (pp. 102–116). RMIT University.
Marton, F. (2006). Sameness and difference in transfer. *Journal of the Learning Sciences, 15*(4), 499–535. https://doi.org/10.1207/s15327809jls1504_3
Marton, F. (2015). *Necessary conditions of learning*. Routledge. https://doi.org/10.4324/9781315816876
Marton, F., & Booth, S. (1997). *Learning and awareness*. Lawrence Erlbaum Associates. https://doi.org/10.4324/9781315816876
Marton, F., & EDB Chinese Language Research Team. (2010). The Chinese learner of tomorrow. In C. K. K. Chan & N. Rao (Eds.), *CERC studies in comparative education, 25: Revisiting the Chinese learner. Changing contexts, changing education* (pp. 133–163). Springer Science+Business Media B.V.
Marton, F., & Pang, M. F. (2006). On some necessary conditions of learning. *The Journal of the Learning Sciences, 15*(2), 193–220. https://doi.org/10.1207/s15327809jls1502_2
Marton, F., & Pang, M. F. (2007). The paradox of pedagogy: The relative contribution of teachers and learners to learning. *Iskolakultura, 1*(1), 1–29.
Marton, F., & Tsui, A. B. M. (2004). *Classroom discourse and the space of learning*. Lawrence Erlbaum. https://doi.org/10.4324/9781410609762
Meletiou-Mavrotheris, M., & Lee, C. (2002). Teaching students the stochastic nature of statistical concepts in an introductory statistics course. *Statistics Education Research Journal, 1*(2), 22–37. https://doi.org/10.52041/serj.v1i2.563
Moore, D. S. (1990). Uncertainty. In L. A. Steen (Ed.), *On the shoulders of giants: New approaches to numeracy* (pp. 95–137). National Academy Press.
Nilsson, P. (2007). Different ways in which students handle chance encounters in the explorative setting of a dice game. *Educational Studies in Mathematics, 66*(3), 293–315. https://doi.org/10.1007/s10649-006-9062-0
Nilsson, P. (2013). Challenges in seeing data as useful evidence in making predictions on the probability of a real-world phenomenon. *Statistics Education Research Journal, 12*(2), 71–83. https://doi.org/10.52041/serj.v12i2.305
Nilsson, P. (2020). Students' informal hypothesis testing in a probability context with concrete random generators. *Statistics Education Research Journal, 19*(3), 53–73. https://doi.org/10.52041/serj.v19i3.56
Noll, J., & Shaughnessy, M. (2012). Aspects of students' reasoning about variation in empirical sampling distributions. *Journal for Research in Mathematics Education, 43*(5), 509–556. https://doi.org/10.5951/jresematheduc.43.5.0509
Pang, M. F. (2010). Boosting financial literacy: Benefits from learning study. *Instructional Science, 38*(6), 659–677. https://doi.org/10.1007/s11251-009-9094-9
Pang, M. F. (2019). Enhancing the generative learning of young people in the domain of financial literacy through learning study. *International Journal for Lesson and Learning Studies, 8*(3), 170–182. https://doi.org/10.1108/IJLLS-09-2018-0065
Paparistodemou, E., & Meletiou-Mavrotheris, M. (2008). Developing young students' informal inference skills in data analysis. *Statistics Education Research Journal, 7*(2), 83–106. https://doi.org/10.52041/serj.v7i2.471
Pfannkuch, M., & Brown, C. M. (1996). Building on and challenging students' intuitions about probability: Can we improve undergraduate learning? *Journal of Statistics Education, 4*(1), 1–22. https://doi.org/10.1080/10691898.1996.11910502
Prodromou, T., & Pratt, D. (2013). Making sense of stochastic variation and causality in a virtual environment. *Technology, knowledge and learning: Learning mathematics, science and the arts in the context of digital technologies, 18*(3), 121–147. https://doi.org/10.1007/s10758-013-9210-4
Reading, C., & Shaughnessy, J. M. (2004). Reasoning about variation. In D. Ben-Zvi & J. Garfield (Eds.), *The challenge of developing statistical literacy, reasoning and thinking* (pp. 201–226). https://doi.org/10.1007/1-4020-2278-6_9.

Ridgway, J. (2015). Implications of the data revolution for statistics education. *International Statistical Review, 84*(3), 528–549. https://doi.org/10.1111/insr.12110

Sánchez, E., García-García, I., & J., & Mercado M. (2018). Determinism and empirical commitment in the probabilistic reasoning of high school students. In C. Batanero & E. Chernoff (Eds.), *Teaching and learning stochastics. ICME-13 monographs*. Springer. https://doi.org/10.1007/978-3-319-72871-1_13

Schrage, G. (1983). (Mis) interpretation of stochastic models. In R. Scholz (Ed.), *Decision making under uncertainty* (pp. 351–361). North-Holland. https://doi.org/10.1016/S0166-4115(08)62207-4

Schulz, L., & Sommerville, J. (2006). God does not play dice: Causal determinism and preschoolers' causal inferences. *Child Development, 77*(2), 427–442. https://doi.org/10.1111/j.1467-8624.2006.00880.x

Schwartz, D. L., & Bransford, D. J. (1998). A time for telling. *Cognition and Instruction, 16*(4), 475–522. https://doi.org/10.1207/s1532690xci1604_4

Schwartz, D., & Martin, T. (2004). Inventing to prepare for future learning: The hidden efficiency of encouraging original student production in statistics instruction. *Cognition and Instruction, 22*(2), 129–184. https://doi.org/10.1207/s1532690xci2202_1

Shaughnessy, J. M. (1977). Misconceptions of probability: An experiment with a small-group, activity-based, model building approach to introductory probability at the college level. *Educational Studies in Mathematics, 8*, 295–316. https://doi.org/10.1007/BF00385927

Shaughnessy, J. M. (1992). Research in probability and statistics: Reflections and directions. In D. A. Grouws (Ed.), *Handbook of research on mathematics teaching and learning: A project of the national council of teachers of mathematics* (pp. 465–494). Macmillan Publishing Co, Inc.

Shaughnessy, J. M. (1997). Missed opportunities in research on the teaching and learning of data and chance. In F. Biddulph & K. Carr (Eds.), *People in mathematics education: Proceedings of the twentieth annual meeting of the mathematics education research group of Australasia* (Vol. 1, pp. 6–22). The University of Waikato Printery.

Shaughnessy, J. M. (2007). Research on statistics learning and reasoning. In F. K. Lester Jr. (Ed.), *Second handbook of research on mathematics teaching and learning* (Vol. 2, pp. 957–1009). Information Age.

Snee, R. D. (1999). Discussion: Development and use of statistical thinking: A new era. *International Statistical Review, 67*(3), 255–258. https://doi.org/10.1111/j.1751-5823.1999.tb00446.x

To, K. K., & Pang, M. F. (2019). A study of variation theory to enhance students' genre awareness and learning of genre features. *International Journal for Lesson and Learning Studies, 8*(3), 183–195. https://doi.org/10.1108/IJLLS-10-2018-0070

Tversky, A., & Kahneman, D. (1973). Availability: A heuristic for judging frequency and probability. *Cognitive Psychology, 5*(2), 207–232. https://doi.org/10.1016/0010-0285(73)90033-9

Tversky, A., & Kahneman, D. (1974). Judgment under uncertainty: Heuristics and biases. *Science, 185*(4157), 1124–1131. https://doi.org/10.1126/science.185.4157.1124

Van Dooren, W., De Bock, D., Depaepe, F., Janssens, D., & Verschaffel, L. (2003). The illusion of linearity: Expanding the evidence towards probabilistic reasoning. *Educational Studies in Mathematics, 53*(2), 113–138. https://doi.org/10.1023/A:1025516816886

Watson, J. (2016). Whither statistics education research? In *Proceedings of the 39th annual conference of the mathematics education research group of Australasia (MERGA), 3–7 July 2016* (pp. 33–58).

Wild, C. J., Pfannkuch, M., Regan, M., & Horton, N. J. (2011). Towards more accessible conceptions of statistical inference. *Journal of the Royal Statistical Society, 174*(2), 247–295. https://doi.org/10.1111/j.1467-985X.2010.00678.x

Wild, C. J., Utts, J. M., & Horton, N. J. (2018). What is statistics? In D. Ben-Zvi, J. Garfield, & K. Makar (Eds.), *The first handbook of research on statistics teaching and learning*. Springer. https://doi.org/10.1007/978-3-319-66195-7_

How Students' Statistics Beliefs Influence Their Attitudes: A Quantitative and a Qualitative Approach

Florian Berens, Kelly Findley, and Sebastian Hobert

Abstract Negative attitudes of students toward statistics in introductory courses are a widespread phenomenon. The reasons lie first and foremost in students' experiences with mathematics, but they are also more diverse than just that. In addition to known reasons, this paper proposes that one's organizing beliefs about the nature and applicability of statistics may explain one's attitudes toward statistics. This study builds on our previous work, from which we developed a statistics beliefs instrument that reflected four broad statistical conceptions found in our data and corroborated in the statistics education literature. For this next stage of work, we surveyed 471 students and interviewed 14 students studying the social sciences in a German university. Quantitative findings show slight correlations between students' beliefs toward statistics and their attitudes toward statistics. In particular, static, rules-based beliefs about statistics have negative impact on the students' attitudes, while more open, investigative beliefs on statistics have a positive impact on all dimensions of attitudes. Qualitative insights show that beliefs work primarily by emphasizing certain parts of students' attitudes and prior experiences, thus shifting their weighting in the composition of the overall attitude.

Keywords Mathematics attitudes · Mixed methods · Attitudes towards statistics · Statistics beliefs · Statistics conceptions

F. Berens (✉) · S. Hobert
University of Goettingen, Goettingen, Germany
e-mail: florian.berens@uni-goettingen.de; sebastian.hobert@uni-goettingen.de

K. Findley
University of Illinois – Urbana-Champaign, Champaign, IL, USA
e-mail: kfindley@illinois.edu

G. F. Burrill et al. (eds.), *Research on Reasoning with Data and Statistical Thinking: International Perspectives*, Advances in Mathematics Education,
https://doi.org/10.1007/978-3-031-29459-4_22

1 Introduction

Students' views about and interest in a discipline can greatly affect their own learning experiences and knowledge acquisition (Van Griethuijsen et al., 2015; Roesken et al., 2011). This relationship has been frequently highlighted in statistics education research. Not only is there a close connection between students' attitudes and their performance in statistics courses (Emmioglu & Capa-Aydin, 2012; Ramirez et al., 2012; Finney & Schraw, 2003), but it is well established that positive attitudes should be a goal of its own for statistics education (Gal et al., 1997).

Statistics education researchers have argued that students' negative attitudes toward statistics may be the result of certain limiting beliefs about statistics (Zieffler et al., 2008; Finney & Schraw, 2003). However, the use of the term *beliefs* in the existing literature varies, reflecting the multi-faceted construct of beliefs (McLeod, 1992). Much of the existing literature that links students' learning of statistics to beliefs (including the research cited previously) focuses on "self-efficacy beliefs"—views that learners have about themselves and their abilities (Finney & Schraw, 2003; Gal et al., 1997). But students may also have disciplinary beliefs that reflect how they see usefulness, structure, and purpose in disciplinary work (Justice et al., 2020; Wild et al., 2018).

Disciplinary beliefs play a critical role in students' motivation for learning (Kloosterman, 2002), how students view disciplinary advancement and knowledge construction (Tsai et al., 2011; Muis, 2004), and how students think about problem solving (Ozturk & Guven, 2016; Bell & Linn, 2002). Even with the same curriculum and the same teacher, students see tasks quite differently depending on the disciplinary beliefs and conceptions they bring to these contexts (Presmeg, 2002).

Unpacking and understanding students' disciplinary beliefs and conceptions has gained much attention in recent years as the field grapples with the growth of data science and the merging of several areas of study. In this chapter, we want to look more carefully at students' disciplinary beliefs about statistics. In addition, we will be relating their beliefs to their attitudes of statistics to see whether disciplinary beliefs may explain negative attitudes among new learners at the college level.

2 Background

2.1 Attitudes About Statistics

One's attitude toward a discipline refers to their disposition toward that discipline and directly stems from their experiences and perceptions of that discipline (Fishbein & Ajzen, 2009). Attitudes are more stable than emotions, which better reflect the in-the-moment responses that students may have in the learning process; they are also more dispositional than beliefs, which better describe students' perceptions of their own abilities, in addition to the cognitive propositions they have constructed (Philipp, 2007; Gal et al., 1997; McLeod, 1992).

Ramirez et al. (2012) reviewed a set of statistics attitudes instruments and identified six attitudinal subcategories that were reflected in the item constructions of these surveys. They described the connection between these subcategories in comprising one's statistics attitudes as follows:

Before devoting the time and energy (Effort) to learn and do statistics, our model indicates that students evaluate their skills (Cognitive Competence) and the Difficulty of statistics and statistics tasks. They choose to expend Effort on statistics tasks and courses that they like (Affect) and are interested in doing (Interest) while they avoid others. They also consider how useful statistics is and will be in their lives (Value). (Ramirez et al., 2012, p. 61).

Ramirez et al. identified the Survey on Attitudes toward Statistics (SATS-36) as best representing these core components of statistics attitudes (Schau, 2003; Schau et al., 1995).

While other authors (e.g., Tempelaar et al., 2007) find negative attitudes among their students, it is not easy to identify the reasons for that. One explanation might be students' previous experiences with mathematics and the role they perceive that mathematics may play in their statistics coursework (Dempster & McCorry, 2009; Nasser, 2004). Ramirez et al. (2012) also posit that other student characteristics, such as gender or race, as well as past experiences with statistics. These factors interplay with students' attitudes and in some ways predict their attitudinal responses. The relationship between these components can be mapped out, as shown in Fig. 1.

In addition to these factors, however, there may be value to understanding how students' previous experiences and background may contribute to their attitudes about statistics. In particular, we ask how their beliefs about the discipline of statistics might explain certain attitudes. We take a look at this construct in more detail next.

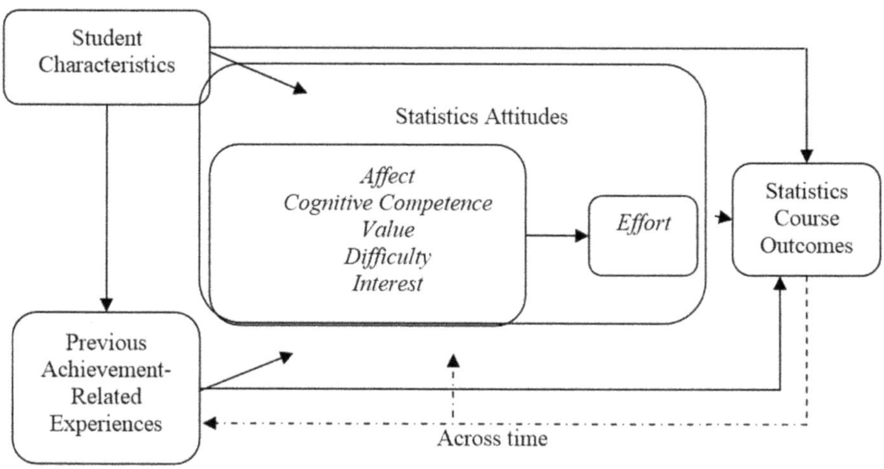

Fig. 1 Students' attitudes toward statistics model. (Ramirez et al., 2012)

2.2 Beliefs About Statistics

Beliefs is arguably a messier construct to define, in part because beliefs can vary widely in their construction and status. On one hand, a belief can be dispositional and deeply embedded in one's affects or desires (e.g., "I believe I can complete this task," "I believe everything happens for a reason"); on the other hand, a belief may be thought of as a conviction held with a varying level of certainty, typically through some type of induction (e.g., "I believe it might rain based on the weather I see") (Philipp, 2007; Southerland et al., 2001; McLeod, 1992). Thus, the usefulness of this construct in research depends on a careful description by the authors of what it means in the context of their study. In our work, we are not specifically concerned with the affective dimensions of a belief, but rather beliefs as representing ideas and constructions. In our case, we are narrowing in on students' beliefs of the discipline of statistics, including how they think about its structure, its application, and the work of experts in the discipline.

Philipp's (2007) use of the term conception may also be a helpful idea to attach to this particular focus on beliefs. According to Philipp, a conception is "a general notion or mental structure encompassing beliefs, meanings, concepts, propositions, rules, mental images, and preferences" (p. 259). Thus, students may have broad disciplinary conceptions that encompass their disciplinary beliefs and organize them into a framework for interacting with the discipline.

Several previous studies have directly investigated students' conceptions of statistics and offered unique framings (Justice et al., 2020; Rolka & Bulmer, 2005; Gordon, 2004; Reid & Petocz, 2002). We found much alignment across the findings in these studies, including a clear novice conception, an expert conception, and two different intermediate conceptions. One of these intermediate conceptions was more theory and method centered, while the other was more data and interpretation centered. These category descriptions are highlighted in Table 1, along with the conception dimensions they align with from each study.

Despite different data collection approaches, there are striking similarities in the conception categories described by all three of these papers. Novice conceptions recognize nothing particularly unique or clearly unifying about statistics (e.g., statistics as disconnected procedures). Intermediate conceptions see growing cohesion and connection to contextual application, either in the direction of its analytical affordances, or in the context of making sense of data more contextually. Expect conceptions add a clear emphasis on meaning-making, with statistics seen not simply as a tool, but also as a way to see and interact with the world differently.

We highlight these conceptions because we see the beliefs we are measuring in our study as indicative of broader conceptions that students have about statistics. By connecting certain belief patterns toward these higher-level conceptions, we hope to more clearly map the potential connection between beliefs and attitudes.

Table 1 Alignment between previous research and statistical conception categories

Conception categories	Research alignment
Novice Conceptions	Gordon Process Mastery Justice et al. "Paint-by-numbers" Reid and Petocz Statistics is individual numeric activities Statistics is using individual techniques Statistics is a collection of techniques Rolka and Bulmer Statistics as a collection of tools and procedures
Intermediate Conceptions (theory centered)	Gordon Tool Justice et al. Step-by-step Painting Class Reid and Petocz Statistics is the analysis and interpretation of data Rolka and Bulmer Statistics as tools with a context
Intermediate Conceptions (data centered)	Justice et al. Realist Perspective Reid and Petocz Statistics is a way of understanding real-life using different statistical models Rolka and Bulmer Statistics as a means of understanding a complex world or as part of a unified picture with understanding data
Expect Conceptions	Gordon Critical Thinking Justice et al. Picasso Reid and Petocz Statistics is an inclusive tool used to make sense of the world and develop personal meanings Rolka and Bulmer Statistics as integrated into the world itself

2.3 *Research Question*

This paper examines whether students' disciplinary beliefs play an important role in the development of statistical attitudes. It is therefore assumed that students have different ideas about the nature of statistics and different personal definitions of the term statistics. These ideas about what statistics means would sensibly play a role in the frame of reference over which negative attitudes are formed. Thus, we wish to answer this question: Is there an association between students' disciplinary beliefs as they relate to statistics and their attitudes toward statistics?

The approach of this paper is to use students' beliefs as explanatory variables. We hypothesize that the more static perspectives characterizing novice conceptions have a negative influence on attitudes, since this view grants students' little autonomy. Additionally, we expect conceptions that focus on statistics as a meaning-making activity to be characterized by higher student autonomy, thus being associated with more positive attitudes.

3 Methods

The data for this study were gathered in an introductory course on statistics for social scientists at the University of Goettingen. Participants come from nine different disciplines in the area of social science, most frequently political science or sociology. Prior to their statistics course students already took a course on qualitative and quantitative empirical methods in social sciences of 14 weeks with eight hours per week.

Data collection took place in connection to the enrolment of the course. Out of 707 students enrolled in the course 471 took part in the survey and completed the items to an extent that the survey could be evaluated. About 50% of the participating students report to be in their second semester of university studies, with an average over all of 4.1 semesters. 61% of the respondents report being female. These characteristics are typical for this course.

For the qualitative part of the study, a random sample of 90 of these students was invited. Of these, 14 students were willing to participate and got interviewed through semi-structured interviews. As part of a bigger interview the following questions were asked (in German) in a standardized manner to answer this paper's question:

1. What do you generally think about statistics?
2. In your studies, you are required to take statistics. What do you think about that?
3. Do you think statistics is useful? Why? Why not?
4. Do you think statistics is difficult compared to other courses you take? Why? Why not?
5. What do you think, how good will you be at statistics?
6. What would you say is statistics? How would you describe its nature? Or could you define it?
7. If you compare statistics to other courses you take, what is special about statistics?
8. When someone does statistical work, what do they do?
9. How do you think you can recognize a statistical expert? What distinguishes him or her as a person and what competencies does he or she have?

In addition, the interviewers asked follow-up questions about what was said.

The interviews were transcribed and pseudonymized. The transcripts were initially analyzed separately by the first author and by an independent person. In the analysis, in-vivo codes were first generated from the interviews related to students' attitudes, justifications for those attitudes, and/or students' disciplinary beliefs. These codes were then inductively grouped into themes. Themes and their occurrences were then coded by consensus in order to collectively conduct further analysis. Regarding students' disciplinary beliefs, consensus assigned one of the four conceptions to be the main conception. For some of the respondents, a second conception was assigned as a minor conception if codes of a second conception appeared, which however was not coded as often as the main conception.

4 Quantitative Instrumentation

As introduced earlier, the SATS 36 instrument, developed by Schau (2003) and widely established in statistics education, was used to survey student attitudes. In the SATS-36 instrument, Schau proposes a six-dimensional structure of statistics attitudes. The dimensions are presented as Affect (I will like statistics.), Cognitive Competence (I will find it difficult to understand statistical concepts.), Value (Statistics is irrelevant in my life.), Difficulty (Statistics is a complicated subject.), Interest (I am interested in learning statistics.), and Effort (I plan to work hard in my statistics course.).

For the theoretical model on statistics beliefs described above, a questionnaire was developed by the first two authors in order to identify different disciplinary beliefs students resonated with. Since this beliefs instrument is not yet published, we first provide some background about the structure of this instrument.

In addition to existing research precluding this project, we have carried out various qualitative preliminary work in our attempts to create an instrument that captures students' disciplinary beliefs about statistics. To create this instrument, the first author collected data through focus group interviews with German students of the social sciences. Data from these interviews were used to describe various student conceptions of statistics. Further work with this concept was carried out with students of mathematics in the USA, so that the authors have developed a theoretical framework that represents students' belief in four categories. Data from both the focus groups and individual student interviews showed high alignment to the conception categories found in previous literature (Findley & Berens, 2020).

The four categories can be seen as a 2x2 matrix in which two data-driven (left column) and two theory-driven (right column) conceptions can be found, as well as two more application-oriented (top row) and two more pure (bottom row) conceptions of statistics (Table 2). The first is a rules-based conception, which understands statistics as a collection of rules and procedures that, when applied correctly, determine solutions to statistical problems. In this conception, however, application is understood only in the sense of computations or algorithmic statistics, not with respect to real-world problems. Data are not necessary either. Secondly, the confirmatory conception describes a picture of statistics in which theories and models can be verified or falsified by comparison with data. Reference to real-world problems arises through the tested theories or models, which can come from a wide variety of

Table 2 Disciplinary conceptions of beliefs instrument

The investigative conception Exploring data with the goal of generating questions and gathering insights	**The confirmation conception** Testing theories and claims using formal methodology in order to come to a conclusion
The descriptive conception Reporting summaries and representations of data in order to share information clearly	**The rules-based conception** Following steps and procedures in order to find correct answers

application fields. They are the starting point for statistics, which then compares data to evaluate the plausibility of the theories or models. The third, descriptive conception views statistics as a large set of tools that can be used to represent data to readers. As in the rules-based conception, statistics is seen here as a toolbox. Likewise, the distance of statistics to real-world problems is a common characteristic. A difference, however, lies in the starting point, which the descriptive conception sees in data. These data can be summarized and depicted objectively by statistics in order to reflect reality dispassionately. The fourth, investigative conception understands statistics as a process of data exploration in which an attempt is made to generate information for decision-making from data by means of a circular procedure. Here, data are again the starting point, but they do not only come from the real world, but it is the task of statistics to investigate this real world and to gain new, valuable insights about it.

For this purpose, nine items were developed for each of the four conceptions on statistics, which were based on the qualitative preliminary work. Three out of nine items were related to the nature of statistics in each conception, three related to the process of doing statistical work, and three described the characteristics of statistical experts and the skills that are most important. The first of the survey's items were tested using principal component analysis and confirmatory factor analysis. All analyses reveal a need for further work on the survey design. Nevertheless, the analyses indicate that the basics of the survey's ideas are reflected in the results. There-fore, the resulting data set can be used for further analyses, even if a revised version of the survey is used in the future.

5 Results

If we first look at the students' ratings broadly, we see that students provide somewhat high ratings for all four conceptions. Nevertheless, students are slightly more inclined towards a descriptive conception (mean of 5.06 out of 7 on a Likert scale), and less inclined towards an investigative conception (4.45). All in all, however, all four conceptions of statistics are relatively close (rules-based 4.89, confirmation 4.77). We also see that the standard deviations are relatively similar, even though here again the investigative conception has the lowest standard deviation. Nevertheless, the standard deviations and different quantiles indicate heterogeneity, which can be used to explain different attitudes.

Table 3 SATS-36 explained by students´ conceptions in 6 linear regressions (one regression per column, p-values in brackets, all attitudes are coded in a way that high values represent positive attitudes)

	Affect	Competence	Value	Difficulty	Interest	Effort
Rules-based	−0.488 (0.000)	−0.410 (0.000)	−0.130 (0.094)	−0.306 (0.000)	−0.160 (0.116)	0.190 (0.042)
Confirmation	0.093 (0.344)	0.125 (0.197)	−0.012 (0.884)	0.082 (0.194)	−0.074 (0.506)	−0.110 (0.266)
Descriptive	0.039 (0.665)	0.070 (0.431)	0.161 (0.039)	−0.044 (0.446)	0.308 (0.003)	0.322 (0.001)
Investigative	0.232 (0.006)	0.187 (0.024)	0.354 (0.000)	0.089 (0.095)	0.618 (0.000)	0.178 (0.039)
Intercept	3.663	4.274	2.775	3.841	1.352	2.150
R^2	0.094	0.066	0.087	0.093	0.145	0.089

5.1 *Quantitative Influence of Beliefs on Attitudes*

To identify the influence of students´ conceptions on their attitudes, six multiple linear regressions are performed. The six dimensions of attitudes about statistics proposed by Schau act as the dependent variables. The four conceptions we identified to organize students' patterns in beliefs are the dependent variable to explain the attitudes. The results of regressions can be found in Table 3.

First and foremost, the effect of the investigative conception on attitudes is particularly remarkable. In all cases, it has a positive effect on students' attitudes, sometimes with effects that can be rated as very large and usually highly significant. So, the investigative conception appears to have a particularly strong and positive effect on students' attitudes.

In contrast, the rules-based conception has a negative effect on students' attitudes. Some of the effects are again quite high. Only the intended effort is positively influenced by this conception. Holding a rules-based conception, therefore, seems to be a big burden for a student decreasing the students´ affect and perceived competence while increasing the perceived difficulty and planned effort in learning.

Regarding the descriptive conception of statistics results give a more mixed picture. The descriptive conception has a positive effect on the personal interest in statistics and the perceived value of statistics. Both are related concepts focusing the usefulness of the application of statistics. The likewise positive effect of a descriptive conception on the planned effort in learning fits well with these findings. However, the pair of perceived difficulty of statistics and the perceived own competence is not related with a descriptive conception. Also, the general affect toward statistics can neither benefit nor is damaged by a descriptive conception.

For the confirmation conception no effects on students' attitudes can be found.

Looking at the R^2 of the six models one can see that the students´ interest in statistics can best be predicted by the students´ conceptions. The other five R^2 are on the same level. It should be noted, however, that all R^2 are not particularly high, indeed quite low. The first thing to say about this is that disciplinary conceptions were never intended to be the strongest predictor of attitudes. They are only one factor in a multicausal model. Even with this in mind, however, the explanatory power is not very high. Looking into qualitative data will therefore provide an explanation of how beliefs influence attitudes and why this influence cannot be measured as high quantitatively here.

5.2 *Insights from the Qualitative Interviews*

Among the 14 interviewees three were classified as mainly holding a rules-based conception of statistics with two of them holding descriptive conceptions as a minor conception. Another three students mainly held a confirmation conception with all three other conceptions appearing as a minor once. Two students mainly expressed an investigative conception of statistics, one with a descriptive conception as a minor. The biggest group of interviewees were six students mainly holding a descriptive conception of statistics. Three of these also held rules-based conceptions as a minor, one holds investigative conceptions as a minor. Overall, then, it appears that the application-oriented beliefs of confirmation and investigative conceptions are relatively rare. Only two of the students only held application-oriented conceptions, another four combine them with rules-based or descriptive conceptions. The attitudes associated with these conceptions are shown in Table 4.

Regarding the reasons for the respondents' attitudes, many different arguments were coded. In particular, the interviewees talked about their previous experiences with mathematics, about the benefits of statistics for further studies, science and personal career perspectives, and about the societal role of statistics. Not every interviewee expressed a direct connection between his or her conception of statistics and these reasons, but for nine interviewees such connections could be coded. Daniel, Nick, Sophie, Ben, and Sam's attitude toward statistics argued directly from the assumption that statistics is mainly math and aims to calculate formulas and algorithms.

Daniel for example started his answer to the first question in the list of questions above saying "Well, I have to be honest, I am not a fan of statistics. Simply because I don't like math very much and statistics is really math for the most part…". Later in the same answer, he said: "that's statistics, that's what you have to expect, it is formulas". Similarly, for question 6, his answer starts "For me, it's really just math, so that's why it's/. I must also say that I really don't know whether statistics is something for me in this case, because it' s just formulas".

Among students who put mathematics and a strong focus on formulas and their calculation in the center of the reasoning, all five held a major or minor rules-based conception. Only two of the remaining nine also held a rules-based conception. This connection, however, does not have to imply negative attitudes. Sophie and Ben reported that they had always liked mathematics and therefore look positively on statistics and formulas. It is striking, however, that interviewees with other

Table 4 Attitudes and conceptions of the 14 interviewees

Pseudonym	Main conception	Minor conception	Attitude
Daniel	Rules-based		Negative
Emma	Rules-based	Descriptive	Rather negative
Nick	Rules-based	Descriptive	Rather positive
Alissa	Descriptive		Neutral
Marc	Descriptive		Rather positive
Chen	Descriptive	Rules-based	Neutral
Sophie	Descriptive	Rules-based	Positive
Ben	Descriptive	Rules-based	Positive
Mia	Descriptive	Investigative	Neutral
Sam	Confirmation	Rules-based	Negative
Selma	Confirmation	Descriptive	Rather positive
Alex	Investigative	Descriptive	Rather positive
Simon	Confirmation	Investigative	Rather positive
Hannah	Investigative		Very positive

conceptions do not draw this direct connection, almost identity, between mathematics and statistics. This means that the influence of previous experiences in mathematics on attitudes towards statistics was also smaller for them. Hannah, for example, reported very negative previous experiences in mathematics, but did not let these flow into her attitudes towards statistics at all. Thus, the influence of a rules-based conception of statistics did not seem to be a direct positive or negative influence on attitudes towards statistics here. Rather, the conception moderates which weighting certain other reasons, such as prior experiences with mathematics, get in the overall evaluation of one's attitudes.

In the cases of Ben, Sophie and Marc, their interviews brought an element of a descriptive conception in direct connection with attitudes towards statistics. In particular, the attitudinal dimension of the value of statistics is addressed by describing statistics as an important tool to describe society and its change. Thereby, statistics is attributed a societal significance, which it unfolds through its use in societal debates.

Marc for example answered the first question saying, "I actually think that's quite important. Especially in the social science field. I think through statistics you can actually reflect how society is developing and also give a certain picture that is true to reality, so to speak, of how people think about certain topics." In his answer to question seven, he added: "[in statistics] you try to give a depiction, but I think that has a certain credibility compared to other things, because you can actually create statistics with it. I think people are also more convinced when you say that this percentage of the population has done this and that…". Once again, it is not the descriptive conception itself that leads to more positive or negative attitudes here, but the descriptive conception focuses on the factual description of (social) reality, which in turn is considered valuable.

Selma, Simon and Hannah also had a focus on the value of statistics in their interviews. They emphasized not so much the importance of statistics for social debates, but rather the use of statistics in science. They described statistics as valuable because it can be used to generate advances in knowledge in research. Simon reflected this idea well: "[It is] important in any case because it reflects data that is then analyzed that disproves or proves certain statements and you need this toolbox of statistics to understand that properly, which I would not have thought before, but that makes absolute sense to learn and grasp statistics" and added "if you have learned statistics, then you also understand whether it is correct and how one has researched. And that really makes sense. I think that I can use this well in the future" answering the third question. A look at the conceptions shows that Selma, Simon and Hannah were three of the five people with more application-oriented conceptions. The conceptions of confirmation or investigation put the use of statistics for research in the foreground and thus justified a high value of statistics. Once again, the conception of statistics thus acted as a moderator in the choice of the importance of other influences on attitudes toward statistics.

6 Conclusion

In summary, the quantitative results show some negative effects of the rules-based conception, as well as slightly positive effects of the descriptive conception and very positive effects of the investigative conception. Some of the effects are certainly not small, but they explain only a small part of the heterogeneity of students' attitudes toward statistics. The qualitative results provide an explanation for this. In the students' arguments, the conceptions of statistics do not appear as direct justifications. Rather, the conceptions moderate which other arguments play a prioritized role in the overall composition of attitudes. Thus, learners' conceptions of statistics play an important role in explaining learners' negative (and positive) attitudes. However, this occurs only slightly directly, but rather as a moderating factor of other variables.

The results of this study should be treated with caution since the results come from only one German university. Thus, it can be concluded from the existing investigation that it is valuable to further investigate the question of the influence of the beliefs on attitudes, especially if analyzed together with other relevant factors as for example attitudes toward mathematics or attitudes toward research. Therefore, further investigation with other student groups is needed. Given the great importance of negative attitudes toward statistics, this further work seems meaningful and promising.

References

Bell, P., & Linn, M. C. (2002). Beliefs about science: How does science instruction contribute? In B. K. Hofer & P. R. Pintrich (Eds.), *Personal epistemology: The psychology of beliefs about knowledge and knowing* (pp. 321–346). Lawrence Erlbaum Associates Publishers.

Dempster, M., & McCorry, N. K. (2009). The role of previous experience and attitudes toward statistics in statistics assessment outcomes among undergraduate psychology students. *Journal of Statistics Education, 17*(2). https://doi.org/10.1080/10691898.2009.11889515

Emmioglu, E., & Capa-Aydin, Y. (2012). Attitudes and achievement in statistics: A meta-analysis study. *Statistics Education Research Journal, 11*(2), 95–102. https://doi.org/10.52041/serj.v11i2.332

Findley, K., & Berens, F. (2020). Assessing the disciplinary perspectives of introductory statistics students. In S. S. Karunakaran, Z. Reed, & A. Higgins (Eds.), *Proceedings of the 23rd annual conference on research in undergraduate mathematics education* (pp. 1099–1104). https://www.researchgate.net/publication/339712352_Assessing_the_Disciplinary_Perspectives_of_Introductory_Statistics_Students

Finney, S. J., & Schraw, G. (2003). Self-efficacy beliefs in college statistics courses. *Contemporary Educational Psychology, 28*(2), 161–186. https://doi.org/10.1016/S0361-476X(02)00015-2

Fishbein, M., & Ajzen, I. (2009). *Predicting and changing behavior: The reasoned action approach*. Psychology Press. https://doi.org/10.4324/9780203838020

Gal, I., Ginsburg, L., & Schau, C. (1997). Monitoring attitudes and beliefs in statistics education. In I. Gal, J. Garfield, & Y. Gal (Eds.), *The assessment challenge in statistics education* (pp. 37–51). IOS Press. https://www.stat.auckland.ac.nz/~iase/publications/assessbkref

Gordon, S. (2004). Understanding students′ experiences of statistics in a service course. *Statistics Education Research Journal, 3*(1), 40–59. https://doi.org/10.52041/serj.v3i1

Justice, N., Morris, S., Henry, V., & Fry, E. B. (2020). Paint-by-number of Picasso? A grounded theory phenomenographical study of students' conceptions of statistics. *Statistics Education Research Journal, 19*(2), 76–102. https://doi.org/10.52041/serj.v19i2.111

Kloosterman, P. (2002). Beliefs about mathematics and mathematics learning in the secondary school: Measurement and implications for motivation. In G. Leder, E. Pehkonen, & G. Törner (Eds.), *Beliefs: A hidden variable in mathematics education?* (pp. 247–269). Springer. https://link.springer.com/chapter/10.1007/0-306-47958-3_15

McLeod, D. B. (1992). Research on affect in mathematics education: A reconceptualization. *Handbook of research on mathematics teaching and learning, 1*, 575–596.

Muis, K. R. (2004). Personal epistemology and mathematics: A critical review and synthesis of research. *Review of Educational Research, 74*(3), 317–377. https://doi.org/10.3102/00346543074003317

Nasser, F. (2004). Structural model of the effects of cognitive and affective factors on the achievement of Arabic-speaking pre-service teachers in introductory statistics. *Journal of Statistics Education, 12*(1). https://doi.org/10.1080/10691898.2004.11910717

Ozturk, T., & Guven, B. (2016). Evaluating students' beliefs in problem solving process: A case study. *Eurasia Journal of Mathematics, Science and Technology Education, 12*(3), 411–429. https://doi.org/10.12973/eurasia.2016.1208a

Philipp, R. A. (2007). Mathematics teachers' beliefs and affect. In F. K. Lester (Ed.), *Second handbook of research on mathematics teaching and learning* (pp. 257–315). National Council of Teachers of Mathematics.

Presmeg, N. (2002). Beliefs about the nature of mathematics in the bridging of everyday and school mathematical practices. In G. Leder, E. Pehkonen, & G. Törner (Eds.), *Beliefs: A hidden variable in mathematics education?* (pp. 293–312). Springer. https://link.springer.com/chapter/10.1007/0-306-47958-3_15

Ramirez, C., Schau, C., & Emmioglu, E. (2012). The importance of attitudes in statistics education. *Statistics Education Research Journal, 11*(2), 57–71. https://doi.org/10.52041/serj.v11i2.329

Reid, A., & Petocz, P. (2002). Students' conceptions of statistics: A phenomenographic study. *Journal of Statistics Education, 10*(2). https://doi.org/10.1080/10691898.2002.11910662

Roesken, B., Hannula, M. S., & Pehkonen, E. (2011). Dimensions of students' views of themselves as learners of mathematics. *ZDM, 43*(4), 497–506. https://doi.org/10.1007/s11858-011-0315-8

Rolka, K., & Bulmer, M. (2005). Picturing student beliefs in statistics. *ZDM, 37*(5), 412–417. https://doi.org/10.1007/s11858-005-0030-4

Schau, C. (2003). Students' attitudes: The "other" important outcome in statistics education. In *Proceedings of the Joint Statistical Meetings, San Francisco* (pp. 3673–3681). http://statlit.org/pdf/2003SchauASA.pdf

Schau, C., Stevens, J., Dauphinee, T. L., & Vecchio, A. D. (1995). The development and validation of the survey of attitudes toward statistics. *Educational and Psychological Measurement, 55*(5), 868–875. https://doi.org/10.1177/0013164495055005022

Southerland, S. A., Sinatra, G. M., & Matthews, M. R. (2001). Belief, knowledge, and science education. *Educational Psychology Review, 13*(4), 325–351. https://doi.org/10.1023/A:1011913813847

Tempelaar, D. T., Schim van der Loeff, S., & Gijselaers, W. H. (2007). A structural equation model analyzing the relationship of students attitudes toward statistics, prior reasoning abilities and course performance. *Statistics Education Research Journal, 6*(2), 78–102. https://doi.org/10.52041/serj.v6i2.486

Tsai, C. C., Ho, H. N. J., Liang, J. C., & Lin, H. M. (2011). Scientific epistemic beliefs, conceptions of learning science and self-efficacy of learning science among high school students. *Learning and Instruction, 21*(6), 757–769. https://doi.org/10.1016/j.learninstruc.2011.05.002

Van Griethuijsen, R. A., van Eijck, M. W., Haste, H., Den Brok, P. J., Skinner, N. C., Mansour, N., et al. (2015). Global patterns in students' views of science and interest in science. *Research in Science Education, 45*(4), 581–603. https://doi.org/10.1007/s11165-014-9438-6

Wild, C., Utts, J. M., & Horton, N. J. (2018). What is statistics? In D. Ben-Zvi, K. Makar, & J. B. Garfield (Eds.), *International handbook of research in statistics education* (pp. 5–36). Springer. https://doi.org/10.1007/978-3-319-66195-7_1

Zieffler, A., Garfield, J., Alt, S., Dupuis, D., Holleque, K., & Chang, B. (2008). What does research suggest about the teaching and learning of introductory statistics at the college level? A review of the literature. *Journal of Statistics Education, 16*(2). https://doi.org/10.1080/10691898.2008.11889566

Algebraization Levels of Activities Linked to Statistical Tables in Spanish Secondary Textbooks

Jocelyn Pallauta, María Gea, Carmen Batanero, and Pedro Arteaga

Abstract In this paper we analyze the algebraization levels involved in the mathematical activity linked to statistical tables in a sample of 18 Spanish secondary school textbooks (12–15-year-old students) from three different publishers. We performed a content analysis based on the classification of the statistical tables suggested by Lahanier-Reuter and on the levels of algebraic reasoning described by Godino and his collaborators. The results show the increase of algebraization levels required to apply both statistical and arithmetic knowledge, as well as algebraic reasoning with school level. The distribution of the types of statistical tables in the textbooks selected shows differences between the selected publishers as the school year progresses, which may go unnoticed by the teacher.

Keywords Statistical tables · Textbooks · Algebraization levels

J. Pallauta
Departamento de Ciencias Exactas, Universidad de los Lagos, Campus Chuyac, Osorno, Chile
e-mail: jocelyn.diaz@ulagos.cl

M. Gea (✉) · C. Batanero
Facultad de Ciencias de la Educación. Departamento de Didáctica de la Matemática. Despacho 321, Campus Universitario de Cartuja, Granada, Spain
e-mail: mmgea@ugr.es; batanero@ugr.es

P. Arteaga
Facultad de Ciencias de la Educación. Departamento de Didáctica de la Matemática. Despacho 363.3, Campus Universitario de Cartuja, Granada, Spain
e-mail: parteaga@ugr.es

G. F. Burrill et al. (eds.), *Research on Reasoning with Data and Statistical Thinking: International Perspectives*, Advances in Mathematics Education, https://doi.org/10.1007/978-3-031-29459-4_23

1 Introduction

Statistical tables are today widely used to summarize and communicate information in the media and in the professional and scientific work (Estrella et al., 2017; Estrella, 2014). Teaching students to build and interpret their own tables will serve them to better generalize and understand other concepts (Burgess, 2002). Feinberg and Wainer (2011) indicate, however, that the learning of these tables is not simple, since interpreting the variables and their categories, as well as the information represented, is involved. It is also possible to raise a variety of tasks related to these tables, such as representing information, reading and interpreting the table, or translating between different representations (Koschat, 2005).

Consequently, the ability to read, to interpret and to build statistical tables is a component of statistical literacy, that all citizens need to successfully face the information society (Gal, 2002; Sharma, 2017; Watson, 2013), which is becoming increasingly relevant, given the amount and variety of information in the media and on the Internet (Engel et al., 2021). In order to fulfill this need, statistical tables are included in the curricular guidelines of several countries (e.g., CCSSI, 2010; NCTM, 2014). In particular, Spain, explicitly incorporates the study of statistical tables from the beginning of primary education (MECD, 2014), where the variety of types and different uses of statistical tables are accompanied by a series of cognitive demands involved in their understanding.

More specifically, the Spanish curriculum for primary education (6 to 11 years-old) propose working with statistical tables throughout all this period to record and classify qualitative and quantitative data, as well as to build absolute and relative frequencies tables (MECD, 2014). In the first and second grades of Secondary Education (MECD, 2015), students are asked to organize data obtained from a population of qualitative or quantitative variables into tables, compute their absolute and relative frequencies, and to represent them graphically, as well as to carry out the opposite process of translating graphs into tables. This work continues in third grade, where cumulative frequencies are introduced.

Research on tables is scarce and some authors argue that the apparent facility of tables is a fallacy (Martí, 2009), because despite that students use tabular representations in different mathematical topics, or in other subjects, the variety of tables implies that each table should be read in a specific way, which is an obstacle for students (Duval, 2003). Moreover, previous research has mainly examined students' competence to build or to read statistical tables (e.g., Díaz-Levicoy et al., 2020; Gabucio et al., 2010; Pallauta et al., 2021a) with little attention paid to the way in which statistical tables are presented in the textbooks.

In this paper we focus on the textbook, which is an important resource for teaching and learning mathematics in the classroom (Alkhateeb, 2019) and receive increasing attention from the research community (Fan et al., 2013). From the official curricular guidelines until the teaching implemented in the classroom, an important step is the written curriculum reflected in textbooks (Herbel-Eisenmann, 2007). According to Weiland (2019), in the case of statistics, the textbook can have a notable influence on the curriculum implemented in the classroom, mainly because teachers have less knowledge of statistics than of other areas of mathematics.

Our aim is analyzing the algebraic activity required in the study of statistical tables in a sample of Spanish secondary textbooks. The study of algebraization levels in the study of probability have been addressed by Burgos et al. (2022). The results of the study will help teacher to be aware of the progressive degree of algebraic activity needed to study and work with different statistical tables and therefore to better plan the introduction of activities with statistical tables, by taking into account the algebraic and statistical previous knowledge of their students. In the next sections, we expose the research background, method and results.

2 Background

2.1 Types of Statistical Tables

We used the classification by Lahanier-Reuter (2003), who describes the following types of tables, considering the specific functions of each table that give them a different meaning:

- *Data table.* They are used to record or represent the values of one or several variables for each element of the population or sample, particularly, when they are collected. They are composed of as many rows as elements in the sample and the heading indicates the variables whose values are collected for each element. The first column contains the list of elements and each cell the data associated to the different variables. Rows and columns can be exchanged.
- *Distribution table of a variable.* These tables display the frequency distribution of a statistical variable, which can be qualitative or quantitative. The first column contains the modalities of the variable represented and the cells contain the ordinary (absolute, relative or percentages) or cumulative frequency, which corresponds to the modality of its row. Sometimes, the last row records the totals.
- *Two-way or contingency table.* It displays the joint distribution of two statistical variables and their modalities are represented in the first row and column, respectively.

2.2 Elementary Algebraization Levels

We base on a framework (Godino et al., 2014, 2015a, b), where Elementary Algebraic Reasoning (EAR) is understood as the system of practices related to the resolution of tasks in which algebraic processes (e.g., symbolization, generalization, modeling) and objects (e.g., variables, unknowns, equations, patterns, relationships) are involved. This model does not describe the pupils' cognitive development, but the mathematical activity carried out to solve a mathematical task. These authors discriminate different levels of EAR by considering the grade of generality of the objects used, the mathematical processes involved, and the type of language used, in the following way:

- *Level 0. Arithmetic reasoning.* This level is characterized by operations with intensive objects of first generality degree such as particular numbers, as well as the use of natural, numerical or iconic languages. The meaning of the equal sign is operational.
- *Level 1. Emerging algebraic reasoning.* The solver recognizes the properties of operations and uses the equal sign to express relationships; the concept of equivalence appears. In functional tasks, a general rule may be identified.
- *Level 2. Intermediate algebraic reasoning.* Symbolic representations are used to represent intensive objects; equations of the type $Ax \pm B = C$ are solved, but there are no operations with variables. In functional tasks, a general rule is recognized.
- *Level 3. Consolidated algebraic reasoning.* Symbols are used analytically. Operations with indeterminates or variables are performed; equations of the type $Ax \pm B = Cx \pm D$ are solved.

Work at each of these levels (Table 1) requires at least one of the proposed conditions (generality, transformations of objects, mathematical language), with no need to use all of them. Among the algebraic objects, binary relations (equivalence/order), operations performed on the elements of a set and their properties, functions and objects involved in the same are considered. These algebraization levels can be identified in activities related to any mathematical content and imply a progressive epistemic and cognitive complexity degree due to the level of generality of mathematical objects, the ostensive representations, and syntactic calculation used (Godino et al., 2015a).

These levels were expanded by Godino et al. (2015b) to analyze the algebraic activity in which parameters intervene. Note that the term parameter is not used here in its statistical meaning (as a characteristic of a population that determine its distribution, such as for example the mean in a normal distribution). Parameters are used in its algebraic meaning that is, a variable that is used with other variables to specify a family of functions or equations. More specifically, in this work we will only consider algebraic level L4 in Godino et al. (2015b), which is characterized by use of equations and or functions, including parameters and coefficients, but when no operations with parameters are carried out.

2.3 Previous Research

There is scarce research analysing statistical tables in textbooks, especially in secondary school education. Evangelista and Guimarães (2019) investigated the different types of tables presented in eight collections of Brazilian textbooks for Primary Education grades one to five. They distinguished data banks (similar to data tables),

Table 1 Characteristics of algebraic levels 0 to 3 (Godino et al., 2014, p. 121)

Ear level	Objects generality	Transformations of objects	Mathematical language
L0	Objects with first degree of generality (particular numbers). Operational meaning of equality.	Arithmetic operations with particular numbers. Properties of operations or natural numbers.	Natural, numerical, iconic, gestural. Value of symbols are obtained by operating with particular numbers.
L1	Objects with second degree of generality (sets, classes or types of numbers). Relational meaning of equality. Variables are used as unknowns.	Operations with first grade of generality objects, applying properties of the natural numbers algebraic structure and equality used as equivalence.	Symbols can be used as operation results, but there are no operations with symbols.
L2	Objects with second degree of generality (sets, classes or types of numbers). Relational meaning of equality. Variables used as unknowns, generalized numbers and changing quantity.	Operations with objects of first grade of generality, applying properties of the natural numbers algebraic structure. Equations of the form: $Ax \pm B = C$ in functional tasks, with no operation with the variables to obtain canonical forms of expression.	Symbolic-literal, although it is associated with context information.
L3	Indeterminate, unknowns, equations, variables and particular functions are used. Intensive objects with second grade generality.	Equations of the form: $Ax \pm B = Cx \pm D$, in structural tasks. Operations with unknowns or variables are made to obtain canonical forms of expression.	Symbolic-literal; symbols are used analytically (without meaning), without considering contextual information.

frequency tables and frameworks (containing auxiliary data that help solving a situation). They found a higher presence of frequency tables (56%), followed by data banks (36.7%) and frameworks (7.8%). García-García et al. (2019) identified different types of tables in a sample of 12 Mexican textbooks for Primary Education grades one to six (6 to 12 years-old). They considered data tables, frequency tables and two-way tables. The data table was present in all the books; there were also many frequency tables, while contingency tables were scarce.

In Chile, Díaz-Levicoy et al. (2015) analysed the statistical tables in textbooks of Basic Education first and second grades (6 and 7 years-old). They considered the type of table (data table, counting table, frequency and two-way table) in two different editorials. Counting tables (a table in which the frequency counts are recorded, through marks or symbols) were 83% of all the tables in one editorial, while in the other, only represented 42%. In the latter, a more balanced distribution was found with frequency tables (50%) and two-way tables (17%), which were not considered by the first editorial. Also, in Chile, Pallauta, Gea, and Arteaga (2021b) analyzed the types of tables and the frequency represented in a sample of 12 school textbooks directed to Basic Education fifth to eighth grades (10 to 13 years-old). The distribution tables of a variable with absolute, relative or percentage frequencies were the most frequent (44.6%), followed by data tables (21.5%), distribution tables of a variable with cumulative frequencies (15.9%), and finally, two-way tables with ordinary frequencies (11.9%).

In a previous paper (Pallauta, Gea, Batanero, & Arteaga, 2021c) we investigated the types of problems, language used, concepts, properties, procedures, and arguments linked to the study of statistical tables in a sample of Spanish secondary school textbooks. The results showed differences between the selected publishers in the type of situations proposed. There were no differences in concepts, properties, and arguments, the treatment of language was similar both by publisher and educational level.

Our research complements the above papers, by analysing the algebraic levels needed to work with different types of tables, which has not been studied by the above authors. The aim is to show that, although visually these tables may seem very similar, different cognitive demands are required to work with each of them (Koschat, 2005). In addition, we analyse the distribution of the types of tables in a sample of Spanish secondary school textbooks.

3 Method

3.1 Description of the Sample

The sample was made of 18 Spanish textbooks aimed at secondary school grades one to four (12–15 years-old). The editorials (Anaya, Edelvives & Santillana) were selected because of their long tradition of dedication to publish mathematics textbooks, as well as by their prestige and wide diffusion in all the country, both in public and private schools. The list of the specific textbooks that have been analyzed is presented in the Appendix. More specifically, six textbooks were analyzed per editorial, since in the first and second grades all students study the same contents, while in third and fourth grades there are two different strands in Spanish secondary education, so that the student can choose between academic mathematics and applied mathematics. This second option includes less contents, which are studied with less detail than in the academic option. For each book in the sample, all the exercises, examples, problems and paragraphs dealing with statistical tables were examined. We will refer as "activities" all of them, independently if they are exercises, problems, examples, or paragraphs dealing with content related to the tables.

The distribution of the activities by grade and publisher is presented in Table 2. We remark the increment in the percentage of activities included in the textbooks analyzed by grade and some variation between editorials.

Table 2 Percentage of activities analyzed in different grades in each publisher

	Editorial			
Grade	Anaya n = 862	Edelvives n = 892	Santillana n = 795	Total n = 2549
1	12.4	9.8	8.9	10.4
2	13.0	16.0	10.8	13.4
3	29.7	30.2	44.8	34.6
4	44.9	44.1	35.5	41.7

3.2 *Analysis*

We performed a content analysis (Neuendorf, 2016) of the statistical tables included in the sample of textbooks to identify the mathematical activity needed to read and/or build each of them. Systematic steps were followed in the content analysis, where the first stage was to identify the topic corresponding to the statistics and probability unit in each textbook. The second step was to select the paragraphs presenting situations in which the statistical table was used (explicitly or implicitly), either as an object of study itself, or as a tool to answer other questions.

The content of these paragraphs was then examined so that, in a cyclical and inductive manner, the tables were studied in order to firstly classify the table according to the categories proposed by Lahanier-Reuter (2003), that is, as a data table, a distribution table of a variable or a two-way table, where some subcategories were used in these two last types of tables, as described in the section of results. Secondly, we identified the algebraic activity involved in each category of table, using the three initial algebraization levels described by Godino et al. (2014), as well as level 4 proposed by Godino et al. (2015b). To ensure the reliability of the coding, continuous revisions of the texts were carried out by the authors and discordant cases were discussed together until agreement was reached. Finally, as a fourth and final step, summary tables of the results were drawn up to facilitate the drawing of conclusions.

4 Results

4.1 *Algebraic Activity Involved in Different Statistical Tables*

In this section we base on Lahanier-Reuter (2003) classification of tables, where we subdivide some of her types, depending on whether cumulative frequencies are represented in the table and whether the data are grouped in intervals or not. Thus, we consider three types of distribution table of a variable: ordinary frequency distribution table; cumulative frequency distribution table; and distribution table with

In this table the height and weight of some students is displayed

Height (cm)	165	172	171	168	174	176	167	173
Weight (kg)	53	67	72	68	80	75	69	64

a. Compute the mean and median of height and weight.
b. Compute the range of each variable.
c. Obtain the mean deviation of each variable.

Fig. 1 Data table translated from a first grade textbook. (Colera et al., 2016, p. 281)

values grouped in intervals. Similarly, we subdivided into two classes the contingency table, depending on whether values grouped in intervals are considered or not. In the following sections, we study the algebraic activity involved for each of these different types of statistical tables.

4.1.1 Data Tables

Data tables present as many rows (or columns) as individuals in the sample under study (See Fig. 1 as an example, where individuals are represented in columns). Generally, some headings are included on the first row (or column) of the table, which describe the variables collected in each individual. Thus, an additional number of rows (or columns) are included to record the data for each individual according to each variable.

In the example of data table in Fig. 1, reproduced from a textbook for grade 1, a data table is used to compute different statistical summaries. Two quantitative variables (height and weight) are represented with different values for each individual in the sample (8 students in total), with the label of each variable in the first column of the table. The data in the table are extensive objects, because each one represents a specific value (a number). The mathematical language in the table is numerical (values) and verbal (labels of the variables and abbreviation of the units of measurement). In these tables, the idea of statistical variable appears but the concepts of frequency or distribution are not explicitly used. However, in case the variable is numerical, as in the example shown in Fig. 1, some summaries could be obtained (the range, minimum and maximum, for example).

When the student obtains information from the table they operate with particular numbers and, consequently, all these objects are used with a first degree of generality (particular numeric values); consequently, the algebraic level required to deal with data table in the EAR model (Godino et al., 2014) is L0, corresponding to arithmetic reasoning.

4.1.2 One Variable Distribution

When the size of the data set increases is not practical to list the values corresponding to each individual in the sample in a data table. Alternatively, the data are classified (data with the same value are assigned to the same class) and the frequency for each different modality of the variable (number of elements in each class) is computed. These values and frequencies are represented in a *one-variable distribution table*, which usually contains several columns. Generally, the type of frequencies represented in the table (absolute, relative, cumulative or percentages) are indicated in the upper part (first row) of the table. In the first column, the different modalities or values that the variable takes in the sample under study are described. Moreover, in each of the following columns, the cells in the table contain the frequency (absolute, relative, etc.) corresponding to that value or modality. Sometimes, the last row records the total values (See example from a book directed to grade one in Fig. 2).

The same algebraic objects involved in a data table appear (variable, values, minimum, maximum, range), while classification of values and equivalence class are new algebraic objects involved in this type of tables; frequency and its different types, as well as distribution are new statistical objects. Note that in these tables, the modalities of the variable are used with a second degree of generality, since each one refers to a class: the class of individuals in the sample with the same value for this modality of the variable and, therefore, we are using a relational meaning of equality. We also remark the introduction of some functions that assigns the frequency to each value of the variable. In the case of ordinary frequency, this function is not given by an algebraic formula, but it is a univocal correspondence from the set of variable values (or modalities) in the set of natural numbers N; this correspondence is represented in the table. There is a second function assigning each ordinary frequency to the relative frequency for each value, which in this case is a linear function defined by $h_i = \frac{f_i}{N}$, where h_i is the relative frequency for the value number

There are four candidates in an election.
Construct a pie chart with the results.

I count and build the table

A→ |||| |||| |||| |

B→ |||| ||||

C→ |||| |||| | |

D→ |||| |||| | |

Candidate	Frequency	Relative Frequency	Percentage
A	16	$\frac{16}{50} = 0{,}32$	32%
B	10	$\frac{10}{50} = 0{,}20$	20%
C	12	$\frac{12}{50} = 0{,}24$	24%
D	12	$\frac{12}{50} = 0{,}24$	24%
Total	50	1,00	100%

Fig. 2 Distribution table with ordinary frequencies translate from a grade one textbook. (Colera et al., 2016, p. 283)

i (x_i) of the variable (x), f_i the absolute frequency for this value and N the sample size. A third linear function appear when transforming relative frequencies to percentages. Therefore, in these tables we start working with the level L1 of EAR, according to Godino et al. (2014). In addition of the natural number set N, the set of decimal numbers (a subset of rational numbers) is involved.

We can distinguish three different types of distribution tables depending on whether cumulative frequencies and class intervals are considered. In each of these subtypes of tables, new mathematical objects appear, thus giving each table different meaning and difficulty for the student.

Ordinary Frequency Distribution Tables

In these tables all or a part of absolute frequencies, relative frequencies or percentage frequencies are displayed for each different modality of the variable. An example in a book directed to first grade is reproduced in Fig. 2. In that example, a counting list has been added, and the formula to compute relative frequencies is remembered to the student. Although this formula is used with particular values, it has a second degree of generality, as it is valid for any value of the variable. We also remark that each modality of the variable is used as an equivalence class; for example, A is the class of the 16 people voting this candidate. As is shown in the example, no symbols are used to refer to the variable values or the different frequencies and the computation formulas are represented using only the particular values in the table, which requires the use of the level L1.

Frequency Distribution Tables with Cumulative Frequencies

In addition to all the statistical and algebraic objects involved in the previous types of tables, we add the cumulative frequencies (that can be absolute, relative or percentage). An example is given in Fig. 3. The cumulative absolute frequency F_i for the i value is computed by the formula $F_i = \sum_{j=1}^{i} f_j$. Therefore, cumulative frequencies are not linear function of the ordinary frequencies, but a function of that frequency and all the frequencies that correspond to all the previous values. From the cumulative absolute frequency, the relative and percentage cumulative frequencies are computed, which are linear function of the cumulative absolute frequency. We also emphasise that constructing or interpreting cumulative frequencies involves understanding and dealing with inequalities. Moreover, from grade two, textbooks incorporate symbols to describe the computation of different frequencies (See example in Fig. 3) or to justify their properties.

Note that, as is shown in the example (Fig. 3), symbols are used to denote the modality of the variable and different types of frequencies as well as their formulas of computation, so that there are some elementary operations with the symbols. Consequently, we work at the EAR level L3, because symbols are used analytically

Absolute frequency (f_i) of a variable value (x_i) is the number of times it repeats.
Relative frequency (h_i) of a variable value (xi) is the quotient between its absolute frequency the total number of data.
Cumulative absolute frequency (F_i) of a variable value (x_i) is the sum of the absolute frequencies of values smaller or equal of that value: $F_i = f_1 + f_2 + f_3 + \ldots + f_i$.
Cumulative relative frequency (H_i) of a variable value (x_i) is the sum of the relative frequency of values smaller or equal of that value: $Hi = h_1 + h_2 + h_3 + \ldots + h_i$.

Example:
Complete the frequency table with the cumulative frequencies.

xi	0	1	2	3	4
fi	6	16	15	10	3

We compute the cumulative frequencies becausse the varible is quantitative.

xi	fi	Fi	hi	Hi
0	6	6	0,12	0,12
1	16	22	0,32	0,44
2	15	37	0,3	0,74
3	10	47	0,2	0,94
4	3	50	0,06	1
	50		1	

The last cumulative absolute frequency is the total of data.

The last cumulative relative frequency is always 1.

The cumulative relative frequency of a variable value (x_i) is equivalent to the quotient between its cumulative absolute frequency and the total number of data;

$$Hi = \frac{f1 + f2 + f3 + \ldots + fi}{N} = \frac{Fi}{N}$$

Fig. 3 Distribution table with cumulative frequencies translated from a grade two textbook. (Almodóvar et al., 2016, p. 277)

without considering contextual information. In the example, arrows are employed to represent iconically the operations that should be performed to complete all the cells and computing the different types of frequencies. Some properties of cumulative frequencies are highlight in notes added to the table.

Distribution Table with Grouped Data

When the set of different values in a distribution table is big, these values are grouped in intervals, before producing a table for the distribution. Tables with grouped data may also contain any type of frequency, both ordinary and cumulative. An example is given in Fig. 4 directed to grade two. We still work at algebraic level L3, because symbols are used to represent the variable values and different types of frequencies, whose formulas include operation with these symbols.

Compute the measures of centre for these data recorded in a frequency table.

x_i	c_i	n_i	$c_i \cdot n_i$	N_i
[13, 14)	13.5	3	40.5	3
[14, 15)	14.5	5	72.5	8
[15, 16)	15.5	4	62	12
[16, 17)	16.5	7	115.5	19
[17, 18)	17.5	8	140	27
Total		27	430.5	

- The modal interval [17, 18) is repeated 8 times.
- The mean is $\bar{x} = \frac{430{,}5}{27} = 15.9$.
- The median interval is [16, 17). It is the first interval whose cumulative absolute frequency exceeds 13.5.

Fig. 4 Distribution table with grouped data. (Adapted from Romero et al., 2016, p. 195)

To all the statistical and algebraic objects described for the previous types of tables, the ideas of interval, its extremes and class mark are added, as well as median and modal interval. Note that semi-open intervals and its mathematical representation are used in Fig. 4. The building of these tables involves a second classification of values in the given intervals; therefore, a second degree of generality of the variable values is used: classes (intervals) of values, each of which is a class of elements in the sample with the same value.

4.1.3 Two-Way Tables

In these tables, the joint distribution for two statistical variables is represented, so that the elements in a sample are classified in a double way, by taking into account the values for each of the two variables in study. Therefore, new statistical objects appear (joint, marginal and conditional distribution; association and independence between the variables). In the joint distribution, the double classification is considered, whereas in the marginal distribution, only the simple classification by one variable is taken into account; in the conditional distribution, a restricted part of the sample (elements with the same value for one variable) are classified for the second variable. All this classification involves working with second degree of generality.

The frequencies of the cells are ordinary (absolute, relative or percentage), each of which can be regarded as joint, marginal or conditional. The last row and the last column are intended to register subtotals, of rows and columns respectively (See use of symbols with two variables in a grade four textbooks in Fig. 5). Given the absolute joint frequency f_{ij} corresponding to a value for each variable (x_i, y_j), the relative frequency is given by

$$h_{ij} = \frac{f_{ij}}{N}$$

x / y	y_1	y_2	...	y_j	Sum
x_1	n_{11}	n_{12}	...	n_{1j}	$n_{11} + n_{12} + \cdots + n_{1j}$
x_2	n_{21}	n_{22}	...	n_{2j}	$n_{21} + n_{22} + \cdots + n_{2j}$
...	...	...	...	...	...
x_i	n_{i1}	n_{i2}	...	n_{ij}	$n_{i1} + n_{i2} + \cdots + n_{ij}$
Sum	$n_{11} + n_{21} + \cdots + n_{i1}$	$n_{12} + n_{12} + \cdots + n_{i2}$	---	$n_{1j} + n_{2j} + \cdots n_{ij}$	N

Marginal distribution for x

Marginal distribution for y

Fig. 5 Use of algebraic symbols in two-way tables translated from a grade four textbook. (García et al., 2016, p. 166)

where N is the sample size. Given the absolute marginal frequencies $f_{i.}$ of value x_i and $f_{.j}$ of value y_j, the marginal relative frequencies are given, respectively, by

$$h_{i.} = \frac{f_{i.}}{N}; h_{.j} = \frac{f_{.j}}{N}$$

And the conditional relative frequency of y_j given x_i by

$$h(y_j \mid x_i) = \frac{f_{ij}}{f_{i.}}$$

Similarly, it is possible to obtain the conditional relative frequency of x_i give n y_j. Again, relative frequency and percentage are linear functions of ordinary frequency, although ordinary frequency is not linear function of the variables, and marginal frequencies are not linear function of joint frequencies. However, conditional frequencies are linear functions of joint frequencies.

Notice that, when working with two-way tables, the joint frequencies are function of two variables x and y and the marginal frequencies are a function only of one variable (either x or y). When considering the conditional frequencies, we only deal with one variable, while the other acts as a parameter, because, for example, when fixing the value of y, we obtain different conditional distributions for x, and vice-versa. Consequently, we are using parameters, although we do not operate with them, and thus we are working at the EAR level L4 defined by Godino et al. (2015b). See use of symbols with two variables in a grade four textbooks in Fig. 5.

Two-Way Table with Categorical Data

We differentiate two types of two-way tables depending on whether class intervals (values grouped in intervals) are considered or not. All the algebraic and statistical objects described above for two-way tables apply and therefore we work at EAR level L4 (Godino et al., 2015b), because of the many new statistical objects used, at least either implicitly or explicitly, such as use of parameters when considering any conditional distributions. One example is presented in Fig. 6.

In a study the nationality and gender of residents in a village were displayed in the following table. If a person from that village is chosen at random, compute the probability of the following events:

	Spanish	French	English	Total
Men	300	150	150	600
Women	200	150	50	400
Total	500	300	200	1000

a. The person is Spanish.
b. The person is a French man.

Fig. 6 Two-way table with categorical data translated from a grade four textbook. (García et al., 2016, p. 187)

Fig. 7 Two-way table with grouped data translated from a grade three textbook. (Colera et al., 2015a, p. 262)

Build a population pyramid with the data in the table.

Age group	Number of men (millions)	Number of women (millions)
0-9	125	111
10-19	122	108
20-29	108	98
30-39	94	88
40-49	74	72
50-59	51	51
60-69	32	32
70-79	15	16

Two-Way Table with Grouped Data

In these tables we consider the grouping of the variable values in intervals, in one or both variables in the two-way tables, for any type of frequency. Consequently, all mathematical and statistical objects described for two-way tables with categorical data take part in the construction or interpretation of these tables (Fig. 7).

To summarise the mathematical activity involved in the different types of tables, we present in Table 3 a synthesis of the analysis performed in the previous sections. We observe the complexity of all the types of tables, in spite of their apparent simplicity, since the student has to deal with many statistical and algebraic objects. Work at the arithmetic level (L0) is restricted to the use of data table, while all the other types of tables involve at least a pre-algebraic level L1, which can raise to upper levels L3, or L4, depending on the operations required from the student with the variables and symbols, as well as presence of parameters. The number and abstractness of mathematical objects also increases from the data table to the two-way table with grouped data, due to the amount of different mathematical objects to be used in each type of table.

Table 3 Statistical and algebraic objects involved in the different types of tables

Mathematical Objects	Data table	One-variable distribution table			Two-way table	
		Ord	Cum	Grouped	Categ	Grouped
Statistical Objects						
Statistical study, population, individual, census sample	x	x	x	x	x	x
Ordinary frequencies (absolute, relative)		x	x	x	x	x
Cumulative frequency (absolute, relative), inequalities			x	x		
Joint frequencies (absolute, relative)					x	x
Median interval, modal interval				x		x
Marginal and conditional frequencies, absolute, relative					x	x
Association, independence					x	x
Expected frequencies in case of independence					x	x
Algebraic Objects						
Variable, values	x	x	x	x	x	x
Variable types and scale of measurement	x	x	x	x	x	x
Parameters					x	x
Range, maximum, minimum (in quantitative variables)	x	x	x	x	x	x
Classification, equivalence class		x	x	x	x	x
Class intervals, extremes, class Centre				x		x
Function (linear and non-linear); proportionality		x	x	x	x	x
Graphs and tables	x	x	x	x	x	x
Dependent / independent variable		x	x	x	x	x
Operational meaning of equality	x	x	x	x	x	x
Relational meaning of equality		x	x	x	x	x

4.2 *Distribution of Statistical Tables in the Textbooks*

Once the semiotic analysis of the different types of tables has been presented, in this section we describe the results of our study about the types of tables in a sample of 18 secondary school grades. First, the distribution of type of tables in the different grades is presented in Table 4. The analysis of the different types of tables in the sample of textbooks by educational level shows that, globally, more than half of the tables correspond to distribution tables (72.8%), mainly those representing ordinary frequencies (43.0%), followed by distribution of grouped data frequencies (22.3%). We have not found related research that analyses statistical tables in secondary school textbooks and for this reason, we now differentiate our research with some studies performed at primary school level.

Table 4 Percentage of type of table by secondary school grade

Type of table		1st n = 265	2nd n = 341	3rd n = 881	4th n = 1062	Total n = 2549
Data table		7.5	5.6	7.3	20.2	12.4
Frequency distribution	Ordinary	74.0	53.4	45.5	29.9	43.0
	Cumulative	0.8	8.5	11.2	5.7	7.5
	Grouped data	10.2	9.4	31.0	22.2	22.3
Two-way	Categorical data	7.5	23.2	2.8	21.5	13.8
	Grouped data	0.0	0.0	2.2	0.5	0.9

Thus, results very similar to those in our study were reported by García-García et al. (2019), although the authors neither considered cumulative frequencies nor class intervals, since they were working at primary school level, where these concepts are not studied. Moreover, our results differ from those by Díaz-Levicoy et al. (2015) in whose research counting tables (frequency is recorded through marks or symbols) that do not appear in our study were predominant, which is explained because in primary school level children still need the support of counting tables as an intermediate step in the construction of frequency tables.

Distribution tables are also more common in our research than in those by Evangelista and Guimarães (2019) and Pallauta, Gea, and Arteaga (2021b). These differences are explained by the fact that those studies were performed with primary education textbooks, while in our case, in secondary education, cumulative frequencies and grouping in intervals are introduced and therefore there is a need of more different data distribution tables to reinforce these new concepts.

Data tables only appear in a small percentage, so that second degree of generality objects are usually involved in the work with statistical tables in secondary school, because these tables imply the concept of variable and the equivalence relationship is performed, with as many different classes of equivalence as modalities takes the one or bi-dimensional variable represented in the one variable distribution table or two-way table, respectively. In addition, the equal symbol is mainly used with relational meaning, when we relate each modality of the variable with its frequency. Thus, the algebraic level required to work with these tables is usually level L1. We also observe the increase of the algebraization levels needed to apply both statistical and arithmetic knowledge and algebraic reasoning as the school degree progresses, because the use of the cumulative frequencies and class intervals mainly in third grade implies the proto algebraic level L2.

We also note the highest number of data tables in fourth grade, where the objects involved have usually first degree of generality (no distribution of the variable is represented), and this implies the level L0. Two-way tables are mainly used in second and fourth grades, usually with algebraic level L4 involving a second degree of generality because the cells contain different frequencies (joint, marginal and conditioned) that implies a great complexity in their interpretation.

Table 5 Percentage of type of table by publisher

Type of table		Anaya n = 862	Edelvives n = 892	Santillana n = 795
Data table		15.2	13.0	8.8
Frequency distribution	Ordinary	37.2	42.3	50.2
	Cumulative	5.1	10.9	6.3
	Grouped data	19.1	19.1	29.3
Two-way	Categorical data	22.2	14.2	4.3
	Grouped data	1.2	0.6	1.1

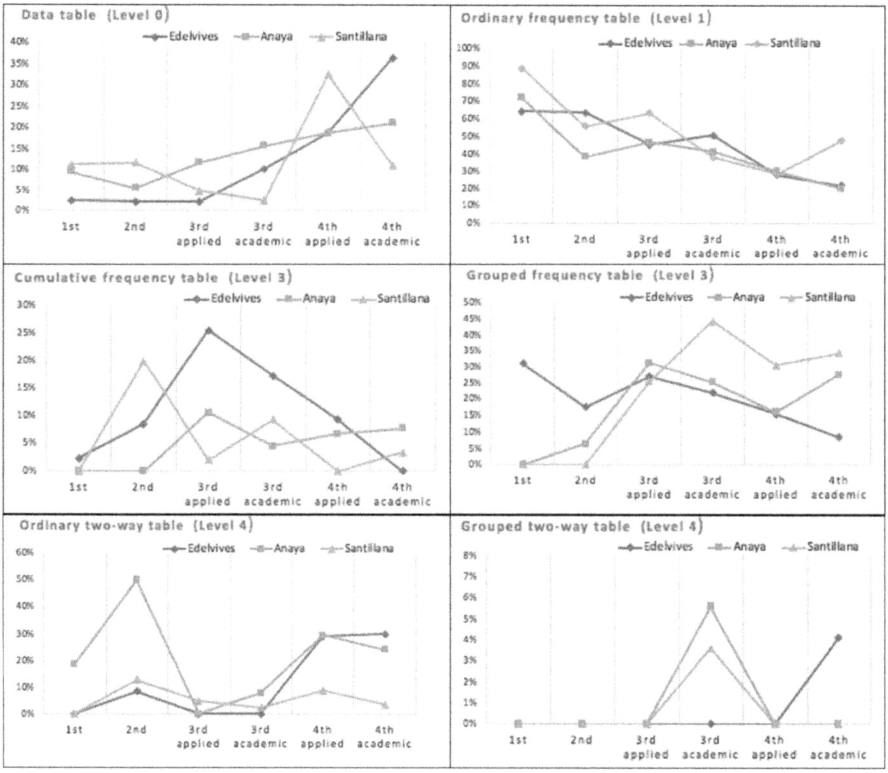

Fig. 8 Distribution of different type of table by educational level and editorial

When analyzing the distribution of types of tables by publisher (Table 5), we observe more data tables and two-way tables in Anaya, while Santillana introduces more one-variable distribution tables with ordinary frequency and grouped data. Edelvives presents a larger number of distribution tables with cumulative frequencies.

In Fig. 8 we represent for each type of table the percentage of tables of this kind that each editorial includes in the different school grades. Data tables are used in all the editorials and grades, because in the upper courses they are used to provide data from which the student later builds two-way tables or computes different statistical summaries. There is an increase of use of data tables by level except in Santillana.

Ordinary frequency distribution tables constitute most of the tables in the first grades, usually presenting distribution of qualitative variables with few different values, such as eye color or gender of the children in a group. Its relevance diminishes by grade to give some space to the introduction and to work with other types of tables, and the tendency is similar in the three editorials analyzed.

There is no clear pattern of use of cumulative frequency tables, as Anaya restricts them to grades three and four, while Edelvives uses a few in the lower grades and Santillana, introduces them in second and third grades. Their frequency is, generally, small only reaching a quarter the tables in grade three in Santillana.

As regards grouped frequency tables, they are limited to grades two and higher in Anaya, grades three and four in Santillana, while Edelvives uses them in all the different grades. In general, books directed to academic specialties contain more complex tables than those in the applied strand.

Categorical two-way tables are presented in all the grades and editorials, although with different frequency, constituting half the tables in grade two in Anaya. This type of table are very scares in third grade in both specialties by the three editorials. We remark that Edelvives only introduces grouped two-way tables in the last grade, and after the students are familiar with two-way tables, while Anaya and Santillana both propose grouped two-way tables in grade three and do not work with these tables in grade four. No rule is followed as regards algebraic level, as data table (Level 0) increase its presence by grade, while ordinary frequency table (Level 1) diminishes in the upper grades. There is, however, more presence of Level 3 and 4 since grade three, with the exception of Anaya that includes a high percentage of two-way tables (Level 4) in grade two.

5 Conclusions

The study of the algebraization level of statistical tables in the activities proposed in the sample of Spanish secondary school textbooks analyzed in this paper contributes to research in statistics education, as this is a topic that has been scarcely addressed in didactic research at this educational level.

We firstly performed a semiotic analysis of the mathematical objects and algebraic reasoning levels (Godino et al., 2014) involved in the different types of statistical tables described by Lahanier-Reuter (2003). This analysis reveals the cognitive complexity of statistical tables, which involves processes of generalization, representation, interpretation and symbolization, as well as understanding a variety of mathematical objects. Each of these objects are used with different levels of generality and formalization, depending on the type of table.

Consequently, working with statistical tables requires the application of statistical and arithmetic knowledge and algebraic reasoning processes, which help working not only with statistics, but also acquiring different mathematical competences, useful in other different contents. A careful introduction and work with the different types of tables, consequently, serves to progressively advance towards higher levels of mathematical generalization, representation and operation (Godino et al., 2015a, 2014).

In this sense, working with statistical tables can contribute to develop the students' progressive levels of EAR, because they are present in the curricula of several countries, in particular Spain, from the first grades of schooling, and the activity with these statistical tables involves the first EAR levels of algebraization.

It is important that the teacher be aware of this possibility and that also takes into account the complexity involved for the student in their learning of tables. For example, previous research (Gea et al., 2016) has reported difficulties in the work with inequalities, where a second level of algebraization (L2) is required, which is also needed in the tables with grouped data.

As suggested by Burgos et al. (2022), the application of the EAR model (Godino et al., 2014) helps to deepen the characterization of mathematics proposed to the students from the epistemological and learning point of view. In particular in this paper serves to reinforce the characterization of statistical tables by Lahanier-Reuter (2003) and to reveal the complexity of different types of tables. This complexity may go unnoticed by the teacher, as well as that the distribution of some types of statistical tables as the school year progresses is not clear, as our study showed.

In this sense, the EAR model has implications for teacher training, both in primary and secondary education. Besides having responsibility to develop the curricular guidelines proposed by the educational authorities, the teacher needs to introduce and develop algebraic reasoning starting from primary school in the different grades he or she teaches. According to Radford (2011) algebraic reasoning is highly sophisticated and was refined along centuries to achieve its current status. Therefore, we should not expect that children develop algebraic reasoning without a careful instruction and without providing them all possible chances to apply this reasoning in different mathematical activities. The EAR is then a useful tool for the teacher to plan the activities to be proposed to their students in order to develop, not only their statistical knowledge, but also their algebraic reasoning.

We finally remark that method of analysis performed in this chapter can be implemented in teacher training, since the analysis of school textbooks is one of the competences that should be included in teacher training (Godino et al., 2014). Moreover, the recognition of tasks that promote different degrees of algebraic reasoning can be useful for teachers, who need to be able to appreciate the complexity of the tasks set in textbooks, compare them and choose appropriately the textbook they will use with their students (Aké & Díaz Godino, 2018).

Acknowledgements Grant PID2019-105601GB-I00 funded by MCIN/ AEI /10.13039/501100011033. Scholarship ANID Folio: 72190280.

Appendix: Textbooks in the sample

Grade	Reference
1st	Colera, J., Gaztelu, I., & Colera R. (2016). *ESO 1 (1st grade Compulsory Secondary School).* Anaya. Mejía D., Romero, R., & Ocaña, J. (2015). *ESO 1 – Matemáticas (1st grade Compulsory Secondary School – Mathematics).* Edelvives. Almodóvar, J., de la Prida, C., Gaztelu, A., González, A., Machín, P., Pérez, C., & Sánchez, D. (2016). *Matemáticas. Serie Resuelve ESO 1 (Mathematics. Solving collection 1st grade Compulsory Secondary School).* Santillana.
2nd	Colera, J., Gaztelu, I., & Colera R. (2017). *ESO 2 – Matemáticas (2nd grade Compulsory Secondary School – Mathematics).* Anaya. Romero, R., Ocaña, J., & Mejía D., (2016). *ESO 2 – Matemáticas (2nd grade Compulsory Secondary School – Mathematics).* Edelvives. Almodóvar, J., Cuadrado A., Díaz, L., Dorce, C., Gámez, J., Marín, S., & Sánchez, D. (2016). *Matemáticas. Serie Resuelve ESO 2 (Mathematics. Solving collection 2nd grade Compulsory Secondary School).* Santillana.
3rd	Colera, J., Oliveira, M., Gaztelu, I., & Colera, R. (2015a). *ESO 3 – Matemáticas orientadas a las Enseñanzas Académicas (3rd grade Compulsory Secondary School – Academic Mathematics Speciality).* Anaya. Ocaña, J., Romero, R., & Mejía, D. (2015). *ESO 3 – Matemáticas orientadas a las Enseñanzas Académicas (3rd grade Compulsory Secondary School – Academic Mathematics Speciality).* Edelvives. De la Prida, C., Gaztelu, A., González, A., Machín, P., Pérez, C., & Sánchez, D. (2016). *Serie Resuelve ESO 3 – Matemáticas orientadas a las Enseñanzas Académicas (Solving collection 3rd grade Compulsory Secondary School – Academic Mathematics Speciality).* Santillana. Colera, J., Oliveira, M., Gaztelu, I., & Colera, R. (2015b). *ESO 3 – Matemáticas orientadas a las Enseñanzas Aplicadas (3rd grade Compulsory Secondary School – Applied Mathematics Speciality).* Anaya. García, M., Municipio, J., Ortega, P., & Villaoslada, E.M. (2015). *ESO 3 – Matemáticas orientadas a las Enseñanzas Aplicadas (3rd grade Compulsory Secondary School – Applied Mathematics Speciality).* Edelvives. De la Prida, C., Gaztelu, A., González, A., Pérez, C., & Sánchez, D. (2016). *Serie Resuelve ESO 3 – Matemáticas orientadas a las Enseñanzas Aplicadas (Solving collection 3rd grade Compulsory Secondary School – Applied Mathematics Speciality).* Santillana.

(continued)

Grade	Reference
4th	Colera, J., Oliveira, M., Gaztelu, I., & Colera, R. (2016a). *ESO 4 – Matemáticas orientadas a las Enseñanzas Académicas (4th grade Compulsory Secondary School – Academic Mathematics Speciality)*. Anaya. Mejía, D., Ocaña, J., & Romero, R (2016). *ESO 4 – Matemáticas orientadas a las Enseñanzas Académicas (4th grade Compulsory Secondary School – Academic Mathematics Speciality)*. Edelvives. Gámez, J., Gaztelu, A., Loysele, F., Marín S., Pérez, C., & Sánchez, D. (2016). *Serie Resuelve ESO 4 – Matemáticas orientadas a las Enseñanzas Académicas (Solving collection 4th grade Compulsory Secondary School – Academic Mathematics Speciality)*. Santillana. Colera, J., Oliveira, M., Gaztelu, I., & Colera, R. (2016b). *ESO 4 – Matemáticas orientadas a las Enseñanzas Aplicadas (4th grade Compulsory Secondary School – Applied Mathematics Speciality)*. Anaya. García, M., Municipio, J., & Ortega, P. (2016). *ESO 4 – Matemáticas orientadas a las Enseñanzas Aplicadas (4th grade Compulsory Secondary School – Applied Mathematics Speciality)*. Edelvives. Pérez, C., Sánchez, D., & Zapata, A. (2016). *Serie Soluciona ESO 4 – Matemáticas orientadas a las Enseñanzas Aplicadas (Solving collection 4th grade Compulsory Secondary School – Applied Mathematics Speciality)*. Santillana.

References

Aké, L. P., & Díaz Godino, J. (2018). Análisis de tareas de un libro de texto de primaria desde la perspectiva de los niveles de algebrización (Task analysis of an elementary school textbook from the algebraization levels perspective). *Educación Matemática, 30*(2), 171–201. https://doi.org/10.24844/EM3002.07

Alkhateeb, M. (2019). The language used in the 8th grade mathematics textbook. *Eurasia Journal of Mathematics, Science and Technology Education, 15*(7), 3–13. https://doi.org/10.29333/ejmste/106111

Burgess, T. (2002). Investigating the "data sense" of preservice teachers. In B. Phillips (Ed.), *Proceedings of the 6th international conference on teaching statistics* (pp. 1–6). International Association for Statistics Education.

Burgos, M., Batanero, C., & Godino, J. D. (2022). Algebraization levels in the study of probability. *Mathematics, 10*(1), 91. https://doi.org/10.3390/math10010091

Common Core State Standards Initiative. (CCSSI). (2010). *Common core state standards for mathematics*. National Governors Association for Best Practices and the Council of Chief State School Officers.

Díaz-Levicoy, D., Morales, R., & López-Martín, M. M. (2015). Tablas estadísticas en libros de texto chilenos de 1° y 2° año de Educación Primaria (Statistical tables in textbook Chilean 1st and 2nd grade of primary school). *Revista Paranaense de Educação Matemática, 4*(7), 10–39. https://doi.org/10.33871/22385800.2015.4.7.10-39

Díaz-Levicoy, D., Morales, R., Arteaga, P., & López-Martín, M. M. (2020). Conocimiento sobre tablas estadísticas por estudiantes chilenos de tercer año de Educación Primaria (Knowledge of statistical tables by Chilean students third year of primary education). *Educación Matemática, 32*(2), 247–277. https://doi.org/10.24844/EM3202.10

Duval, R. (2003). Comment analyser le fonctionnement représentationnel des tableaux et leur diversité? (How to analyse the representational functioning of the tables and their diversity?). *Spirale-Revue de recherches en éducation, 32*(32), 7–31. https://doi.org/10.3406/spira.2003.1377

Engel, J., Ridgway, J., & Weber, F. (2021). Educación estadística, democracia y empoderamiento de los ciudadanos (Statistics education, democracy and citizen's empowerment). *Paradigma, 42*(Extra-1), 1–31. https://doi.org/10.37618/PARADIGMA.1011-2251.2021.p01-31.id1016

Estrella, S. (2014). El formato tabular: una revisión de literatura (Tabular format: a review of literature). *Revista Actualidades Investigativas en Educación, 14*(2), 1–23. https://doi.org/10.15517/AIE.V14I2.14817

Estrella, S., Mena-Lorca, A., & Olfos-Ayarza, R. (2017). Naturaleza del objeto matemático Tabla (Nature of the mathematical object table). *MAGIS, 10*(20), 105–122. https://doi.org/10.11144/Javeriana.m10-20.nomt

Evangelista, B., & Guimarães, G. (2019). Análise de atividades sobre tabelas em livros didáticos brasileiros dos anos iniciais do ensino fundamental (Analysis of activities about tables in the primary school Brazilian textbooks). In J. M. Contreras, M. M. Gea, M. M. López-Martín, & E. Molina-Portillo (Eds.), *Actas del Tercer Congreso Internacional Virtual de Educación Estadística*. www.ugr.es/local/fqm126/civeest.html

Fan, L., Zhu, Y., & Miao, Z. (2013). Textbook research in mathematics education: Development status and directions. *Zentralblatt für Didaktik der Mathematik, 45*(5), 633–646. https://doi.org/10.1007/s11858-013-0539-x

Feinberg, R., & Wainer, H. (2011). Extracting sunbeams from cucumbers. *Journal of Computational and Graphical Statistics, 20*(4), 793–810. https://doi.org/10.1198/jcgs.2011.204a

Gabucio, F., Martí, E., Enfedaque, J., Gilabert, S., & Konstantinidou, A. (2010). Niveles de comprensión de las tablas en alumnos de primaria y secundaria (Levels of graph comprehension in primary and secondary school students). *Cultura y Educación, 22*(2), 183–197. https://doi.org/10.1174/113564010791304528

Gal, I. (2002). Adults' statistical literacy: Meanings, components, responsibilities. *International Statistical Review, 70*(1), 1–25. https://doi.org/10.1111/j.1751-5823.2002.tb00336.x

García-García, J., Díaz-Levicoy, D., Vidal, H., & Arredondo, E. (2019). Las tablas estadísticas en libros de texto de educación primaria en México (Statistical tables in primary education textbooks in Mexico). *Paradigma, 40*(2), 153–175. https://doi.org/10.37618/PARADIGMA.1011-2251.2019.p153-175.id754

Gea, M., Batanero, C., Fernandes, J., & Arteaga, P. (2016). Interpretación de resúmenes estadísticos por futuros profesores de educación secundaria (Interpretation of summaries on statistics addressed to prospective teachers in middle school). *REDIMAT, 5*(2), 135–157. https://doi.org/10.17583/redimat.2016.1902

Godino, J., Aké, L., Gonzato, M., & Wilhelmi, M. (2014). Niveles de algebrización de la actividad matemática escolar. Implicaciones para la formación de maestros (Algebrization levels of school mathematics activity. Implication for primary school teacher education). *Enseñanza de las Ciencias, 32*(1), 199–219. https://doi.org/10.5565/rev/ensciencias.965

Godino, J., Neto, T., Wilhelmi, M., Aké, L., Etchegaray, S., & Lasa, A. (2015a). Niveles de algebrización de las prácticas matemáticas escolares. Articulación de las perspectivas ontosemiótica y antropológica (Algebrization levels of school mathematics practices. Networking of the Onto-semiotic and Anthropological perspectives). *Avances de Investigación en Educación Matemática, 8*, 117–142. https://doi.org/10.35763/aiem.v1i8.105

Godino, J. D., Neto, T., Wilhelmi, M. R., Aké, L., Etchegaray, S., & Lasa, A. (2015b). Algebraic reasoning levels in primary and secondary education. In K. Krainer & N. Vondrová (Eds.), *Proceedings of the ninth congress of the European society for research in mathematics education* (CERME 9) (pp. 426–432). ERME.

Herbel-Eisenmann, B. A. (2007). From intended curriculum to written curriculum: Examining the voice of a mathematics textbook. *Journal for Research in Mathematics Education, 38*(4), 344–369. https://doi.org/10.2307/30034878

Koschat, M. (2005). A case for simple tables. *The American Statistician, 59*(1), 31–40. https://doi.org/10.1198/000313005X21429

Lahanier-Reuter, D. (2003). Différents types de tableaux dans l'enseignement des statistiques (Different types of tables in statistics education). *Spirale-Revue de recherches en éducation, 32*(32), 143–154. https://doi.org/10.3406/spira.2003.1386

Martí, E. (2009). Tables as cognitive tools in primary education. In C. Andersen, N. Scheuer, M. P. Pérez Echeverría, & E. Teubal (Eds.), *Representational systems and practices as learning tools in different fields of learning* (pp. 133–148). Sense Publishers. https://doi.org/10.1163/9789087905286_009

MECD. (2014). *Real Decreto 126/2014, de 28 de febrero, por el que se establece el currículo básico de la educación primaria* (Royal Decree 126/2014, 28th February, establishing the basic curriculum guidelines for primary education). Ministerio de Educación, Cultura y Deportes.

MECD. (2015). *Real Decreto 1105/2014, de 26 de diciembre, por el que se establece el currículo básico de la Educación Secundaria Obligatoria y del Bachillerato* (Royal Decree 1105/2014, 26th December, establishing the basic curriculum guidelines for Compulsory Secondary Education and High school). Ministerio de Educación, Cultura y Deportes.

National Council of Teachers of Mathematics (NCTM). (2014). *Principles to actions: Ensuring mathematical success for all*. NCTM.

Neuendorf, K. (2016). *The content analysis guidebook*. Sage.

Pallauta, J., Arteaga, P., & Garzón-Guerrero, J. A. (2021a). Secondary school students' construction and interpretation of statistical tables. *Mathematics, 9*(24), 3197. https://doi.org/10.3390/math9243197

Pallauta, J., Gea, M. M., & Arteaga, P. (2021b). Caracterización de las tareas propuestas sobre tablas estadísticas en libros de texto chilenos de educación básica (Characterization of tasks related to statistical tables in chilean basic education textbooks). *Paradigma, 40*(1), 32–60. https://doi.org/10.37618/PARADIGMA.1011-2251.2021.p32-60.id1017

Pallauta, J., Gea, M. M., Batanero, C., & Arteaga, P. (2021c). Significado de la tabla estadística en libros de texto españoles de educación secundaria (Meaning of the statistical table in Spanish secondary education textbooks). *Bolema: Boletim de Educação Matemática, 35*, 1803–1824. https://doi.org/10.1590/1980-4415v35n71a26

Radford, L. (2011). Grade 2 students' non-symbolic algebraic thinking. In J. Cai & E. Knuth (Eds.), *Early algebraization* (Advances in mathematics education) (pp. 303–322). SpringerVerlag.

Sharma, S. (2017). Definitions and models of statistical literacy: A literature review. *Open Review of Educational Research, 4*(1), 118–133. https://doi.org/10.1080/23265507.2017.1354313

Watson, J. M. (2013). *Statistical literacy at school: Growth and goals*. Routledge. https://doi.org/10.4324/9780203053898

Weiland, T. (2019). The contextualized situations constructed for the use of statistics by school mathematics textbooks. *Statistics Education Research Journal, 18*(2), 18–38. https://doi.org/10.52041/serj.v18i2.13

References

Abrahamson, D. (2009). Embodied design: Constructing means for constructing meaning. *Educational Studies in Mathematics, 70*, 20–47. https://doi-org.proxy.library.uu.nl/10.1007/s10649-008-9137-1

Abrahamson, D. (2019). A new world: Educational research on the sensorimotor roots of mathematical reasoning. In A. Shvarts (Ed.), *Proceedings of the annual meeting of the Russian chapter of the International Group for the Psychology of Mathematics Education (PME) & Yandex* (pp. 48–68). HSE Publishing House. https://www.igpme.org/wp-content/uploads/2020/01/PMEYandex2019Final.pdf

Abrahamson, D., & Sánchez-García, R. (2016). Learning is moving in new ways: The ecological dynamics of mathematics education. *Journal of the Learning Sciences, 25*(2), 203–239. https://doi.org/10.1080/10508406.2016.1143370

Abrahamson, D., Nathan, M. J., Williams-Pierce, C., Walkington, C., Ottmar, E. R., Soto, H., & Alibali, M. W. (2020). The future of embodied design for mathematics teaching and learning. *Frontiers in Education, 5*, 147. https://doi.org/10.3389/feduc.2020.00147

Abrahamson, D., Dutton, E., & Bakker, A. (2021). Towards an enactivist mathematics pedagogy. In S. A. Stolz (Ed.), *The body, embodiment, and education: An interdisciplinary approach*. Routledge. https://doi.org/10.4324/9781003142010

Ainley, J., Gould, R., & Pratt, D. (2015). Learning to reason from samples: Commentary from the perspectives of task design and the emergence of "big data.". *Educational Studies in Mathematics, 88*(3), 405–412. https://doi.org/10.1007/s10649-015-9592-4

Aké, L. P., & Díaz Godino, J. (2018). Análisis de tareas de un libro de texto de primaria desde la perspectiva de los niveles de algebrización (Task analysis of an elementary school textbook from the algebraization levels perspective). *Educación Matemática, 30*(2), 171–201. https://doi.org/10.24844/EM3002.07

Alberto, R. A., Shvarts, A., Drijvers, P., & Bakker, A. (2022). Action-based embodied design for mathematics learning: A decade of variations on a theme. *International Journal of Child-Computer Interaction, 32*, 100419. https://doi.org/10.1016/j.ijcci.2021.100419

Alkhateeb, M. (2019). The language used in the 8th grade mathematics textbook. *Eurasia Journal of Mathematics, Science and Technology Education, 15*(7), 3–13. https://doi.org/10.29333/ejmste/106111

Alrø, H., Blomhøj, M., Skovsmose, O., & Skånstrøm, M. (2000). Farlige små tal : almendannelse i et risikosamfund (Dangerous small numbers : Allgemeinbildung in a risk society). *Kvan – et tidsskrift for læreruddannelsen og folkeskolen, 20*(56), 17–27.

American Statistical Association (ASA) [website]. (2022). https://www.amstat.org/

G. F. Burrill et al. (eds.), *Research on Reasoning with Data and Statistical Thinking: International Perspectives*, Advances in Mathematics Education, https://doi.org/10.1007/978-3-031-29459-4

Andersen, M. W., & Weng, P. (2019). Dannelse gennem meningsfulde oplevelser med matematik (Bildung through meaningful experiences with mathematics). In I. J. Hansen, M. Rønø, S. E. Soneff, & A. H. Yates (Eds.), *Dannelse i alle fag* (pp. 85–100). Dafolo.

Anshari, M., Alas, Y., & Guan, L. S. (2015). Developing online learning resources: Big data, social networks, and cloud computing to support pervasive knowledge. *Education and Information Technologies, 21*(6), 1663–1677. https://doi.org/10.1007/s10639-015-9407-3

Arbeitskreis Stochastik der Gesellschaft für Didaktik der Mathematik. (2003). Empfehlungen zu Zielen und zur Gestaltung des Stochastikunterrichts [Recommendations on objectives and the design of stochastics lessons]. *Stochastik in der Schule, 23*(3), 21–26.

Arnold, P. (2022). *Statistical investigations Te Tūhuratanga Tauanga*. To be published. NZCER Press.

Arnold, P., & Franklin, C. (2021). What makes a good statistical question? *Journal of Statistics and Data Science Education, 29*(1), 122–130. https://doi.org/10.1080/26939169.2021.1877582

Arnold, P., Pfannkuch, M., Wild, C. J., Regan, M., & Budgett, S. (2011). Enhancing students' inferential reasoning: From hands-on to "movies". *Journal of Statistics Education, 19*(2), 1–32. https://doi.org/10.1023/A:1009854103737

Arnold, P., Confrey, J., Jones, R. S., Lee, H., & Pfannkuch, M. (2018). Statistics learning trajectories. In D. Ben-Zvi, K. Makar, & J. Garfield (Eds.), *International handbook of research in statistics education* (pp. 295–326). Springer International Publishing AG.

Artigue, M. (2007). Digital technologies: A window on theoretical issues in mathematics education. In D. Pitta-Pantazi & G. Philippou (Eds.), *Proceedings of the 5th congress of the European society for research in mathematics education* (pp. 68–82). Cyprus University. http://erme.site/wp-content/uploads/CERME5/plenaries.pdf

Australian Curriculum, Assessment, and Reporting Authority. (2012). *The Australian curriculum: Mathematics, Sydney, Australia: Author*.

Baig, M. I., Shuib, L., & Yadegaridehkordi, E. (2020). Big data in education: A state of the art, limitations, and future research directions. *International Journal of Educational Technology in Higher Education, 17*(1). https://doi.org/10.1186/s41239-020-00223-0

Bakker, A. (2004a). *Design research in statistics education – On symbolizing and computer tools* [Dissertation, University of Utrecht].

Bakker, A. (2004b). Reasoning about shape as a pattern in variability. *Statistics Education Research Journal, 3*(2), 64–83. https://doi.org/10.52041/serj.v3i2.552

Bakker, A. (2018). *Design research in education. A practical guide for early career researchers*. Routledge. https://doi.org/10.4324/9780203701010

Bakker, A., & Derry, J. (2011). Lessons from inferentialism for statistics education. *Mathematical Thinking and Learning, 13*(1–2), 5–26. https://doi.org/10.1080/10986065.2011.538293

Bakker, A., & Gravemeijer, K. (2004). Learning to reason about distribution. In D. Ben-Zvi & J. Garfield (Eds.), *The challenge of developing statistical literacy, reasoning and thinking* (pp. 147–168). Kluwer Academic Publishers. https://doi.org/10.1007/1-4020-2278-6_7

Bakker, A., & Hoffmann, M. H. (2005). Diagrammatic reasoning as the basis for developing concepts: A semiotic analysis of students' learning about statistical distribution. *Educational Studies in Mathematics, 60*(3), 333–358. https://doi-org.proxy.library.uu.nl/10.1007/s10649-005-5536-8

Bakker, A., & van Eerde, D. (2015). An introduction to design-based research with and example from statistics education. In A. Bikner-Ahsbahs, C. Knipping, & N. Presmeg (Eds.), *Approaches to qualitative research in mathematics education. Examples of methodology and methods* (pp. 429–466). Springer. https://doi.org/10.1007/978-94-017-9181-6_16

Ball, D. L., Thames, M. H., & Phelps, G. (2008). Content knowledge for teaching: What makes it special? *Journal of Teacher Education, 59*(5), 389–407. https://doi.org/10.1177/0022487108324554

Bansilal, S., & Lephoto, T. (2022). Exploring particular learner factors associated with South African mathematics learners' achievement: Gender gap or not. *African Journal of Research in Mathematics, Science and Technology education*. https://doi.org/10.1080/18117295.2022.2057730

Bargagliotti, A., & Franklin, C. (2021). *Statistics and data science for teachers*. American Statistical Association. https://www.amstat.org/docs/default-source/amstat-documents/gaisei-iprek-12_full.pdf

Bargagliotti, A., Franklin, C., Arnold, P., Gould, R., Johnson, S., Perez, L., & Spangler, D. (2020a). *Pre-K-12 guidelines for assessment and instruction in statistics education (GAISE) report II*. American Statistical Association and National Council of Teachers of Mathematics.

Bargagliotti, A., Franklin, C., Arnold, P., Gould, R., Johnson, S., Perez, L., & Spangler, D. A. (2020b). *Pre-K-12 guidelines for assessment and instruction in statistics education II (GAISE II)*. American Statistical Association and National Council of Teachers of Mathematics.

Bargagliotti, A., Franklin, C., Arnold, P., Johnson, S., Perez, L., & Spangler, D. A. (2020c). *Pre-K-12 guidelines for assessment and instruction in statistics education II: (GAISE II): A framework for statistics and data science education*. American Statistical Association.

Bargagliotti, A., Franklin, C., Arnold, P., Gould, R., Johnson, S., Perez, L., & Spangler, D. (2020e). *Pre-K–12 guidelines for assessment and instruction in statistics education II (GAISE II): A framework for statistics and data science education*. American Statistical Association. https://www.amstat.org/asa/files/pdfs/GAISE/GAISEIIPreK-12_Full.pdf

Bargagliotti, A., Franklin, C., Arnold, P., Gould, R., Johnson, S., Perez, L., & Spangler, D. (2020f). *Pre-K-12 guidelines for assessment and instruction in statistics education (GAISE) II*. American Statistical Association and National Council of Teachers of Mathematics.

Barrett, J. E., Clements, D. H., Klanderman, D., Pennisi, S.-J., & Plaki, M. V. (2006). Students' coordination of geometric reasoning and measuring strategies on a fixed perimeter task: Developing mathematical understanding of linear measurement. *Journal for Research in Mathematics Education, 37*(3), 187–221. https://doi.org/10.2307/30035058

Batanero, C., & Díaz, C. (2010). Training teachers to teach statistics: What can we learn from research? *Statistique et Enseignement, 1*(1), 5–20.

Batanero, C., Godino, J. D., Vallecillos, A., Green, D. R., & Holmes, P. (1994). Errors and difficulties in understanding elementary statistical concepts. *International Journal of Mathematics Education in Science and Technology, 25*(4), 527–547. https://doi.org/10.1080/0020739940250406

Batanero, C., Estepa, A., Godino, J. D., & Green, D. R. (1996). Intuitive strategies and preconceptions about association in contingency tables. *Journal for Research in Mathematics Education, 27*, 151–169. https://doi.org/10.2307/749598

Batanero, C., Godino, J., & Estepa, A. (1998). Building the meaning of statistical association through data analysis activities. In A. Olivier & K. Newstead (Eds.), *Proceedings of the 22nd conference of the international group for the psychology of mathematics education* (pp. 221–236). University of Stellenbosch.

Batanero, C., Tauber, L. M., & Sánchez, V. (2004). Students' reasoning about the normal distribution. In D. Ben-Zvi & J. Garfield (Eds.), *The challenge of developing statistical literacy, reasoning and thinking* (pp. 257–276). Springer. https://doi-org.proxy.library.uu.nl/10.1007/1-4020-2278-6_11

Battista, M. T. (1999). The importance of spatial structure in geometric reasoning. *Teaching Children Mathematics, 6*(3), 170–177. https://doi.org/10.5951/TCM.6.3.0170

Batur, A., & Baki, A. (2022). Examination of the relationship between statistical literacy levels and statistical literacy self-efficacy of high school students. *Eğitim ve Bilim, 47*(209), 171–205.

Batur, A., Özmen, Z. M., Topan, B., Akoğlu, K., & Güven, B. (2021). A cross-national comparison of statistics curricula. *Turkish Journal of Computer and Mathematics Education, 12*(1), 290–319.

Behar, R. (2021). *El histograma como un instrumento para la comprensión de las funciones de densidad de probabilidad* [The histogram as a tool for understanding probability density functions]. Project description on researchgate. https://www.researchgate.net/project/El-histograma-como-un-instrumento-para-la-comprension-de-las-funciones-de-densidad-de-probabilidad

Bell, P., & Linn, M. C. (2002). Beliefs about science: How does science instruction contribute? In B. K. Hofer & P. R. Pintrich (Eds.), *Personal epistemology: The psychology of beliefs about knowledge and knowing* (pp. 321–346). Lawrence Erlbaum Associates Publishers.

Bender, P., Beyer, D., Brück-Binninger, U., Kowallek, R., Schmidt, S., Sorger, P., et al. (1999). Überlegungen zur fachmathematischen Ausbildung der angehenden Grundschullehrerinnen und -lehrer [reflections on the mathematics education of prospective elementary school teachers]. *Journal für Mathematikdidaktik, 20*, 301–310. https://doi.org/10.1007/BF03338903

Ben-Zvi, D. (2006a, July 2–7). Scaffolding students' informal inference and argumentation. In A. Rossman. & B. Chance (Eds.), *Proceedings of the 7th international conference on teaching of statistics (CD-ROM), Salvador.*

Ben-Zvi, D. (2006b). Scaffolding students' informal inference and argumentation. In A. Rossman & B. Chance (Eds.) *Proceedings of the 7th international conference on teaching statistics.* International Statistical Institute. Retrieved from http://www.stat.auckland.ac.nz/~iase/publications/17/2D1_BENZ.pdf

Ben-Zvi, D. (2016). Tres paradigmas en el Desarrollo del razonamiento estadístico de los estudiantes *[*Three paradigms in developing students' statistical reasoning*]*. In S. Estrella et al. (Eds.), *XX Actas de las Jornadas Nacionales de Educación Matemática* (pp. 13–22). SOCHIEM.

Ben-Zvi, D. (2018). Foreword. In A. Leavy, M. Meletiou-Mavrotheris, & E. Paparistodemou (Eds.), *Statistics in early childhood and primary education* (pp. vii–viii). Springer.

Ben-Zvi, D., & Garfield, J. (2004). Statistical literacy, reasoning, and thinking: Goals, definitions, and challenges. In D. Ben-Zvi & J. Garfield (Eds.), *The challenges of developing statistical literacy, reasoning, and thinking* (pp. 3–15). Kluwer. https://doi.org/10.1007/1-4020-2278-6_1

Ben-Zvi, D., & Garfield, J. (2008). Introducing the emerging discipline of statistics education. *School Science and Mathematics, 108*(8), 355–361. https://doi.org/10.1111/j.1949-8594.2008.tb17850.x

Ben-Zvi, D., & Makar, K. (2016). International perspectives on the teaching and learning of statistics. In D. Ben-Zvi & K. Makar (Eds.), *The teaching and learning of statistics* (pp. 1–10). Springer. https://doi.org/10.1007/978-3-319-23470-0_1

Ben-Zvi, D., Aridor, K., Makar, K., & Bakker, A. (2012). Students' emergent articulations of uncertainty while making informal statistical inferences. *ZDM, 44*(7), 913–925. https://doi.org/10.1007/s11858-012-0420-3

Ben-Zvi, D., Bakker, A., & Makar, K. (2015). Learning to reason from samples. *Educational Studies in Mathematics, 88*(3), 291–303. https://doi.org/10.1007/s10649-015-9593-3

Ben-Zvi, D., Makar, K., & Garfield, J. (Eds.). (2018a). *International handbook of research in statistics education.* Springer Cham. https://doi.org/10.1007/978-3-319-66195-7

Ben-Zvi, D., Gravemeijer, K., & Ainley, J. (2018b). Design of statistics learning environments. In D. Ben-Zvi, K. Makar, & J. Garfield (Eds.), *International handbook on research in statistics education* (pp. 473–502). Springer Cham. https://doi.org/10.1007/978-3-319-66195-7_16

Bernstein, A. N. (1967). *The coordination and regulation of movements.* Pergamon press.

Beswick, K., & Goos, M. (2018). Mathematics teacher educator knowledge: What do we know and where to from here? *Journal of Mathematics Teacher Education, 21*(5), 417–427. https://doi.org/10.1007/s10857-018-9416-4

Beuving, J., & de Vries, G. (2020). Teaching qualitative research in adverse times. *Learning and Teaching, 13*(1), 42–66. https://doi.org/10.3167/latiss.2020.130104

Bhargava, R., Kadouaki, R., Bhargava, E., Castro, G., & D'Ignazio, C. (2016). Data murals: Using the arts to build data literacy. *The Journal of Community Informatics, 12*(3). https://doi.org/10.15353/joci.v12i3.3285

Biehler, R. (1989). Educational perspectives on exploratory data analysis. In R. Morris (Ed.), *Studies in mathematics education: The teaching of statistics* (Vol. 7, pp. 185–201). UNESCO.

Biehler, R. (1997). Software for learning and for doing statistics. *International Statistical Review, 65*(2), 167–189. https://doi.org/10.1111/j.1751-5823.1997.tb00399.x

Biehler, R. (2006). Leitidee "Daten und Zufall" in der didaktischen Konzeption und im Unterrichtsexperiment [guiding principle "data and chance" in the didactic conception and in the teaching experiment]. In J. Meyer (Ed.), *Anregungen zum Stochastikunterricht* (Vol. 3). Franzbecker.

Biehler, R. (2007a). Denken in Verteilungen – Vergleichen von Verteilungen (Thinking in distributions – comparing distributions). *Der Mathematikunterricht, 53*(3), 3–11.

Biehler, R. (2007b). Students' strategies of comparing distributions in an exploratory data analysis context. In *56th session of the International Statistical Institute*. https://www.stat.auckland.ac.nz/~iase/publications/isi56/IPM37_Biehler.pdf

Biehler, R. (2019). Allgemeinbildung, mathematical literacy, and competence orientation. In H. N. Jahnke & L. Hefendehl-Hebeker (Eds.), *Traditions in German-speaking mathematics education research* (pp. 141–170). Springer International Publishing. https://doi.org/10.1007/978-3-030-11069-7_6

Biehler, R., & Fleischer, Y. (2021). Introducing students to machine learning with decision trees using CODAP and Jupyter notebooks. *Teaching Statistics, 43*(S1). https://doi.org/10.1111/test.12279

Biehler, R., & Hartung, R. (2006). Leitidee Daten und Zufall [guiding principle data and chance]. In W. Blum, C. Drüke-Noe, R. Hartung, & O. Köller (Eds.), *Bildungsstandards Mathematik: konkret. Sekundarstufe I: Aufgabenbeispiele, Unterrichtsanregungen, Fortbildungsideen* (pp. 51–80). Cornelson Scriptor.

Biehler, R., Hofmann, T., Maxara, C., & Prömmel, A. (2011). *Daten und Zufall mit fathom - Unterrichtsideen für die SI und SII mit software-Einführung [data and chance with fathom - teaching ideas for SI and SII with software introduction]*. Schroedel.

Biehler, R., Ben-Zvi, D., Bakker, A., & Makar, K. (2013). Technology for enhancing statistical reasoning at the school level. In M. A. Clements, A. J. Bishop, C. Keitel, J. Kilpatrick, & F. K. S. Leung (Eds.), *Third international handbook of mathematics education* (pp. 643–689). Springer Science and Business Media. https://doi.org/10.1007/978-1-4614-4684-2_21

Biehler, R., Frischemeier, D., & Podworny, S. (2018a). Elementary preservice teachers' reasoning about statistical modeling in a civic statistics context. *ZDM, 50*(7), 1237–1251. https://doi.org/10.1007/s11858-018-1001-x

Biehler, R., Frischemeier, D., Reading, C., & Shaughnessy, M. (2018b). Reasoning about data. In D. Ben-Zvi, K. Makar, & J. Garfield (Eds.), *International handbook of research in statistics education* (pp. 139–192). Springer International. https://doi.org/10.1007/978-3-319-66195-7_5

Biggs, J. B., & Collis, K. F. (1982). *Evaluating the quality of learning: The Solo taxonomy*. Academic.

Biggs, J. B., & Collis, K. F. (2014). *Evaluating the quality of learning: The SOLO taxonomy (structure of the observed learning outcome)*. Academic.

Bilgin, A., & Petocz, P. (2013). Students' experience of becoming a statistical consultant. In *Proceedings of the Joint IASE/IAOS Satellite Conference Statistics Education for Progress* (pp. 1–6).

Birel, G. K. (2017). The investigation of pre-service elementary mathematics teachers' subject matter knowledge about probability. *Mersin Üniversitesi Eğitim Fakültesi Dergisi, 13*(1), 348–362.

Blanton, M. L., & Kaput, J. J. (2011). Functional thinking as a route into algebra in the elementary grades. In J. Cai & E. Knuth (Eds.), *Early algebraization: A global dialogue from multiple perspectives* (pp. 5–23). Springer. https://doi.org/10.1007/978-3-642-17735-4_2

Blomhøj, M. (2001). Hvorfor matematikundervisning? : Matematik og almendannelse i et højteknologisk samfund (why mathematics education? Mathematics and Allgemeinbildung in a high-tech society). *Centre for Research in Learning Mathematics, 24*, 218–246.

Bock, L. (2017). *Design, durchführung und evaluation einer unterrichtseinheit zur förderung des lesens und interpretierens von eindimensionalen streudiagrammen Für lernschwache kinder einer vierten klasse unter verwendung kooperativer lernformen* (Design, implementation and evaluation of a teaching unit to promote the reading and interpretation of one-dimensional scatter plots for children with learning disabilities in a fourth-grade class using cooperative learning methods). (Bachelor of Education), University of Paderborn.

Boels, L., Bakker, A., Van Dooren, W., & Drijvers, P. (2019a). Conceptual difficulties when interpreting histograms: A review. *Educational Research Review, 28*, 100291. https://doi.org/10.1016/j.edurev.2019.100291

Boels, L., Bakker, A., & Drijvers, P. (2019b). Eye tracking secondary school students' strategies when interpreting statistical graphs. In M. Graven, H. Venskat, A. A. Esien, & P. Vale (Eds.), *Proceedings of the forty-third psychology of mathematics education conference* (pp. 113–120). PME. https://www.igpme.org/wp-content/uploads/2019/07/PME43-proceedings.zip

Boels, L., Bakker, A., & Drijvers, P. (2019c). Unravelling teachers' strategies when interpreting histograms: An eye-tracking study. In U. T. Jankvist, M. Van den Heuvel-Panhuizen, & M. Veldhuis (Eds.), *Proceedings of the 11th congress of the European society for research in mathematics education* (pp. 888–895). Freudenthal Group & Freudenthal Institute, Utrecht University & ERME. https://hal.archives-ouvertes.fr/hal-02411575/document

Boels, L., Bakker, A., Van Dooren, W., & Drijvers, P. (2022). *Secondary school students' strategies when interpreting histograms and case-value plots: An eye-tracking study.* [submitted] Freudenthal Institute, Utrecht University.

Børne- og Undervisningsministeriet. (2022, 2/2/2019). *Folkeskolens historie: Et kort rids over folkeskolens lange historie* (The history of the Danish school: A short outline of a long history). https://emu.dk/grundskole/uddannelsens-formaal-og-historie/folkeskolens-historie

Boyd, D., & Crawford, K. (2012). Critical questions for big data: Provocations for a cultural, technological, and scholarly phenomenon. *Information, Communication & Society, 15*(5), 662–679. https://doi.org/10.1080/1369118X.2012.678878

Braham, H. M., & Ben-Zvi, D. (2017). Students' emergent articulations of statistical models and modeling in making informal statistical inferences. *Statistics Education Research Journal, 16*(2), 116–143. https://doi.org/10.52041/serj.v16i2.187

Bransford, J. D., & Brown, A. L. (1999). In R. R. Cocking (Ed.), *How people learn: Brain, mind, experience, and school*. National Academy Press.

Brasil. Ministério da Educação. (2018). *Base Nacional Comum Curricular – BNCC*. MEC/SEF.

Brasil. Ministério da Educação e do Desporto. (1998). *Parâmetros curriculares nacionais*. MEC/SEF.

Bromage, A., Pierce, S., Reader, T., & Compton, L. (2022). Teaching statistics to non-specialists: Challenges and strategies for success. *Journal of Further and Higher Education, 46*(1), 46–61. https://doi.org/10.1080/0309877X.2021.1879744

Budgett, S., & Pfannkuch, M. (2010). Assessing students' statistical literacy. In P. Bidgood, N. Hunt, & F. Jolliffe (Eds.), *Assessment methods in statistical education: An international perspective* (pp. 103–121). John Wiley & Sons Ltd..

Budgett, S., & Rose, D. (2017). Developing statistical literacy in the final school year. *Statistics Education Research Journal, 16*(1), 139–162. https://doi.org/10.52041/serj.v16i1.221

Burgess, T. (2002). Investigating the "data sense" of preservice teachers. In B. Phillips (Ed.), *Proceedings of the 6th international conference on teaching statistics* (pp. 1–6). International Association for Statistics Education.

Burgos, M., Batanero, C., & Godino, J. D. (2022). Algebraization levels in the study of probability. *Mathematics, 10*(1), 91. https://doi.org/10.3390/math10010091

Burrill, G. (2018). Concept images and statistical thinking: The role of interactive dynamic technology. In M. A. Sorto (Ed.). *Proceedings of the tenth international congress on teaching statistics*. Kyoto, Japan. https://iaseweb.org/Conference_Proceedings.php?p=ICOTS_10_2018

Burrill, G. (2020a). Statistical literacy and quantitative reasoning: Rethinking the curriculum. In P. Arnold (Ed.), *Proceedings of the roundtable conference of the International Association for Statistical Education (IASE)*.

Burrill, G. (2020b, July 6–12). Statistical literacy and quantitative reasoning: Rethinking the curriculum. In P. Arnold (Ed.), *New skills in the changing world of statistics education: Proceedings of the roundtable conference of the International Association for Statistical Education (IASE)*, Held online.

Burrill, G., & Biehler, R. (2011). Fundamental statistical ideas in the school curriculum and in training teachers. In C. Batanero, G. Burrill, & C. Reading (Eds.), *Teaching statistics in school mathematics-challenges for teaching and teacher education* (pp. 57–69). Springer. https://doi.org/10.1007/978-94-007-1131-0_10

Cairney, P. (2016). *The politics of evidence-based policy making*. Palgrave Pivot. https://doi.org/10.1057/978-1-137-51781-4

Caldas, M. S., & Silva, E. C. (2016). Fundamentos e aplicação do Big Data: como tratar informações em uma sociedade de yottabytes [Fundamentals and application of Big Data: How to handle information in a yottabyte society]. *University Libraries: Research, Experiences, and Perspectives, 3*(1), 65–85. https://periodicos.ufmg.br/index.php/revistarbu/article/view/3086

Campos, C. R. (2016a). *La educación estadística y la educación crítica* [Statistical education and critical education]. Segundo Encuentro Colombiano de Educación Estocástica (2 ECEE).

Campos, C. R. (2016b). *Towards critical statistics education. Theory and practice*. Lambert Academic Publishing.

Campos, C.R., Wodewotzki, M.L., & Jacobini, O. R. (2013). *Educação estatística: Teoria e prática em ambientes de modelagem matemática* [Statistical education: Theory and practice in mathematical modeling environments] (2 edn). Autêntica Editora.

Carmi, E., Yates, S. J., Lockley, E., & Pawluczuk, A. (2020). Data citizenship: Rethinking data literacy in the age of disinformation, misinformation, and malinformation. *Internet Policy Review, 9*(2). https://doi.org/10.14763/2020.2.1481

Carter, J., Brown, M., & Simpson, K. (2017). From the classroom to the workplace: How social science students are learning to do data analysis for real. *Statistics Education Research Journal, 16*(1), 80–101. https://doi.org/10.52041/serj.v16i1.218

Carzola, I., & Giordano, C. (2021). O papel do letramento estatístico na implementação dos temas contemporâneos transversais da BNCC [the role of statistical literacy in the implementation of contemporary transversal themes of BNCC]. In C. Moteiro & L. Carvalho. (Eds.), *Temas emergentes em letramento estatístico [recurso eletrônico]* [emerging themes in statistical literacy] (pp. 88–111). Ed. UFPE, 2021.

Casey, S. A. (2008). *Subject matter knowledge for teaching statistical association*. Doctoral Dissertation, Illinois State University.

Casey, S. A. (2014). Teachers' knowledge of students' conceptions and their development when learning linear regression. In K. Makar, B. de Sousa, & R. Gould (Eds.), *Sustainability in statistics education: Proceedings of the ninth international conference on teaching statistics, Flagstaff*. International Statistical Institute.

Casey, S. A. (2015). Examining student conceptions of covariation: A focus on the line of best fit. *Journal of Statistics Education, 23*(1). https://doi.org/10.1080/10691898.2015.11889722

Casey, S. A., & Wasserman, N. H. (2015). Teachers' knowledge about informal line of best fit. *Statistics Education Research Journal, 14*(1), 8–35. https://doi.org/10.52041/serj.v14i1.267

Castro Sotos, A. E. C., Vanhoof, S., Van den Noortgate, W., & Onghena, P. (2007). Students' misconceptions of statistical inference: A review of the empirical evidence from research on statistics education. *Educational Research Review, 2*(2), 98–113. https://doi.org/10.1016/j.edurev.2007.04.001

Castro, A., Vanhoof, S., Van den Noortgate, W., & Onghena, P. (2007). Students' misconceptions of statistical inference: A review of the empirical evidence from research on statistics education. *Educational Research Review, 2*(2), 98–113. https://doi.org/10.1016/j.edurev.2007.04.001

Çatman-Aksoy, E., & Işıksal-Bostan, M. (2021). Seventh graders' statistical literacy: An investigation on bar and line graphs. *International Journal of Science and Mathematics Education, 19*, 397–418.

CBMS. (2012). *Conference board of mathematical sciences*.

Chance, B., & Rossman, A. (2006). Using simulation to teach and learn statistics. In A. Rossman & B. Chance (Eds.), *Working cooperatively in statistics education: Proceedings of the seventh international conference on teaching statistics* (pp. 1–6). Salvador. https://iase-web.org/Conference_Proceedings.php?p=ICOTS_7_2006

Chance, B., Ben-Zvi, D., Garfield, J., & Medina, E. (2007). The role of technology in improving student learning of statistics. *Technology Innovations in Statistics Education, 1*(1), 1–24. https://doi.org/10.5070/T511000026

Change the Equation Analysis of 2015 National Assessment of Educational Progress. (2017). *Presentation at the National Council of Teachers of Mathematics Annual Meeting, Chicago IL.* https://schools.saisd.net/upload/page/0061/docs/NCTM_2017_Chicago_share.pdf

Clement, J. (2000). Analysis of clinical interview: Foundations and model viability. In A. E. Kelly & R. A. Lesh (Eds.), *Handbook of research design in mathematics and science education* (pp. 547–589). Lawrence Erlbaum Associates Publishers.

Cobb, P. (1999). Individual and collective mathematical development: The case of statistical data analysis. *Mathematical Thinking and Learning, 1*(1), 5–43. https://doi.org/10.1207/s15327833mtl0101_1

Cobb, G. (2007). The introductory statistics course: A Ptolemaic curriculum? *Technology Innovations in Statistics Education, 1*(1), 1–15. https://doi.org/10.5070/T511000028

Cobb, G. W. (2015). Mere renovation is too little, too late: We need to rethink the undergraduate curriculum from the ground up. *The American Statistician, 69*(4), 266–282. https://doi.org/10.1080/00031305.2015.1093029

Cobb, P., & McClain, K. (2004). Principles of instructional design for supporting the development of students' statistical reasoning. In D. Ben-Zvi & J. Garfield (Eds.), *The challenge of developing statistical literacy, reasoning, and thinking* (pp. 375–395). Kluwer. https://doi.org/10.1007/1-4020-2278-6_16

Cobb, G. W., & Moore, D. S. (1997). Mathematics, statistics, and teaching. *The American Mathematical Monthly, 104*(9), 801–823. https://doi.org/10.1080/00029890.1997.11990723

Cobb, P., Confrey, J., diSessa, A., Lehrer, R., & Schauble, L. (2003). Design experiments in educational research. *Educational Researcher, 32*(1), 9–13. https://doi.org/10.3102/0013189x032001009

CODAP (Common Online Data Analysis Platform) [Computer software]. (2022). Retrieved from https://codap.concord.org

Coffey, A., & Atkinson, P. (2003). *Encontrar el sentido a los datos cualitativos. Estrategias complementarias de investigación* [Finding the meaning of the qualitative data. Complementary research strategies]. Editorial Universidad de Antioquia.

Common Core State Standards Initiative (CCSSI). (2010). *Common core state standards for mathematics*. Author.

Common Core State Standards Initiative. (CCSSI). (2010). *Common core state standards for mathematics*. National Governors Association for Best Practices and the Council of Chief State School Officers.

Common Online Data Analysis Platform [Computer software]. (2014). *The Concord Consortium.* https://codap.concord.org/app/static/dg/en/cert/index.html

Conference Board of the Mathematical Sciences. (2012). *The mathematical education of teachers.* Providence.

Conner, A. (2008). Expanded Toulmin diagrams: A tool for investigating complex activity in classrooms. In O. Figueras, J. L. Cortina, S. Alatorre, T. Rojano, & A. Sepulveda (Eds.), *Proceedings of the joint meeting of the international group for the psychology of mathematics education 32 and the North American chapter of the international group for the psychology of mathematics education XXX* (Vol. 2, pp. 361–368). Cinvestav-UMSNH.

Conner, A., Singletary, L. M., Smith, R. C., Wagner, P. A., & Francisco, R. T. (2014a). Identifying kinds of reasoning in collective argumentation. *Mathematical Thinking and Learning, 16*(3), 181–200. https://doi.org/10.1080/10986065.2014.921131

Conner, A., Singletary, L. M., Smith, R. C., Wagner, P. A., & Francisco, R. T. (2014b). Teacher support for collective argumentation: A framework for examining how teachers support students' engagement in mathematical activities. *Educational Studies in Mathematics, 86*(3), 401–429. https://doi.org/10.1007/s10649-014-9532-8

Conner, A., Tabach, M., & Rasmussen, C. (2022). Collectively engaging with others' reasoning: Building intuition through argumentation in a paradoxical situation. In *International Journal of Research in Undergraduate Mathematics Education*. Advance online publication. https://doi.org/10.1007/s40753-022-00168-x

Consortium for the Advancement of Undergraduate Statistics Education (CAUSE) [website]. (2022). https://www.causeweb.org/cause/

Cooper, L. L., & Shore, F. S. (2008). Students' misconceptions in interpreting center and variability of data represented via histograms and stem-and-leaf plots. *Journal of Statistics Education, 16*(2), 1. https://doi.org/10.1080/10691898.2008.11889559

Cooper, L. L., & Shore, F. S. (2010). The effects of data and graph type on concepts and visualizations of variability. *Journal of Statistics Education, 18*(2). http://jse.amstat.org/v18n2/cooper.pdf

Cope, B., & Kalantzis, M. (2016). Big data comes to school: Implications for learning, assessment, and research. *AERA Open, 2*(2), 2332858416641907. https://doi.org/10.1177/2332858416641907

Crooks, N., Bartel, A., & Albali, M. (2019). Conceptual knowledge of confidence intervals in psychology undergraduate and graduate students. *Statistics Education Research Journal, 18*(1), 46–62. https://doi.org/10.52041/serj.v18i1.149

Cui, H., & Zhang, D. (2018). Strategies on teacher professional development in big data era. In *2018 4th International Conference on Education Technology, Management and Humanities Science (ETMHS 2018)*. Atlantis Press.

Cyrino, M., & Grando, R. (2022). (Des)construção curricular necessária: Resistir, (re)existir, possibilidades insubordinadas criativamente. [Necessary curricular (de)construction: Resist, (re) exist, creatively insubordinate possibilities]. *Revista de Educação Matemática, 19* (edição especial), 1–25. https://www.revistasbemsp.com.br/index.php/REMat-SP/article/view/728

D'Ignazio, C. (2017). Creative data literacy: Bridging the gap between the data-haves and data-have nots. *Information Design Journal, 23*(1), 6–18. https://doi.org/10.1075/idj.23.1.03dig

Da Valle, S., & Osti, S. (2016). Statistica enigmista: An ISTAT puzzle magazine to introduce. In J. Engel (Ed.), *Proceedings of the roundtable conference of the International Association of Statistics Education (IASE).*

Dai, Q., Chen, Y., & Hua, G. (2017). Relationship between big data and teaching evaluation. In *Proceedings of the 3rd International Conference on Education and Social Development – ICESD 2017* (pp. 8–9).

Dai, H., Tao, Y., & Shi, T. W. (2018). Research on mobile learning and micro course in the big data environment. In *Proceedings of the 2nd international conference on e-education, e-business and e-technology* (pp. 48–51).

Darling-Hammond, L., Hyler, M., & Gardner, M. (2017). *Effective teacher professional development*. Learning Policy Institute. https://files.eric.ed.gov/fulltext/ED606743.pdf

David, I., & Maligalig, D. (2006). Are we teaching statistics correctly to our youth? *The Philippine Statistician, 55*(3–4), 1–28.

de Vetten, A., Schoonenboom, J., Keijzer, R., & van Oers, B. (2018). The development of informal statistical inference content knowledge of pre-service primary school teachers during a teacher college intervention. *Educational Studies in Mathematics, 99*(2), 217–234. https://doi.org/10.1007/s10649-018-9823-6

del Mas, R. C. (2004). A comparison of mathematical and statistical reasoning. In D. Ben-Zvi & J. Garfield (Eds.), *The challenge of developing statistical literacy, reasoning, and thinking* (pp. 79–95). Kluwer. https://doi.org/10.1007/1-4020-2278-6_4

del Mas, R., Garfield, J., & Chance, B. (1999). A model of classroom research in action: Developing simulation activities to improve students' statistical reasoning. *Journal of Statistics Education, 7*(3). https://doi.org/10.1080/10691898.1999.12131279

Dempster, M., & McCorry, N. K. (2009). The role of previous experience and attitudes toward statistics in statistics assessment outcomes among undergraduate psychology students. *Journal of Statistics Education, 17*(2). https://doi.org/10.1080/10691898.2009.11889515

Deng, M. (2017). Analysis on value evolution of higher vocational teachers and its development paths under the big data background. In *3rd International Conference on Arts, Design and Contemporary Education, ICADCE 2017* (pp. 738–740). Atlantis Press.

Denzin, N. K., & Lincoln, Y. (2012). *El campo de la investigación cualitativa.* (The field of qualitative research). *Manual de investigación cualitativa, 1.* Gedisa.

Department for International Development. (2015). *UK aid: Tackling global challenges in the national interest.*

Department of Basic Education (DBE), Republic of South Africa. (2011). *Curriculum and assessment policy statement grades 4–6: Life skills.* DBE, Republic of South Africa.

Department of Education. (1997). *Curriculum 2005 lifelong learning for the 21st century.* National Department of Education.

Department of Education. (2002). *Revised national curriculum statement grades R-9 (schools): Mathematics.* Department of Education.

Department of Education. (2013). *The K to 12 curriculum guide for mathematics.* Department of Education. Retrieved from http://www.deped.gov.ph/

Department of Education. (2020, July 3). *Basic education learning continuity plan.* Retrieved from Department of Education. https://www.deped.gov.ph/wp-content/uploads/2020/07/DepEd_LCP_July3.pdf

Department of Education. (2021). *Department of Education Education Management Information System Division.* Retrieved from Number of enrollment in all sector senior high school, from AY 2017–18 to 2020–21. https://www.deped.gov.ph/alternative-learning-system/resources/facts-and-figures/datasets/

Derouet, C., & Parzysz, B. (2016). How can histograms be useful for introducing continuous probability distributions? *ZDM Mathematics Education, 48*(6), 757–773. https://doi-org.proxy.library.uu.nl/10.1007/s11858-016-0769-9

Díaz-Levicoy, D., Morales, R., & López-Martín, M. M. (2015). Tablas estadísticas en libros de texto chilenos de 1° y 2° año de Educación Primaria (Statistical tables in textbook Chilean 1st and 2nd grade of primary school). *Revista Paranaense de Educação Matemática, 4*(7), 10–39. https://doi.org/10.33871/22385800.2015.4.7.10-39

Díaz-Levicoy, D., Morales, R., Arteaga, P., & López-Martín, M. M. (2020). Conocimiento sobre tablas estadísticas por estudiantes chilenos de tercer año de Educación Primaria (Knowledge of statistical tables by Chilean students third year of primary education). *Educación Matemática, 32*(2), 247–277. https://doi.org/10.24844/EM3202.10

Dierdorp, A., Bakker, A., Ben-Zvi, D., & Makar, K. (2017). Secondary students' considerations of variability in measurement activities based on authentic practices. *Statistics Education Research Journal, 16*(2), 397–418. https://doi.org/10.52041/serj.v16i2.198

Doerr, H. M., & English, L. D. (2003). A modeling perspective on students' mathematical reasoning about data. *Journal for Research in Mathematics Education, 34*(2), 110–136. https://doi.org/10.2307/30034902

Dreyfus, T., Kouropatov, A., & Ron, K. (2021). Research as a resource in a high-school calculus curriculum. *ZDM – Mathematics Education, 53*, 679–693. https://doi.org/10.1007/s11858-021-01236-3

Drijvers, P. (2012). Teachers transforming resources into orchestrations. In G. Gueudet, B. Pepin, & L. Trouche (Eds.), *From text to 'lived' resources: Mathematics curriculum materials and teacher development* (pp. 265–281). Springer.

Drijvers, P. (2015). Digital technology in mathematics education: Why it works (or doesn't). In S. J. Cho (Ed.), *Selected regular lectures from the 12th international congress on mathematical education* (pp. 135–151). Springer. https://doi.org/10.1007/978-3-319-17187-6_8

Drijvers, P. (2019). Embodied instrumentation: Combining different views on using digital technology in mathematics education. In U. T. Jankvist, M. van den Heuvel-Panhuizen, & M. Veldhuis (Eds.), *Proceedings of the 11th congress of the European society for research in mathematics education* (pp. 8–28). Freudenthal Group & Freudenthal Institute, Utrecht University & ERME. https://hal.archives-ouvertes.fr/hal-02436279v1

Drozda, Z., Johnstone, D., & Van Horne, B. (2022). Previewing the national landscape of K-12 data science implementation. Paper commissioned for the *Workshop on foundations of data science for students in grades K-12*. Board on science education, the board on mathematical sciences and analytics, the computer science and telecommunications board, and the National Academy of Sciences, engineering, and medicine.

Duval, R. (2003). Comment analyser le fonctionnement représentationnel des tableaux et leur diversité? (How to analyse the representational functioning of the tables and their diversity?). *Spirale-Revue de recherches en éducation, 32*(32), 7–31. https://doi.org/10.3406/spira.2003.1377

Dvir, M., & Ben-Zvi, D. (2021). Informal statistical models and modeling. *Mathematical Thinking and Learning, 1–21*, 79–99. https://doi.org/10.1080/10986065.2021.1925842

Eichler, A., & Vogel, M. (2013). *Leitidee Daten und Zufall. Von konkreten Beispielen zur Didaktik der Stochastik [guiding principle data and chance. From concrete examples to the didactics of stochastics]*. Springer Spektrum.

Ekol, G. (2015). Exploring foundation concepts in introductory statistics using dynamic data points. *International Journal of Education in Mathematics, Science and Technology, 3*(3), 230–241. https://doi.org/10.18404/ijemst.15371

Emmioglu, E., & Capa-Aydin, Y. (2012). Attitudes and achievement in statistics: A meta-analysis study. *Statistics Education Research Journal, 11*(2), 95–102. https://doi.org/10.52041/serj.v11i2.332

Engel, J. (2017). Statistical literacy for active citizenship: A call for data science education. *Statistics Education Research Journal, 16*(1), 44–49. https://doi.org/10.52041/serj.v16i1.213

Engel, J., & Sedlmeier, P. (2011). Correlation and regression in the training of teachers. In *Teaching statistics in school mathematics-challenges for teaching and teacher education* (pp. 247–258). Springer.

Engel, J., Sedlmeier, P., & Wörn, C. (2008). Modeling scatterplot data and the signal-noise metaphor: Towards statistical literacy for pre-service teachers. In C. Batanero, G. Burrill, C. Reading, & A. Rossman (Eds.), *Proceedings of the ICMI study 18 and IASE round table conference*. International Commission on Mathematics Instruction and International Association for Statistical Education.

Engel, J., Ridgway, J., & Weber, F. (2021). Educación estadística, democracia y empoderamiento de los ciudadanos (Statistics education, democracy and citizen's empowerment). *Paradigma, 42*(Extra-1), 1–31. https://doi.org/10.37618/PARADIGMA.1011-2251.2021.p01-31.id1016

English, L. D. (2010). Young children's early modelling with data. *Mathematics Education Research Journal, 22*(2), 24–47. https://doi.org/10.1007/BF03217564

English, L. D. (2012). Data modelling with first-grade students. *Educational Studies in Mathematics, 81*(1), 15–30. https://doi.org/10.1007/s10649-011-9377-3

English, L. D., & Watson, J. M. (2015). Exploring variation in measurement as a foundation for statistical thinking in the elementary school. *International Journal of STEM Education, 2*(1), 1–20. https://doi.org/10.1186/s40594-015-0016-x

English, L. D., & Watson, J. M. (2016). Development of probabilistic understanding in fourth grade. *Journal for Research in Mathematics Education, 47*(1), 28–62. https://doi.org/10.5951/jresematheduc.47.1.0028

English, L. D., & Watson, J. (2018). Modelling with authentic data in sixth grade. *ZDM, 50*(1), 103–115. https://doi.org/10.1007/s11858-017-0896-y

Ernest, P. (2015). The social outcomes of learning mathematics: Standard, unintended or visionary? *International Journal of Education in Mathematics, Science and Technology, 3*(3), 187–192.

Ertaş, G., & Aslan-Tutak, F. (2021). Mathematics teacher education in Turkey through the lens of international TEDS-M study. *REDIMAT–Journal of Research in Mathematics Education, 10*(2), 152–174.

Estepa, A., & Sánchez Cobo, F. T. (2001). Empirical research on the understanding of association and implications for the training of researchers. In C. Batanero (Ed.), *Training researchers in the use of statistics* (pp. 37–51). International Association for Statistical Education and International Statistical Institute.

Estrella, S. (2014). El formato tabular: una revisión de literatura (Tabular format: a review of literature). *Revista Actualidades Investigativas en Educación, 14*(2), 1–23. https://doi.org/10.15517/AIE.V14I2.14817

Estrella, S. (2018). Data representations in early statistics: Data sense, meta-representational competence and transnumeration. In A. Leavy, M. Meletiou-Mavrotheris, & E. Paparistodemou (Eds.), *Statistics in early childhood and primary education – Supporting early statistical and probabilistic thinking* (pp. 239–256). Springer. https://doi.org/10.1007/978-981-13-1044-7_14

Estrella, S., Mena-Lorca, A., & Olfos-Ayarza, R. (2017). Naturaleza del objeto matemático Tabla (Nature of the mathematical object table). *MAGIS, 10*(20), 105–122. https://doi.org/10.11144/Javeriana.m10-20.nomt

Estrella, S., Mena, A., & Olfos, R. (2018). Lesson study in Chile: A very promising but still uncertain path. In M. Quaresma, C. Winsløw, S. Clivaz, J. da Ponte, A. Ní Shúilleabháin, & A. Takahashi (Eds.), *Mathematics lesson study around the world: Theoretical and methodological issues* (pp. 105–122). Springer. https://doi.org/10.1007/978-3-319-75696-7

Estrella, S., Zakaryan, D., Olfos, R., & Espinoza, G. (2020). How teachers learn to maintain the cognitive demand of tasks through lesson study. *Journal of Mathematics Teacher Education, 23*, 293–310. https://doi.org/10.1007/s10857-018-09423-y

Estrella, S., Vergara, A., & González, O. (2021). Developing data sense: Making inferences from variability in tsunamis at primary school. *Statistics Education Research Journal, 20*(2), 16. https://doi.org/10.52041/serj.v20i2.413

Estrella, S., Méndez-Reina, M., Olfos, R., & Aguilera, J. (2022). Early statistics in kindergarten: Analysis of an educator's pedagogical content knowledge in lessons promoting informal inferential reasoning. *International Journal for Lesson and Learning Studies, 11*(1), 1–13. https://doi.org/10.1108/IJLLS-07-2021-0061

Evangelista, B., & Guimarães, G. (2019). Análise de atividades sobre tabelas em livros didáticos brasileiros dos anos iniciais do ensino fundamental (Analysis of activities about tables in the primary school Brazilian textbooks). In J. M. Contreras, M. M. Gea, M. M. López-Martín, & E. Molina-Portillo (Eds.), *Actas del Tercer Congreso Internacional Virtual de Educación Estadística*. www.ugr.es/local/fqm126/civeest.html

Fan, L., Zhu, Y., & Miao, Z. (2013). Textbook research in mathematics education: Development status and directions. *Zentralblatt für Didaktik der Mathematik, 45*(5), 633–646. https://doi.org/10.1007/s11858-013-0539-x

Fathom [Computer software]. (2022). Retrieved from https://fathom.concord.org/

Fazekas, M., & Kocsis, G. (2020). Uncovering high-level corruption: Cross-national objective corruption risk indicators using public procurement data. *British Journal of Political Science, 50*(1), 155–164. https://doi.org/10.1017/S0007123417000461

Feinberg, R., & Wainer, H. (2011). Extracting sunbeams from cucumbers. *Journal of Computational and Graphical Statistics, 20*(4), 793–810. https://doi.org/10.1198/jcgs.2011.204a

Fidler, F. (2006). Should psychology abandon *p*-values and teach CIs instead? Evidence-based reforms in statistics education. In A. Rossman & B. Chance (Eds.), *Working cooperatively in statistics education: Proceedings of the seventh international conference on teaching statistics (ICOTS-7)*. Salvador. https://iase-web.org/documents/papers/icots7/5E4_FIDL.pdf

Fidler, F., & Loftus, G. R. (2009). Why figures with error bars should replace *p*-values: Some conceptual arguments and empirical demonstrations. *Zeitschrift für Psychologie/Journal of Psychology, 217*(1), 27–37.

Fielding-Wells, J. (2014). Where's your evidence? Challenging young students' equiprobability bias through argumentation. In K. Makar, B. DeSousa, & R. Gould (Eds.), S*ustainability in statistics education. Proceedings of the ninth international conference on teaching statistics (ICOTS9, July, 2014), Flagstaff*. International Statistical Institute. https://iase-web.org/icots/9/proceedings/pdfs/ICOTS9_2B2_FIELDINGWELLS.pdf

Fielding-Wells, J. (2018). Dot plots and hat plots: Supporting young students emerging understandings of distribution, center and variability through modeling. *ZDM, 50*(7), 1125–1138. https://doi.org/10.1007/s11858-018-0961-1

Fielding-Wells, J., & Makar, K. (2015). Inferring to a model: Using inquiry-based argumentation to challenge young children's expectations of equally likely outcomes. In A. Zieffler & E. Fry (Eds.), *Reasoning about uncertainty: Learning and teaching informal inferential reasoning* (pp. 1–27). Catalyst Press.

Findley, K., & Berens, F. (2020). Assessing the disciplinary perspectives of introductory statistics students. In S. S. Karunakaran, Z. Reed, & A. Higgins (Eds.), *Proceedings of the 23rd annual conference on research in undergraduate mathematics education* (pp. 1099–1104). https://www.researchgate.net/publication/339712352_Assessing_the_Disciplinary_Perspectives_of_Introductory_Statistics_Students

Finney, S. J., & Schraw, G. (2003). Self-efficacy beliefs in college statistics courses. *Contemporary Educational Psychology, 28*(2), 161–186. https://doi.org/10.1016/S0361-476X(02)00015-2

Fiofanova, O. A. (2020). New literacy and data-future in education: Advanced technology smart big-data. *Revista Inclusiones, 7*, 174–180. http://revistainclusiones.org/index.php/inclu/article/view/1276

Fishbein, M., & Ajzen, I. (2009). *Predicting and changing behavior: The reasoned action approach*. Psychology Press. https://doi.org/10.4324/9780203838020

Foster, J. K., Zhuang, Y., Conner, A., Park, H., & Singletary, L. (2020). One Teacher's Analysis of Her Questioning in Support of Collective Argumentation. In *Mathematics education across cultures: Proceedings of the 42nd annual meeting of the North American chapter of the international group for the psychology of mathematics education* (pp. 2067–2071). Cinvestav/AMIUTEM/PME-NA. https://doi.org/10.51272/pmena.42.2020

François, K., Monteiro, C., & Allo, P. (2020). Big data literacy as new vocation for statistical literacy. *Statistics Education Research Journal, 19*(1), 194–205. https://doi.org/10.52041/serj.v19i1.130

Frank, M., Walker, J., Attard, J., & Tygel, A. (2016). Data literacy-what is it and how can we make it happen? *The Journal of Community Informatics, 12*(3), 4–8. https://doi.org/10.15353/joci.v12i3.3274

Franklin, C. (2021). As Covid makes clear, statistics education is a must. *Significance, 18*(2), 35.

Franklin, C., & Mewborn, D. (2006). The statistical education of grades pre-K-2 teachers: A shared responsibility. In G. Burrill (Ed.), *NCTM 2006 yearbook: Thinking and reasoning with data and Chance* (pp. 335–344). NCTM.

Franklin, C., Kader, G., Mewborn, D. S., Moreno, J., Peck, R., Perry, M., & Scheaffer, R. (2005a). *Guidelines for assessment and instruction in statistics education (GAISE) report: A pre-K–12 curriculum framework*. American Statistical Association.

Franklin, C., Kader, G., Mewborn, D., Moreno, J., Peck, R., Perry, M., & Scheaffer, R. (2005b). *Guidelines for assessment and instruction in statistics education (GAISE) report*. American Statistical Association.

Franklin, C., Kader, G., Mewborn, D., Moreno, J., Peck, R., Perry, M., & Scheaffer, R. (2007a). *Guidelines for assessment and instruction in statistics education (GAISE) report: A pre-K-12 curriculum framework*. American Statistical Association. https://www.amstat.org/docs/default-source/amstat-documents/gaiseprek-12_full.pdf

Franklin, C., Kader, G., Mewborn, D., Moreno, J., Peck, R., Perry, M., et al. (2007b). *Guidelines for assessment and instruction in statistics education (GAISE) report: A preK-12 curriculum framework*. American Statistical Association.

Franklin, C., Kader, G., Mewborn, D., Moreno, J., Peck, R., Perry, M., & Scheaffer, R. (2007c). *Guidelines for assessment and instruction in statistics education (GAISE) report*. American Statistical Association.

Franklin, C., Bargagliotti, A., Case, C., Kader, G., Scheaffer, R., & Spangler, D. A. (2015a). *The statistical education of teachers*. American Statistical Association. https://www.amstat.org/docs/default-source/amstat-documents/edu-set.pdf

Franklin, C., Bargagliotti, A. E., Case, C. A., Kader, G. D., Schaeffer, R. L., & Spangler, D. A. (2015b). *The statistical education of teachers*. American Statistical Association.

Freedman, D., Pisani, R., & Purves, R. (1978). *Statistics*. W. W. Norton & Co.

Friel, S. N., Curcio, F. R., & Bright, G. W. (2001). Making sense of graphs: Critical factors influencing comprehension and instructional implications. *Journal for Research in Mathematics Education, 32*(2), 124–158. https://doi.org/10.2307/749671

Frischemeier, D. (2017). *Statistisch denken und forschen lernen mit der software TinkerPlots [learning to think and research statistically with the software TinkerPlots]*. Springer Spektrum.

Frischemeier, D. (2018). Design, implementation, and evaluation of an instructional sequence to lead primary school students to comparing groups in statistical projects. In A. Leavy, M. Meletiou-Mavrotheris, & E. Paparistodemou (Eds.), *Statistics in early childhood and primary education* (pp. 217–238). Springer Nature. https://doi.org/10.1007/978-981-13-1044-7_13

Frischemeier, D. (2019). Primary school students' reasoning when comparing groups using modal clumps, medians, and hatplots. *Mathematics Education Research Journal, 31*(4), 485–505. https://doi.org/10.1007/s13394-019-00261-6

Frischemeier, D., & Schnell, S. (2021). Statistical investigations in primary school–the role of contextual expectations for data analysis. *Mathematics Education Research Journal*, 1–26. https://doi.org/10.1007/s13394-021-00396-5

Frischemeier, D., Biehler, R., Podworny, S., & Budde, L. (2021). A first introduction to data science education in secondary schools: Teaching and learning about data exploration with CODAP using survey data. *Teaching Statistics, 43*(S1). https://doi.org/10.1111/test.12283

Fujita, T., Kazak, S., Turmo, M., & Mansour, N. (2018). *Strategic partnership for innovative in data analytics in schools (SPIDAS)*. State of the Art Review.

Fülöp, É. (2019). *Learning to solve problems that you have not learned to solve: Strategies in mathematical problem solving*. Doctoral Theses from University of Gothenburg. http://hdl.handle.net/2077/60464

Gabucio, F., Martí, E., Enfedaque, J., Gilabert, S., & Konstantinidou, A. (2010). Niveles de comprensión de las tablas en alumnos de primaria y secundaria (Levels of graph comprehension in primary and secondary school students). *Cultura y Educación, 22*(2), 183–197. https://doi.org/10.1174/113564010791304528

GAISE College Report ASA Revision Committee. (2016). *Guidelines for assessment and instruction in statistics education college report*. http://www.amstat.org/education/gaise

Gal, I. (2002a). Adults' statistical literacy: Meanings, components, and responsibilities. *International Statistical Review, 70*(1), 1–25.

Gal, I. (2002b). Adults' Statistical Literacy: Meanings, components, responsibilities. *International Statistical Review / Revue Internationale de Statistique, 70*(1), 1. https://doi.org/10.2307/1403713

Gal, I., Ginsburg, L., & Schau, C. (1997). Monitoring attitudes and beliefs in statistics education. In I. Gal, J. Garfield, & Y. Gal (Eds.), *The assessment challenge in statistics education* (pp. 37–51). IOS Press. https://www.stat.auckland.ac.nz/~iase/publications/assessbkref

Galotti, K. (1989). Approaches to studying formal and everyday reasoning. *Psychological Bulletin, 105*(3), 331–351. https://doi.org/10.1037/0033-2909.105.3.331

García-García, J., Díaz-Levicoy, D., Vidal, H., & Arredondo, E. (2019). Las tablas estadísticas en libros de texto de educación primaria en México (Statistical tables in primary education textbooks in Mexico). *Paradigma, 40*(2), 153–175. https://doi.org/10.37618/PARADIGMA.1011-2251.2019.p153-175.id754

García-Pérez, M. A., & Alcalá-Quintana, R. (2016). The interpretation of scholars' interpretations of confidence intervals: Criticism, replication, and extension of Hoekstra et al. (2014). *Frontiers in Psychology, 7*, 1–12. https://doi.org/10.3389/fpsyg.2016.01042

Garfield, J. (2002). The challenge of developing statistical reasoning. *Journal of Statistics Education, 10*(3). https://doi.org/10.1080/10691898.2002.11910676

Garfield, J., & Ben Zvi, D. (2004). Research on statistical literacy, reasoning, and thinking: Issues, challenges, and implications. In D. Ben Zvi & J. Garfield (Eds.), *The challenge of developing statistical literacy, reasoning and thinking* (pp. 397–409). Springer.

Garfield, J., & Ben-Zvi, D. (2008a). *Developing students' statistical reasoning*. Springer.

Garfield, J., & Ben-Zvi, D. (2008b). *Developing students' statistical reasoning: Connecting research and teaching practice*. Springer. https://doi.org/10.1007/978-1-4020-8383-9

Garfield, J., & Ben-Zvi, D. (2009). Helping students develop statistical reasoning: Implementing a statistical reasoning learning environment. *Teaching Statistics, 31*(3), 72–77. https://doi.org/10.1111/j.1467-9639.2009.00363.x

Garfield, J., Le, L., Zieffler, A., & Ben-Zvi, D. (2015). Developing students' reasoning about samples and sampling variability as a path to expert statistical thinking. *Educational Studies in Mathematics, 88*(3), 327–342. https://doi.org/10.1007/s10649-014-9541-7

Gea, M., Batanero, C., Fernandes, J., & Arteaga, P. (2016). Interpretación de resúmenes estadísticos por futuros profesores de educación secundaria (Interpretation of summaries on statistics addressed to prospective teachers in middle school). *REDIMAT, 5*(2), 135–157. https://doi.org/10.17583/redimat.2016.1902

Gilmartin, K., & Rex, K. (2000). *Student toolkit: More charts, graphs and tables*. Open University. https://ahpo.net/assets/more-charts-graphs-and-tables-toolkit.pdf

Giroux, H. (2006). *La escuela y la lucha por la ciudadanía. Pedagogía crítica de la época moderna* [The school and the struggle for citizenship. Critical pedagogy of the modern age] (4th ed.). Siglo XXI.

Glencross, M. J., & Binyavanga, K. W. (1997). The role of technology in statistics education: A view from a developing region. In J. Garfield & G. Burrill (Eds.), *Research on the role of technology in teaching and learning statistics* (pp. 301–308). International Statistical Institute.

Gnanadesikan, M., & Scheaffer, R. (1987). The art and technique of simulation. Pearson Learning.

Godino, J., Aké, L., Gonzato, M., & Wilhelmi, M. (2014). Niveles de algebrización de la actividad matemática escolar. Implicaciones para la formación de maestros (Algebrization levels of school mathematics activity. Implication for primary school teacher education). *Enseñanza de las Ciencias, 32*(1), 199–219. https://doi.org/10.5565/rev/ensciencias.965

Godino, J., Neto, T., Wilhelmi, M., Aké, L., Etchegaray, S., & Lasa, A. (2015a). Niveles de algebrización de las prácticas matemáticas escolares. Articulación de las perspectivas ontosemiótica y antropológica (Algebrization levels of school mathematics practices. Networking of the Onto-semiotic and Anthropological perspectives). *Avances de Investigación en Educación Matemática, 8*, 117–142. https://doi.org/10.35763/aiem.v1i8.105

Godino, J. D., Neto, T., Wilhelmi, M. R., Aké, L., Etchegaray, S., & Lasa, A. (2015b). Algebraic reasoning levels in primary and secondary education. In K. Krainer & N. Vondrová (Eds.), *Proceedings of the ninth congress of the European society for research in mathematics education* (CERME 9) (pp. 426–432). ERME.

Gökce, R. (2019). *Ortaokul matematik öğretmenlerinin istatistiksel akıl yürütmeye ilişkin alan ve pedagojik alan bilgilerinin incelenmesi* [*Examining middle school mathematics teachers' content and pedagogical content knowledge of statistical reasoning*] (Unpublished doctoral dissertation). Pamukkale University.

Goldin, G. A. (2000). A scientific perspective on structured, task-based interviews in mathematics education research. In A. E. Kelly & R. A. Lesh (Eds.), *Handbook of research design in mathematics and science education*. Lawrence Erlbaum Associates.

Gomez Marchant, C. N., Park, H., Zhuang, Y., Foster, J., & Conner, A. (2021). Theory to practice: Prospective mathematics teachers' recontextualizing discourses surrounding collective argumentation. *Journal of Mathematics Teacher Education, 24*, 1–29. https://doi.org/10.1007/s10857-021-09500-9

Gómez-Blancarte, A., & Tobías-Lara, M. G. (2018). Using the Toulmin model of argumentation to validate students' inferential reasoning. In M. A. Sorto, A. White, & L. Guyot (Eds.), *Looking back, looking forward. Proceedings of the tenth international conference on teaching statistics (ICOTS20, July, 2018), Kyoto*. International Statistical Institute. https://iase-web.org/icots/10/proceedings/pdfs/ICOTS10_8C1.pdf

González, M. T., Espinel, M. C., & Ainley, J. (2011). Teachers' graphical competence. In C. Batanero, G. Burrill, & C. Reading (Eds.), *Teaching statistics in school mathematics-challenges for teaching and teacher education* (pp. 187–197). Springer. https://doi.org/10.1007/978-94-007-1131-0_20

Gordon, S. (2004). Understanding students′ experiences of statistics in a service course. *Statistics Education Research Journal, 3*(1), 40–59. https://doi.org/10.52041/serj.v3i1

Gould, R. (2010). *Statistics and the modern student*. Department of Statistics, UCLA. Retrieved from https://escholarship.org/uc/item/9p97w3zf

Grant, T., & Nathan, M. (2008). *Students' conceptual metaphors influence their statistical reasoning about confidence intervals*. WCER Working Paper No. 2008-5. Wisconsin: Wisconsin Center for Education Research. [Online: https://wcer.wisc.edu/docs/working-papers/Working_Paper_No_2008_05.pdf]

Gratzer, W., & Carpenter, J. E. (2008/2009). The histogram-area connection. *The Mathematics Teacher, 102*(5), 226–340.

Gravetter, F. J., & Wallnau, L. B. (2013). Introduction to regression. In *Statistics for the behavioral sciences* (pp. 558–572). Wadsworth.

Green, D. (1993). Data analysis: What research do we need? In L. Pereira-Mendoza. (Ed.), Introducing data analysis in the schools: Who should teach it and how? *Proceedings of the international statistical institute round table conference, August 10–14, 1992*.

Groth, R. E. (2007). Toward a conceptualization of statistical knowledge for teaching. *Journal for Research in Mathematics Education, 38*(5), 427–437.

Groth, R. E. (2009). Characteristics of teachers' conversations about mean, median, and mode. *Teaching and Teacher Education, 25*, 707–716. https://doi.org/10.1016/jtate.2008.11.005

Groth, R. E. (2013). Characterizing key developmental understandings and pedagogically powerful ideas within a statistical knowledge for teaching framework. *Mathematical Thinking and Learning, 15*(2), 121–145. https://doi.org/10.1080/10986065.2013.770718

Guerrero, O. (2008). *Educación matemática crítica: Influencias teóricas y aportes*. [Critical mathematics education: Theoretical influences and contributions]. *Evaluación e Investigación, 3*(1), 63–78.

Gutiérrez, R. (2009). Embracing the inherent tensions in teaching mathematics from an equity stance. *Democracy & Education, 18*(3), 9–16.

Gutiérrez, R. (2013). The sociopolitical turn in mathematics education. *Journal for Research in Mathematics Education, 44*(1), 37–68.

Gutstein, E. (2003). Teaching and learning mathematics for social justice in an urban, Latino school. *Journal for Research in Mathematics Education, 34*(1), 37–73.

Gutstein, E. (2006). *Reading and writing the world with mathematics: Toward a pedagogy for social justice*. Routledge.

Gutstein, E. (2007). Possibilities and challenges in teaching mathematics for social justice. In *Third annual National Research Symposium of the Maryland Institute for Minority Achievement and Urban Education*. University of Illinois-Chicago.

Harel, G. (2013). DNR-based curricula: The case of complex numbers. *Journal of Humanistic Mathematics, 3*(2), 2–61. https://doi.org/10.5642/jhummath.201302.03

Harradine, A., & Konold, C. (2006). How representational medium affects the data displays students make. *Seventh international conference on teaching statistics, Salvador.*

Hasemann, K., & Mirwald, E. (2012). Daten, häufigkeit und wahrscheinlichkeit (Data, frequency and chance). In G. Walther, M. van den Heuvel-Panhuizen, D. Granzer, & O. Köller (Eds.), *Bildungsstandards für die grundschule: Mathematik konkret* (Educational standards for elementary school: Mathematics in concrete terms) (pp. 141–161). Cornelsen Scriptor.

Heitele, D. (1975). An epistomological view on fundamental stochastic ideas. *Educational Studies in Mathematics, 6*(2), 187–205.

Henriques, A. (2016). Students' difficulties in understanding of confidence intervals. In D. Ben-Zvi & K. Makar (Eds.), *The teaching and learning of statistics* (pp. 129–138). Springer. https://doi.org/10.1007/978-3-319-23470-0_18

Henriques, A., & Oliverira, H. M. (2016). Students' expressions of uncertainty in making informal inference when engaged in a statistical investigation using Tinkerplots. *Statistics Education Research Journal, 15*(2), 62–80. https://doi.org/10.52041/serj.v15i2.24115(2)

Herbel-Eisenmann, B. A. (2007). From intended curriculum to written curriculum: Examining the voice of a mathematics textbook. *Journal for Research in Mathematics Education, 38*(4), 344–369. https://doi.org/10.2307/30034878

Hill, H. C., Ball, D. L., & Schilling, S. G. (2008). Unpacking pedagogical content knowledge: Conceptualizing and measuring teachers' topic-specific knowledge of students. *Journal for Research in Mathematics Education, 39*(4), 372–400. https://doi.org/10.5951/jresematheduc.39.4.0372

Hoekstra, R., Kiers, H., & Johnson, A. (2012). Are assumptions of well-known statistical techniques checked, and why (not)? *Frontiers in Psychology, 3*. https://doi.org/10.3389/fpsyg.2012.00137

Hoekstra, R., Morey, R., Rouder, J., & Wagenmakers, E. (2014). Robust misinterpretation of confidence intervals. *Psychonomic Bulletin & Review, 21*(5), 1157–1164. https://doi.org/10.3758/s13423-013-0572-3

Hoel, T., Chen, W., & Lu, Y. (2020). Teachers' perceptions of data management as educational resource: A comparative case study from China and Norway. *Nordic Journal of Digital Literacy, 15*(3), 178–189. https://doi.org/10.18261/issn.1891-943x-2020-03-04

Holmqvist, M., & Lindgren, G. (2009). Students learning English as second language: An applied linguistics learning study. *Problems of Education in the 21st Century, 2009*(18), 86–96.

Holmqvist, M., & Selin, P. (2019). What makes the difference? An empirical comparison of critical aspects identified in phenomenographic and variation theory analyses. *Palgrave Communications, 5*(71), 1–8. https://doi.org/10.1057/s41599-019-0284-z

Holmqvist, M., Gustavsson, L., & Wernberg, A. (2007). Generative learning: Learning beyond the learning situation. *Educational Action Research, 15*(2), 181–208. https://doi.org/10.1080/09650790701314684

Huang, L., Wei, Y., Zamboni, A., Zhang, J., & Xu, H. (2015). Big data analysis in a social learning platform. In *4th International conference on computer, mechatronics, control and electronic engineering* (pp. 1467–1470). Atlantis Press.

Huck, S. W. (2016). *Statistical misconceptions. Classic edition*. Routledge.

Huda, M., Maseleno, A., Shahrill, M., Jasmi, K. A., Mustari, I., & Basiron, B. (2017). Exploring adaptive teaching competencies in big data era. *International Journal of Emerging Technologies in Learning (IJET), 12*(03), 68. https://doi.org/10.3991/ijet.v12i03.6434

Illustrative Mathematics. https://curriculum.illustrativemathematics.org/HS/index.html

Inglis, M., Mejia-Ramos, J. P., & Simpson, A. (2007). Modelling mathematical argumentation: The importance of qualification. *Educational Studies in Mathematics, 66*(1), 3–21. https://doi.org/10.1007/s10649-006-9059-8

Innabi, H. (2007). Factors considered by secondary students when judging the validity of a given statistical generalization. *International electronic journal of mathematics education*. Special Issue. Emerging Research in Statistics Education, *3*(2), 168–186.

Innabi, H., & Emanuelsson, J. (2021). Enrichment in school principals' ways of seeing mathematics. *International Journal of Mathematical Education in Science and Technology, 52*(10), 1508–1539. https://doi.org/10.1080/0020739X.2020.1782496

International Association for Statistical Education (IASE) [website]. (2022). https://iase-web.org/

International Data Science in Schools Project (IDSSP). (2019). *Curriculum frameworks for introductory data science*. http://idssp.org/files/IDSSP_Frameworks_1.0.pdf

Investigations in Number, Data and Space®. (2017). (3rd edition). Pearson.

Isoda, M., & Olfos, R. (2009). *El enfoque de resolución de problemas en la enseñanza de la matemática a partir del estudio de clases* (*The problem-solving approach in the teaching of mathematics from the lesson study*). Ediciones Universitarias de Valparaíso.

Isoda, M., & Olfos, R. (2021). *Teaching multiplication with lesson study*. Springer. https://doi.org/10.1007/978-3-030-28561-6

Isoda, M., Olfos, R., Estrella, S., & Baldin, Y. (2022). Two contributions of japanese lesson study for the mathematics teacher education: The effective terminology for designing lessons and as a driving force to promote sustainable study groups. *Educação Matemática Em Revista, 1*(23), 98–112. https://doi.org/10.37001/EMR-RS.v.2.n.23.2022.p.98-112

Jahnke, H. N. (2019). Mathematics and Bildung 1810 to 1850. In H. N. Jahnke & L. Hefendehl-Hebeker (Eds.), *Traditions in German-speaking mathematics education research* (pp. 115–140). Springer International Publishing. https://doi.org/10.1007/978-3-030-11069-7_5

Jaramillo, D. (2003). *(Re)constituição do ideário de futuros professores de Matemática num contexto de investigação sobre a prática pedagógica.* [(Re)constitution of the ideas of prospective mathematics teachers in a context of research on pedagogical practice]. Tesis Doctoral. Universidade Estadual de Campinas, Brasil.

Justice, N., Morris, S., Henry, V., & Fry, E. B. (2020). Paint-by-number of Picasso? A grounded theory phenomenographical study of students' conceptions of statistics. *Statistics Education Research Journal, 19*(2), 76–102. https://doi.org/10.52041/serj.v19i2.111

Kahneman, D. (2011). *Thinking, Fast and Slow*. Farrar, Straus & Giroux.

Kalinowski, P., Lai, J., & Cumming, G. (2018). A cross-sectional analysis of students' intuitions when interpreting CIs. *Frontiers in Psychology, 9*(112). https://doi.org/10.3389/fpsyg.2018.00112

Kalobo, L. (2016). Teachers' perceptions of learners' proficiency in statistical literacy, reasoning and thinking. *African journal of research in mathematics, Science and Technology Education, 20*(3), 225–233. https://www.tandfonline.com/doi/full/10.1080/18117295.2016.1215965

Kaput, J. J. (2008). What is algebra? What is algebraic reasoning? In J. J. Kaput, D. W. Carraher, & M. L. Blanton (Eds.), *Algebra in the early grades* (pp. 5–18). Lawrence Erlbaum Associates.

Kazak, S., Fujita, T., & Turmo, M. P. (2021). Students' informal statistical inferences through data modeling with a large multivariate dataset. *Mathematical Thinking and Learning, 25*, 23–43. https://doi.org/10.1080/10986065.2021.1922857

Kelly, A., & Lesh, R. (2000). *Handbook of research design in mathematics and science education*. Routledge. https://doi.org/10.4324/9781410602725

Kelly, A. E., Sloane, F., & Whittaker, A. (1997). Simple approaches to assessing underlying understanding of statistical concepts. In I. Gal & J. B. Garfield (Eds.), *The assessment challenge in statistics education* (pp. 85–90). IOS Press.

Kenett, R., & Thyregod, P. (2006). Aspects of statistical consulting not taught by academia. *Statistica Neerlandica*, 396–411. https://doi.org/10.1111/j.1467-9574.2006.00327.x

Kloosterman, P. (2002). Beliefs about mathematics and mathematics learning in the secondary school: Measurement and implications for motivation. In G. Leder, E. Pehkonen, & G. Törner (Eds.), *Beliefs: A hidden variable in mathematics education?* (pp. 247–269). Springer. https://link.springer.com/chapter/10.1007/0-306-47958-3_15

Konold, C. (2002). Alternatives to scatterplots. *Paper presented at the sixth international conference on teaching statistics, Cape Town.*

Konold, C. (2006). Designing a data analysis tool for learners. In M. Lovett & P. Shah (Eds.), *Thinking with data: The 33rd annual Carnegie symposium on cognition*. Lawrence Erlbaum Associates.

Konold, C., & Higgins, T. L. (2003). Reasoning about data. In J. Kilpatrick, W. G. Martin, & D. Schifter (Eds.), *A research companion to principles and standards for school mathematics* (pp. 193–215). National Council of Teachers of Mathematics.

Konold, C., & Miller, C. D. (2005). *TinkerPlots: Dynamic data explorations*. Key Curriculum Press.

Konold, C., & Pollatsek, A. (2002). Data analysis as the search for signals in noisy processes. *Journal for Research in Mathematics Education, 33*(4), 259–289. https://doi.org/10.2307/749741

Konold, C., & Pollatsek, A. (2004). Conceptualizing an average as a stable feature of a noisy process. In D. Ben Zvi & J. Garfield (Eds.), *The challenge of developing statistical literacy, reasoning and thinking* (pp. 169–199). Springer.

Konold, C., Robinson, A., Khalil, K., Pollatsek, A., Well, A., Wing, R., & Mayr, S. (2002). Students' use of modal clumps to summarize data. *Paper presented at the sixth international conference on teaching statistics, Cape Town.*

Konold, C., Higgins, T., Russell, S. J., & Khalil, K. (2015). Data seen through different lenses. *Educational Studies in Mathematics, 88*(3), 305–325. https://doi.org/10.1007/s10649-013-9529-8

Koremenos, B. (2020/29/1). https://www.yardbarker.com/nba/articles/major_concern_or_blip_dissecting_james_hardens_recent_slump/s1_13132_31170638

Koschat, M. (2005). A case for simple tables. *The American Statistician, 59*(1), 31–40. https://doi.org/10.1198/000313005X21429

Kotz, S., Balakrishnan, N., Read, C. B., & Vidakovic, B. (2005). *Encyclopedia of statistical sciences* (Vol. 1, 2nd ed.). John Wiley and Sons.

Krüger, K., Sill, H. D., & Sikora, C. (2015). *Didaktik der Stochastik in der Sekundarstufe I [didactics of stochastics in secondary school]*. Springer.

Krummenauer, J., & Kuntze, S. (2018). Primary students' data-based argumentation—An empirical reanalysis. In E. Bergqvist, M. Osterholm, C. Granberg, & L. Sumpter (Eds.), *Proceedings of the 42nd international group for the psychology of mathematics education* (Vol. 3, pp. 251–258). PME.

Krummheuer, G. (1995). The ethnography of argumentation. In P. Cobb & H. Bauersfeld (Eds.), *The emergence of mathematical meaning: Interaction in classroom cultures* (pp. 229–269). Erlbaum.

Kullberg, A., Runesson Kempe, U., & Marton, F. (2017). What is made possible to learn when using the variation theory of learning in teaching mathematics? *ZDM Mathematics Education, 49*, 559–569. https://doi.org/10.1007/s11858-017-0858-4

Kultusministerkonferenz. (2005). *Bildungsstandards im Fach Mathematik für den Primarbereich [educational standards in mathematics for the primary level]*. Luchterhand.

Kultusministerkonferenz. (2012). *Bildungsstandards im Fach Mathematik für die Allgemeine Hochschulreife [educational standards in mathematics for the general qualification for university entrance]*. Wolters Kluwer.

Lahanier-Reuter, D. (2003). Différents types de tableaux dans l'enseignement des statistiques (Different types of tables in statistics education). *Spirale-Revue de recherches en éducation, 32*(32), 143–154. https://doi.org/10.3406/spira.2003.1386

Lampen, E. (2015). Teacher narratives in making sense of the statistical mean algorithm. *Pythagoras, 36*(1). https://doi.org/10.4102/pythagoras.v36i1.281

Landwehr, J., Swift, J., & Watkins, A. (1987). *Exploring surveys and information from samples*. Pearson Learning.

Lane, D., & Peres, S. (2006). Interactive simulations in the teaching of statistics: Promise and pitfalls. In A. Rossman & B. Chance (Eds.), *Working cooperatively in statistics education: Proceedings of the seventh international conference on the teaching of statistics*. Salvador. https://iase-web.org/Conference_Proceedings.php?p=ICOTS_7_2006

Langrall, C., Nisbet, S., Mooney, E., & Jansem, S. (2011). The role of context expertise when comparing groups. *Mathematical Thinking and Learning, 13*(1–2), 47–67. https://doi.org/10.1080/10986065.2011.538620

Leavy, A. (2008). An examination of the role of statistical investigation in supporting the development of young children's statistical reasoning. In O. Saracho & B. Spodek (Eds.), *Contemporary perspectives on mathematics education in early childhood* (pp. 215–232). Information Age Publishing.

Leavy, A., & Hourigan, M. (2018). Inscriptional capacities and representations of young children engaged in data collection during a statistical investigation. In A. Leavy, M. Meletiou-Mavrotheris, & E. Paparistodemou (Eds.), *Statistics in early childhood and primary education* (pp. 89–108). Springer Nature. https://doi.org/10.1007/978-981-13-1044-7_6

Leavy, A., Meletiou-Mavrotheris, M., & Paparistodemou, E. (Eds.). (2018). *Statistics in early childhood and primary education: Supporting early statistical and probabilistic thinking*. Springer Nature. https://doi.org/10.1007/978-981-13-1044-7

Lee, J. T. (1999). It's all in the area. *Mathematics Teacher, 92*(8), 670–672. https://www.jstor.org/stable/27971168

Lee, J. T., & Lee H. S. (2014). Visual representations of empirical probability distributions when using the granular density metaphor. Invited paper. In K. Makar (Ed.) *Proceedings of the ninth international conference on teaching statistics, Flagstaff.*

Lee, L., & Tan, S. (2020). Teacher learning in lesson study: Affordances, disturbances, contradictions, and implications. *Teaching and Teacher Education, 89*, 102986. https://doi.org/10.1016/j.tate.2019.102986

Lehrer, R., & English, L. (2018). Introducing children to modeling variability. In D. Ben-Zvi, K. Makar, & J. Garfield (Eds.), *International handbook of research in statistics education* (pp. 229–260). Springer International Publishing. https://doi.org/10.1007/978-3-319-66195-7_7

Lehrer, R., & Schauble, L. (2000). Inventing data structures for representational purposes: Elementary grade students' classification models. *Mathematical Thinking and Learning, 2*(1–2), 51–74. https://doi.org/10.1207/S15327833MTL0202_3

Lehrer, R., & Schauble, L. (2004). Modeling natural variation through distribution. *American Educational Research Journal, 41*(3), 635–679. https://doi.org/10.3102/00028312041003635

LeMire, S. D. (2010). An argument framework for the application of null hypothesis statistical testing in support of research. *Journal of Statistics Education, 18*(2). https://doi.org/10.1080/10691898.2010.11889492

Leontyev, A. N. (2009). Activity and consciousness. *Marxist internet archive*. http://www.marxists.org/archive/leontev/works/activity-consciousness.pdf

Lesser, L. M. (2007). Critical values and transforming data: Teaching statistics with social justice. *Journal of Statistics Education, 15*(1), 1–21.

Lesser, L. M. (2014). Teaching statistics for engagement beyond classroom walls. In K. Makar, B. de Sousa, & R. Gould (Eds.), *Sustainability in statistics education. Proceedings of the ninth international conference on teaching statistics (ICOTS 9)*. International Statistical Institute.

Lewis, C., Perry, R., & Murata, A. (2006). How should research contribute to instructional improvement? The case of lesson study. *Educational Researcher, 35*(3), 3–14. https://doi.org/10.3102/0013189X035003003

Li, J., Yang, Q., & Zou, X. (2019). Big data and higher vocational and technical education: Green food and its industry orientation. In *Proceedings of the 2019 International Conference on Big Data and Education* (pp. 118–123). https://doi.org/10.1145/3322134.3322150

Lipson, K. (2002). The role of computer-based technology in developing understanding of the concept of sampling distribution. In *Proceedings of the 6th International Conference on Teaching Statistics*.

Liu, Y., & Thompson, P. W. (2009). Mathematics teachers' understandings of proto-hypothesis testing. *Pedagogies, 4*(2), 126–138. https://doi.org/10.1080/15544800902741564

Liu, F., & Zhang, Q. (2021). A new reciprocal teaching approach for information literacy education under the background of big data. *International Journal of Emerging Technologies in Learning (IJET), 16*(03), 246. https://doi.org/10.3991/ijet.v16i03.20459

Lo, M. L. (2012). *Variation theory and the improvement of teaching and learning*. Göteborgs Universitet, Acta Universitatis Gothoburgensis. http://hdl.handle.net/2077/29645.

Lo, M. L., & Marton, F. (2012). Towards a science of the art of teaching: Using variation theory as a guiding principle of pedagogical design. *International Journal for Lesson and Learning Studies, 1*(1), 7–22. https://doi.org/10.1108/20468251211179678

Local Report Turkey. (2018). *Strategic partnership for innovative in data analytics in schools project*. https://spidasproject.org.uk/about/research

Logica, B., & Magdalena, R. (2015). Using big data in the academic environment. *Procedia Economics and Finance, 33*, 277–286. https://doi.org/10.1016/S2212-5671(15)01712-8

Lopes, C. E., & de Oliveira Souza, L. (2016). Aspectos filosóficos, psicológicos e políticos no estudo da probabilidade e da estatística na Educação básica. [Philosophical, psychological and political features while studying probability and statistics in basic education]. Educação Matemática Pesquisa: *Revista do Programa de Estudos Pós-Graduados em Educação Matemática, 18*(3). https://revistas.pucsp.br/emp/article/view/31494

López-Belmonte, J., Pozo-Sánchez, S., Fuentes-Cabrera, A., & Trujillo-Torres, J.-M. (2019). Analytical competences of teachers in big data in the era of digitalized learning. *Education Sciences, 9*(3), 177. https://doi.org/10.3390/educsci9030177

Lovett, J. N., & Lee, H. S. (2017). New standards require teaching more statistics: Are preservice secondary mathematics teachers ready? *Journal of Teacher Education, 68*(3), 299–311. https://doi.org/10.1177/0022487117697918

Loya, H. (2008). Los modelos pedagógicos en la formación de profesores [Pedagogical models in teacher education]. *Revista Iberoamericana de Educación, 3*(46), 1–8.

Makar, K. (2014). Young children's explorations of average through informal inferential reasoning. *Educational Studies in Mathematics, 86*(1), 61–78. https://doi.org/10.1007/s10649-013-9526-y

Makar, K. (2018). Theorising links between context and structure to introduce powerful statistical ideas in early years. In A. Leavy, M. Meletiou-Mavrotheris, & E. Paparistodemou (Eds.), *Statistics in early childhood and primary education* (pp. 3–20). Springer Nature. https://doi.org/10.1007/978-981-13-1044-7_1

Makar, K., & Allmond, S. (2018). Statistical modelling and repeatable structures: Purpose, process and prediction. *ZDM, 50*, 1–12. https://doi.org/10.1007/s11858-018-0956-y

Makar, K., & Confrey, J. (2002). Comparing two distributions: Investigating secondary teachers' statistical thinking. *Sixth international conference on teaching statistics, Cape Town.*

Makar, K., & Rubin, A. (2009). A framework for thinking about informal statistical inference. *Statistics Education Research Journal, 8*(1), 82–105. https://doi.org/10.52041/serj.v8i1.457

Makar, K., & Rubin, A. (2018). Learning about statistical inference. In D. Ben-Zvi, K. Makar, & J. Garfield (Eds.), *International handbook of research in statistics education* (pp. 261–294). Springer. https://doi.org/10.1007/978-3-319-66195-7_8

Makar, K., Bakker, A., & Ben-Zvi, D. (2011). The reasoning behind informal statistical inference. *Mathematical Thinking and Learning, 13*((1–2)), 152–173. https://doi.org/10.1080/10986065.2011.538301

Martí, E. (2009). Tables as cognitive tools in primary education. In C. Andersen, N. Scheuer, M. P. Pérez Echeverría, & E. Teubal (Eds.), *Representational systems and practices as learning tools in different fields of learning* (pp. 133–148). Sense Publishers. https://doi.org/10.1163/9789087905286_009

Marton, F. (1981). Phenomenography - describing conceptions of the world around us. *Instructional Science, 10*, 177–200. https://doi.org/10.1007/BF00132516

Marton, F. (1986). Phenomenography – A research approach to investigating different understandings of reality. *J. Thought, 21*, 28–49. http://www.jstor.org/stable/42589189

Marton, F. (2000). The structure of awareness. In J. Bowden & E. Walsh (Eds.), *Phenomenography* (pp. 102–116). RMIT University.

Marton, F. (2006). Sameness and difference in transfer. *Journal of the Learning Sciences, 15*(4), 499–535. https://doi.org/10.1207/s15327809jls1504_3

Marton, F. (2015). *Necessary conditions of learning*. Routledge. https://doi.org/10.4324/9781315816876

Marton, F., & Booth, S. (1997). *Learning and awareness*. Lawrence Erlbaum Associates. https://doi.org/10.4324/9781315816876

Marton, F., & EDB Chinese Language Research Team. (2010). The Chinese learner of tomorrow. In C. K. K. Chan & N. Rao (Eds.), *CERC studies in comparative education, 25: Revisiting the Chinese learner. Changing contexts, changing education* (pp. 133–163). Springer Science+Business Media B.V.

Marton, F., & Pang, M. F. (2006). On some necessary conditions of learning. *The Journal of the Learning Sciences, 15*(2), 193–220. https://doi.org/10.1207/s15327809jls1502_2

Marton, F., & Pang, M. F. (2007). The paradox of pedagogy: The relative contribution of teachers and learners to learning. *Iskolakultura, 1*(1), 1–29.

Marton, F., & Tsui, A. B. M. (2004). *Classroom discourse and the space of learning*. Lawrence Erlbaum. https://doi.org/10.4324/9781410609762

Mathematical Association of America & National Council of Teachers of Mathematics. (2017). *The role of calculus in the transition from high school to college mathematics: Report of the workshop held at the MAA Carriage House, Washington, DC, march 17–19, 2016*. Mathematical Association of America. https://www.maa.org/sites/default/files/RoleOfCalc_rev.pdf

Mayring, P. (2015). Qualitative content analysis: Theoretical background and procedures. In A. Bikner-Ahsbahs, C. Knipping, & N. Presmeg (Eds.), *Approaches to qualitative research in mathematics education* (pp. 365–380). Springer.

Mbombo, A. B., & Cavus, N. (2021). Smart university: A university in the technological age. *TEM Journal, 10*(1), 13–17.

McGatha, M., Cobb, P., & McClain, K. (2002). An analysis of students' initial statistical understandings: Developing a conjectured learning trajectory. *The Journal of Mathematical Behavior, 21*(3), 339–355. https://doi.org/10.1016/S0732-3123(02)00133-5

McGowan, B. S. (2020). OpenStreetMap mapathons support critical data and visual literacy instruction. *Journal of the Medical Library Association: JMLA, 108*(4), 649. https://doi.org/10.5195/jmla.2020.1070

McLeod, D. B. (1992). Research on affect in mathematics education: A reconceptualization. *Handbook of research on mathematics teaching and learning, 1*, 575–596.

MEB. (2018a). *Matematik Öğretim Programı (İlkokul ve Ortaokul 1, 2, 3, 4, 5, 6, 7 ve 8. Sınıflar) [Mathematics Curriculum (Primary and Middle Schools Grades 1–8)]*. MEB Yayınları.

MEB. (2018b). *Matematik Öğretim Programı (9, 10, 11 ve 12. Sınıflar) [Mathematics curriculum (Grades 9–12)]*. MEB Yayınları.

MECD. (2014). *Real Decreto 126/2014, de 28 de febrero, por el que se establece el currículo básico de la educación primaria* (Royal Decree 126/2014, 28th February, establishing the basic curriculum guidelines for primary education). Ministerio de Educación, Cultura y Deportes.

MECD. (2015). *Real Decreto 1105/2014, de 26 de diciembre, por el que se establece el currículo básico de la Educación Secundaria Obligatoria y del Bachillerato* (Royal Decree 1105/2014, 26th December, establishing the basic curriculum guidelines for Compulsory Secondary Education and High school). Ministerio de Educación, Cultura y Deportes.

Meletiou-Mavrotheris, M., & Lee, C. (2002). Teaching students the stochastic nature of statistical concepts in an introductory statistics course. *Statistics Education Research Journal, 1*(2), 22–37. https://doi.org/10.52041/serj.v1i2.563

Meletiou-Mavrotheris, M., & Paparistodemou, E. (2015). Developing students' reasoning about samples and sampling in the context of informal inferences. *Educational Studies in Mathematics, 88*(3), 385–404. https://doi.org/10.1007/s10649-014-9551-5

Merriam, S. B., & Tisdell, E. J. (2015). *Qualitative research: A guide to design and implementation*. John Wiley & Sons.

Mezhennaya, N. M., & Pugachev, O. V. (2019). Advantages of using the CAS mathematica in a study of supplementary chapters of probability theory. *European Journal of Contemporary Education, 8*(1). https://doi.org/10.13187/ejced.2019.1.4

Michigan Department of Education. (2016). *M-STEP Final Reports Webcast*.

Ministry of Education. (1992). *Mathematics in the New Zealand curriculum*. Learning Media.

Ministry of Education. (2007). *The New Zealand curriculum*. Learning Media.

Mohammed, A., Kumar, S., Singh, S. P., & Sharma, R. P. (2018). Enhancing teaching and learning in educational institutes using the concept of big data technology. In *2018 International conference on computing, power and communication technologies (gucon)* (pp. 1038–1041). IEEE.

Mokros, J., & Russell, S. J. (1995). Children's concepts of average and representativeness. *Journal for Research in Mathematics Education, 26*(1), 20–39. https://doi.org/10.2307/749226

Monteiro, C. E. F. (2021). Letramento estatístico e big data: Uma revisão integrativa da literature [Statistical literacy and big data: An integrative literature review]. In C. E. F. Monteiro & L. M. T. L. Carvalho (Eds.), *Temas emergentes em letramento estatístico [Emerging themes in statistical literacy]* (pp. 158–181). UFPE. https://editora.ufpe.br/books/catalog/view/666/677/2080

Moore, D. S. (1988). Should mathematicians teach statistics? *College Mathematics Journal., 19*(1), 3–7.

Moore, D. S. (1990). Uncertainty. In L. A. Steen (Ed.), *On the shoulders of giants: New approaches to numeracy* (pp. 95–137). National Academy Press.

Moore, D. S. (1991). *Statistics: Concepts and controversies* (3rd ed.). W. H. Freeman.
Moritz, J. (2004). Reasoning about covariation. In D. Ben Zvi & J. Garfield (Eds.), *The challenge of developing statistical literacy, reasoning and thinking* (pp. 227–255). Springer.
Muis, K. R. (2004). Personal epistemology and mathematics: A critical review and synthesis of research. *Review of Educational Research, 74*(3), 317–377. https://doi.org/10.3102/00346543074003317
Murata, A. (2011). Introduction: Conceptual overview of lesson study. In L. Hart, A. Alston, & A. Murata (Eds.), *Lesson study research and practice in mathematics education* (pp. 1–12). Springer. https://doi.org/10.1007/978-90-481-9941-9
Mutodi, P., & Ngirande, H. (2014). The influence of students' perceptions on mathematics performance. A case of a selected high school in South Africa. *Mediterranean Journal of Social Sciences, 5*(3), 431. https://doi.org/10.5901/mjss.2014.v5n3p431
Naidoo, J., & Mkhabela, N. (2017). Teaching data handling in foundation phase: Teachers' experiences. *Research in Education, 97*(1), 95–111. https://doi.org/10.1177/0034523717697513
Nasser, F. (2004). Structural model of the effects of cognitive and affective factors on the achievement of Arabic-speaking pre-service teachers in introductory statistics. *Journal of Statistics Education, 12*(1). https://doi.org/10.1080/10691898.2004.11910717
Nathan, M. J., & Koedinger, K. R. (2000). Teachers' and researchers' beliefs about the development of algebraic reasoning. *Journal for Research in Mathematics Education, 31*(2), 168–190. https://doi.org/10.2307/749750
National Council of Teachers of Mathematics. (2009a). *Navigating through data analysis and probability in prekindergarten-grade 2* (Vol. 1). Author.
National Council of Teachers of Mathematics. (2009b). *Focus in high school mathematics: Reasoning and sense making*. Author.
National Council of Teachers of Mathematics. (2018). *Catalyzing change in high school mathematics: Initiating critical conversations*. The Council.
National Council of Teachers of Mathematics (NCTM). (1989). *Curriculum and evaluation standards for school mathematics*. Reston.
National Council of Teachers of Mathematics (NCTM). (2000). *Principles and standards for school mathematics*. NCTM.
National Council of Teachers of Mathematics (NCTM). (2014). *Principles to actions: Ensuring mathematical success for all*. NCTM.
National Council of Teachers of Mathematics (NCTM). (2018). *Catalyzing change in high school mathematics: Initiating critical conversations. National Council of teachers of mathematics*. Reston. https://www.nctm.org/change/
National Governor's Association Center for Best Practices & Council of Chief State School Officers. (2010). *Common Core state standards in mathematics*. Authors.
National Governors Association Center for Best Practice & Council of Chief State School Officers. (2010). *Common core state standards for mathematics*. Author.
National Governors Association Center for Best Practices and Council of Chief State School Officers. (2010). *Common core state standards for mathematics.*
Neubrand, M. (2015). Bildungstheoretische Grundlagen des Mathematikunterrichts (education theory foundations of mathematics teaching). In R. Bruder, L. Hefendehl-Hebeker, B. Schmidt-Thieme, & H.-G. Weigand (Eds.), *Handbuch der Mathematikdidaktik (handbook of mathematics education)* (pp. 51–73). Springer. https://doi.org/10.1007/978-3-642-35119-8_3
Neuendorf, K. (2016). *The content analysis guidebook*. Sage.
Nilsson, P. (2007). Different ways in which students handle chance encounters in the explorative setting of a dice game. *Educational Studies in Mathematics, 66*(3), 293–315. https://doi.org/10.1007/s10649-006-9062-0
Nilsson, P. (2013). Challenges in seeing data as useful evidence in making predictions on the probability of a real-world phenomenon. *Statistics Education Research Journal, 12*(2), 71–83. https://doi.org/10.52041/serj.v12i2.305

Nilsson, P. (2020). Students' informal hypothesis testing in a probability context with concrete random generators. *Statistics Education Research Journal, 19*(3), 53–73. https://doi.org/10.52041/serj.v19i3.56

Niss, M. (2000). Gymnasiets opgave, almen dannelse og kompetencer (the task of upper secondary school, Allgemeinbildung and competencies). *Uddannelse: Undervisningsministeriets tidsskrift, 33*(2), 23–33.

Niss, M. (2010). Modeling a crucial aspect of students' mathematical modeling. In R. Lesh, P. L. Galbraith, C. R. Haines, & A. Hurford (Eds.), *Modeling students' mathematical modeling competencies: ICTMA 13* (pp. 43–59). Springer US. https://doi.org/10.1007/978-1-4419-0561-1_4

Niss, M., & Blum, W. (2020). *The learning and teaching of mathematical modelling*. Routledge. https://doi.org/10.4324/9781315189314

Noll, J., & Shaughnessy, M. (2012). Aspects of students' reasoning about variation in empirical sampling distribution. *Journal for Research in Mathematics Education, 43*(5), 509–556. https://doi.org/10.5951/jresematheduc.43.5.0509

North, D., Gal, I., & Zewotir, T. (2014). Building capacity for developing statistical literacy in a developing country: Lessons learned from an intervention. *Statistics Education Research Journal, 13*(2), 15–27.

Oehrtman, M. (2008). Layers of abstraction: Theory and design for the instruction of limit concepts. In M. Carlson & C. Rasmussen (Eds.), *Making the connection: Research and teaching in undergraduate mathematics education*. http://hub.mspnet.org//index.cfm/19688

Organization for Economic Co-operation and Development (OECD). (2018). *PISA 2022 mathematics framework (draft)*. Author. https://pisa2022-maths.oecd.org/files/PISA%202022%20Mathematics%20Framework%20Draft.pdf

Osana, H. P., Leath, E. P., & Thompson, S. E. (2004). Improving evidential argumentation through statistical sampling: Evaluating the effects of a classroom intervention for at-risk 7th-graders. *Journal of Mathematical Behavior, 23*, 351–370. https://doi.org/10.1016/jmathb.2004.06.005

Oslington, G., Mulligan, J., & Van Bergen, P. (2020). Third-graders' predictive reasoning strategies. *Educational Studies in Mathematics, 104*(1), 5–24. https://doi.org/10.1007/s10649-020-09949-0

Otani, H. (2019). Comparing structures of statistical hypothesis testing with proof by contradiction: In terms of argument. *Hiroshima Journal of Mathematics Education, 12*, 1–12.

Ozturk, T., & Guven, B. (2016). Evaluating students' beliefs in problem solving process: A case study. *Eurasia Journal of Mathematics, Science and Technology Education, 12*(3), 411–429. https://doi.org/10.12973/eurasia.2016.1208a

Packer, T. (2021). *AP statistics exam 2021 results*. https://allaccess.collegeboard.org/ap-statistics-exam-2021-results

Pallauta, J., Arteaga, P., & Garzón-Guerrero, J. A. (2021a). Secondary school students' construction and interpretation of statistical tables. *Mathematics, 9*(24), 3197. https://doi.org/10.3390/math9243197

Pallauta, J., Gea, M. M., & Arteaga, P. (2021b). Caracterización de las tareas propuestas sobre tablas estadísticas en libros de texto chilenos de educación básica (Characterization of tasks related to statistical tables in chilean basic education textbooks). *Paradigma, 40*(1), 32–60. https://doi.org/10.37618/PARADIGMA.1011-2251.2021.p32-60.id1017

Pallauta, J., Gea, M. M., Batanero, C., & Arteaga, P. (2021c). Significado de la tabla estadística en libros de texto españoles de educación secundaria (Meaning of the statistical table in Spanish secondary education textbooks). *Bolema: Boletim de Educação Matemática, 35*, 1803–1824. https://doi.org/10.1590/1980-4415v35n71a26

Pang, M. F. (2010). Boosting financial literacy: Benefits from learning study. *Instructional Science, 38*(6), 659–677. https://doi.org/10.1007/s11251-009-9094-9

Pang, M. F. (2019). Enhancing the generative learning of young people in the domain of financial literacy through learning study. *International Journal for Lesson and Learning Studies, 8*(3), 170–182. https://doi.org/10.1108/IJLLS-09-2018-0065

Papacharissi, Z. (2015). The unbearable lightness of information and the impossible gravitas of knowledge: Big data and the makings of a digital orality. *Media, Culture & Society, 37*(7), 1095–1100. https://doi.org/10.1177/0163443715594103

Paparistodemou, E., & Meletiou-Mavrotheris, M. (2008). Developing young students' informal inference skills in data analysis. *Statistics Education Research Journal, 7*(2), 83–106. https://doi.org/10.52041/serj.v7i2.471

Park, Y. E. (2020). Uncovering trend-based research insights on teaching and learning in big data. *Journal of Big Data, 7*(1). https://doi.org/10.1186/s40537-020-00368-9

Park, H., Conner, A., Foster, J. K., Singletary, L., & Zhuang, Y. (2020). One Teacher's Learning to Facilitate Argumentation: Focus on the Use of Repeating. In *Mathematics education across cultures: Proceedings of the 42nd annual meeting of the North American chapter of the international group for the psychology of mathematics education* (pp. 1961–1962). Cinvestav/AMIUTEM/PME-NA. https://doi.org/10.51272/pmena.42.2020

Parzysz, B. (2018). Solving probabilistic problems with technologies in middle and high school: The French case. In N. Amado et al. (Eds.), *Broadening the scope of research on mathematical problem solving* (Research in mathematics education). Springer.

Patel, A., & Pfannkuch, M. (2018). Developing a statistical modeling framework to characterize year 7 students' reasoning. *ZDM, 50*(7), 1197–1212. https://doi.org/10.1007/s11858-018-0960-2

Pawluczuk, A. (2020). Digital youth inclusion and the big data divide: Examining the Scottish perspective. *Internet Policy Review, 9*(2). https://doi.org/10.14763/2020.2.1480

Peirce, C. S. (1956). Sixth paper: Deduction, induction, and hypothesis. In M. R. Cohen (Ed.), *Chance, love, and logic: Philosophical essays* (pp. 131–153). G. Braziller. (Original work published 1878).

Peñaloza-Figueroa, J. L., & Vargas-Perez, C. (2017). Big-data and the challenges for statistical inference and economics teaching and learning. *Multidisciplinary Journal for Education, Social and Technological Sciences, 4*(1), 64. https://doi.org/10.4995/muse.2017.6350

Pfannkuch, M. (2006, July 2–7). Informal inferential reasoning. In A. Rossman & B. Chance (Eds.), *Proceedings of the 7th international conference on teaching of statistics (CD-ROM), Salvador.*

Pfannkuch, M. (2008). Building sampling concepts for statistical inference: A case study. In *Proceedings of the eleventh international congress of mathematics education* (ICOTS 11). Monterrey, Mexico. Online: http://tsg.icme11.org/tsg/show/15

Pfannkuch, M. (2011). The role of context in developing informal statistical inferential reasoning: A classroom study. *Mathematical Thinking and Learning, 13*(1–2), 27–46. https://doi.org/10.1080/10986065.2011.538302

Pfannkuch, M. (2018). Reimagining curriculum approaches. In D. Ben-Zvi, K. Makar, & J. Garfield (Eds.), *International handbook of research in statistics education* (pp. 387–413). Springer. https://doi.org/10.1007/978-3-319-66195-7_12

Pfannkuch, M., & Ben-Zvi, D. (2011). Developing teachers' statistical thinking. In C. Batanero, G. Burrill, & C. Reading (Eds.), *Teaching statistics in school mathematics-challenges for teaching and teacher education. A joint ICMI/IASE study: The 18th ICMI study* (pp. 323–333). Springer.

Pfannkuch, M., & Brown, C. M. (1996). Building on and challenging students' intuitions about probability: Can we improve undergraduate learning? *Journal of Statistics Education, 4*(1), 1–22. https://doi.org/10.1080/10691898.1996.11910502

Pfannkuch, M., & Budgett, S. (2014). Constructing inferential concepts through bootstrap and randomization-test simulations: A case study. In K. Makar, B. de Sousa, & R. Gould (Eds.). Sustainability in statistics education. *Proceedings of the ninth international conference on teaching statistics* (ICOTS9) Flagstaff, Arizona, USA.

Pfannkuch, M., & Wild, C. (2004). Towards an understanding of statistical thinking. In D. BenZvi & J. Garfield (Eds.), *The challenge of developing statistical literacy, reasoning and thinking* (pp. 17–46). Kluwer Academic Publishers.

Pfannkuch, M., Wild, C. J., & Parsonage, R. (2012a). A conceptual pathway to *confidence intervals*. *ZDM – The International Journal on Mathematics Education, 44*(7), 899–911. https://doi.org/10.1007/s11858-012-0446-6

Pfannkuch, M., Wild, C., & Parsonage, R. (2012b). A conceptual pathway to confidence intervals. *ZDM, 44*(7), 899–911. https://doi.org/10.1007/s11858-012-0446-6

Pfannkuch, M., Arnold, P., & Wild, C. (2015). What I see is not quite the way it really is: Students' emergent reasoning about sampling variability. *Educational Studies in Mathematics., 88*, 343–360. https://doi.org/10.1007/s10649-014-9539-1

Pfannkuch, M., Wild, C., Arnold, P., & Budgett, S. (2020). Reflections on a 20-year research journey, 1999–2019. *SET, 1*, 27–33.

Philipp, R. A. (2007). Mathematics teachers' beliefs and affect. In F. K. Lester (Ed.), *Second handbook of research on mathematics teaching and learning* (pp. 257–315). National Council of Teachers of Mathematics.

Pittard, V. (2018). *The integration of data science in the primary and secondary curriculum*. Final Report to The Royal Society Advisory Committee on Mathematics Education's (ACME https://royalsociety.org/-/media/policy/Publications/2018/2018-07-16-integration-of-data-science-primary-secondary-curriculum.pdf

Podworny, S. (2019). *Simulationen und Randomisierungstests mit der software TinkerPlots [simulations and randomization tests with the software TinkerPlots]*. Springer Spektrum.

Porciúncula, M., Schreiber, K. P., & Almeida, R. L. (2019). Statistical literacy: A strategy to promote social justice. *Revista Internacional de Pesquisa em Educação Matemática, RIPEM, 9*(1), 25–44.

Prado, M. M., & Gravoso, R. S. (2011). Improving high school students' statistical reasoning skills: A case of applying anchored instruction. *The Asia-Pacific Education Researcher, 20*(1), 61–72.

Prediger, S., & Zwetzschler, L. (2013). Topic-specific design research with a focus on learning processes: The case of understanding algebraic equivalence in grade 8. In T. Plomp & N. Nieveen (Eds.), *Educational design research: Illustrative cases* (pp. 407–424). SLO.

Presmeg, N. (2002). Beliefs about the nature of mathematics in the bridging of everyday and school mathematical practices. In G. Leder, E. Pehkonen, & G. Törner (Eds.), *Beliefs: A hidden variable in mathematics education?* (pp. 293–312). Springer. https://link.springer.com/chapter/10.1007/0-306-47958-3_15

Prodromou, T., & Pratt, D. (2013). Making sense of stochastic variation and causality in a virtual environment. *Technology, knowledge and learning: Learning mathematics, science and the arts in the context of digital technologies, 18*(3), 121–147. https://doi.org/10.1007/s10758-013-9210-4

Radford, L. (2003). On the epistemological limits of language: Mathematical knowledge and social practice during the renaissance. *Educational Studies in Mathematics, 52*(2), 123–150. http://www.jstor.org/stable/20749442

Radford, L. (2010). The eye as a theoretician: Seeing structures in generalizing activities. *For the Learning of Mathematics, 30*(2), 2–7.

Radford, L. (2011). Grade 2 students' non-symbolic algebraic thinking. In J. Cai & E. Knuth (Eds.), *Early algebraization* (Advances in mathematics education) (pp. 303–322). SpringerVerlag.

Ramirez, C., Schau, C., & Emmioglu, E. (2012). The importance of attitudes in statistics education. *Statistics Education Research Journal, 11*(2), 57–71. https://doi.org/10.52041/serj.v11i2.329

Ramos, M. A., & Gonçalves, R. E. (1996). As narrativas autobiográficas do professor Como estratégia de desenvolvimento e a prática da supervisao [the teacher's autobiographical narratives as a development strategy and the practice of supervision]. In I. Alarcão (Ed.), *Formação reflexiva de profesores. Estratégias de Supervisão* (pp. 123–150).

Rasmussen, C. L., & Stephan, M. (2008). A methodology for documenting collective activity. In A. Kelly, R. Lesh, & J. Baek (Eds.), *Handbook of design research methods in education: Innovations in science, technology, engineering, and mathematics teaching and learning*. Routledge. https://doi.org/10.4324/9781315759593.ch10

Reading, C., & Canada, D. (2011). Teachers' knowledge of distribution. In C. Batanero, G. Burrill, & C. Reading (Eds.), *Teaching statistics in school mathematics. Challenges for teaching and teacher education* (New ICMI Study Series, 14). Springer. https://doi.org/10.1007/978-94-007-1131-0_23

Reading, C., & Reid, J. (2006). An emerging hierarchy of reasoning about distribution from a variation perspective. *Statistics Education Research Journal, 5*(2), 46–68. https://doi.org/10.52041/serj.v5i2.500

Reading, C., & Shaughnessy, J. M. (2004). Reasoning about variation. In D. Ben-Zvi & J. Garfield (Eds.), *The challenge of developing statistical literacy, reasoning and thinking* (pp. 201–226). https://doi.org/10.1007/1-4020-2278-6_9.

Reddy, V., Winnaar, L., Juan, A., Arends, F., Harvey, J., Hannan, S., et al. (2020). TIMSS 2019: Highlights of south African grade 9 results in mathematics and science. In *Achievement and achievement gaps. Pretoria: Department of Basic Education.*

Reid, A., & Petocz, P. (2002). Students' conceptions of statistics: A phenomenographic study. *Journal of Statistics Education, 10*(2). https://doi.org/10.1080/10691898.2002.11910662

Republic of Korea Ministry of Education, Science and Technology. Common Curriculum Mathematics (2011).

Reston, E., & Bersales, L. G. (2011). Reform efforts in training mathematics teachers to teach statistics: Challenges and prospects. In C. Batanero, G. Burrill, & C. Reading (Eds.), *Teaching statistics in school mathematics - challenges for teaching and teacher education: A joint study ICMI/IASE study book*. Springer.

Reston, E., & Cañizares, M. (2019). Needs assessment of teachers' knowledge bases, pedagogical approaches and self-efficacy in implementing the K to 12 science and mathematics curriculum. *International Journal of Research Studies in Education, 8*(2), 29–45. Retrieved from https://www.academia.edu/37702617/

Reston, E., & Krishnan, S. (2014). Statistics education research in Malaysia and The Philippines: A comparative analysis for future directions. *Statistics Education Research Journal, 13*(2), 218–231.

Ridgway, J. (2015). Implications of the data revolution for statistics education. *International Statistical Review, 84*(3), 528–549. https://doi.org/10.1111/insr.12110

Ridgway, J. (2016). Implications of the data revolution for statistics education. *International Statistical Review, 84*(3), 528–549. https://doi.org/10.1111/insr.12110

Ridgway, J., McCusker, S., & Nicholson, J. (2003). *Reasoning with evidence: Development of a scale.*

Roditi, E. (2009). L'histogramme : à la recherche du savoir à enseigner. [The histogram: In search of knowing how to teach it]. Spirale. *Revue de Recherches en Éducation, 43*, 129–138. https://halshs.archives-ouvertes.fr/halshs-00609704

Roesken, B., Hannula, M. S., & Pehkonen, E. (2011). Dimensions of students' views of themselves as learners of mathematics. *ZDM, 43*(4), 497–506. https://doi.org/10.1007/s11858-011-0315-8

Rojas, T., & Salinas, R. (2020). *Una secuencia de aprendizaje que desarrolla el razonamiento inferencial estadístico informal, diseñada en un estudio de clases para una enseñanza escolar online.* [A learning sequence that develops informal statistical inferential reasoning, designed in a lesson study for online school teaching]. [Unpublished undergraduate thesis, Pontificia Universidad Católica de Valparaíso].

Rolka, K., & Bulmer, M. (2005). Picturing student beliefs in statistics. *ZDM, 37*(5), 412–417. https://doi.org/10.1007/s11858-005-0030-4

Royal Society Te Apārangi. (2021). *Pāngarau mathematics and tauanga statistics in Aotearoa New Zealand: Advice on refreshing the English-medium mathematics and statistics learning area of the New Zealand curriculum*. Author.

Rubin, A. (2021). What to consider when we consider data. *Teaching Statistics, 43*(1), 23–33. https://doi.org/10.1111/test.12275

Ruiz-Palmero, J., Colomo-Magaña, E., Ríos-Ariza, J. M., & Gómez-García, M. (2020). Big data in education: Perception of training advisors on its use in the educational system. *Social Sciences, 9*(4), 53. https://doi.org/10.3390/socsci9040053

Rumsey, D. (2022). *Statistics II. For dummies* (2nd ed.). John Wiley & Sons.

Russell, S. J., & Corwin, R. B. (1989). *Statistics: The shape of the data*. Dale Seymour Publications.

Sacristan, A., Calder, N., Rojano, T., Santos-Trigo, M., Friedlander, A., & Meissner, H. (2010). The influence and shaping of digital technologies on the learning – And learning trajectories - of mathematical concepts. In C. Hoyles & J. Lagrange (Eds.), *Mathematics education and technology - rethinking the terrain* (The 17th ICMI Study) (pp. 179–226). Springer. https://doi.org/10.1007/978-1-4419-0146-0_9

Saldanha, L. A. (2003). "Is this sample unusual?" *An investigation of students exploring connections between sampling distributions and statistical inference*. Unpublished doctoral dissertation, Vanderbilt University, Nashville, TN.

Saldanha, L., & Thompson, P. (2014). Conceptual issues in understanding the inner logic of statistical inference: Insights from two teaching experiments. *Journal of Mathematical Behavior, 35*, 1–30. https://doi.org/10.1016/j.jmathb.2014.03.001

Sánchez, S. G. (1998). Fundamentos Para la investigación educativa. Presupuestos epistemológicos que orientan al investigador [Foundations for educational research. Epistemological assumptions that guide the researcher]. *Cooperativa editorial Magisterio*.

Sánchez, E., García-García, I., & J., & Mercado M. (2018). Determinism and empirical commitment in the probabilistic reasoning of high school students. In C. Batanero & E. Chernoff (Eds.), *Teaching and learning stochastics. ICME-13 monographs*. Springer. https://doi.org/10.1007/978-3-319-72871-1_13

Sander, I. (2020). What is critical big data literacy and how can it be implemented? *Internet Policy Review, 9*(2). https://doi.org/10.14763/2020.2.1479

Savard, A. (2014). Developing probabilistic thinking: What about people's conceptions? In E. J. Chernoff & B. Sriraman (Eds.), *Probabilistic thinking: Presenting plural perspectives* (pp. 283–298). Springer. https://doi.org/10.1007/978-94-007-7155-0

Schau, C. (2003). Students' attitudes: The "other" important outcome in statistics education. In *Proceedings of the Joint Statistical Meetings, San Francisco* (pp. 3673–3681). http://statlit.org/pdf/2003SchauASA.pdf

Schau, C., Stevens, J., Dauphinee, T. L., & Vecchio, A. D. (1995). The development and validation of the survey of attitudes toward statistics. *Educational and Psychological Measurement, 55*(5), 868–875. https://doi.org/10.1177/0013164495055005022

Scheaffer, R. (1990). The ASA-NCTM Quantitative Literacy Project: An overview. In the *Proceedings of the Third International Congress on Mathematics Education*. https://iase-web.org/documents/papers/icots3/BOOK1/A1-2.pdf?1402524941

Scheaffer, R. (2006a). Statistics and mathematics: On making a happy marriage. In G. Burrill (Ed.), *Thinking and reasoning with data and chance, 68th NCTM yearbook (2006)* (pp. 309–321). National Council of Teachers of Mathematics.

Scheaffer, R. L. (2006b). Statistics and mathematics: On making a happy marriage. In G. F. Burrill (Ed.), *Thinking and reasoning with data and chance: Sixty-eighth yearbook* (pp. 309–321). National Council of Teachers of Mathematics.

Scheaffer, R. L., & Jacobbe, T. (2014). Statistics education in the K-12 schools of the United States: A brief history. *Journal of Statistics Education, 22*(2).

Scheaffer, R., Gnanadesikan, M., Watkins, A., & Witmer, J. (1996). *Activity-based statistics*. Springer.

Schildkamp, K. (2019). Data-based decision-making for school improvement: Research insights and gaps. *Educational Research, 61*(3), 257–273. https://doi.org/10.1080/00131881.2019.1625716

Schnell, S. (2014). Types of arguments when dealing with chance experiments. In C. Nicol, S. Oesterle, P. Liljedahl, & D. Allan (Eds.), *Proceedings of the joint meeting of PME 38 and PME-NA 36* (Vol. 5, pp. 113–120). PME.

Schouten, G. (2017). On meeting students where they are: Teacher judgment and the use of data in higher education. *Theory and Research in Education, 15*(3), 321–338. https://doi.org/10.1177/1477878517734452

Schrage, G. (1983). (Mis) interpretation of stochastic models. In R. Scholz (Ed.), *Decision making under uncertainty* (pp. 351–361). North-Holland. https://doi.org/10.1016/S0166-4115(08)62207-4

Schulz, L., & Sommerville, J. (2006). God does not play dice: Causal determinism and preschoolers' causal inferences. *Child Development, 77*(2), 427–442. https://doi.org/10.1111/j.1467-8624.2006.00880.x

Schwartz, D. L., & Bransford, D. J. (1998). A time for telling. *Cognition and Instruction, 16*(4), 475–522. https://doi.org/10.1207/s1532690xci1604_4

Schwartz, D., & Martin, T. (2004). Inventing to prepare for future learning: The hidden efficiency of encouraging original student production in statistics instruction. *Cognition and Instruction, 22*(2), 129–184. https://doi.org/10.1207/s1532690xci2202_1

Sharma, S. (2017). Definitions and models of statistical literacy: A literature review. *Open Review of Educational Research, 4*(1), 118–133. https://doi.org/10.1080/23265507.2017.1354313

Shaughnessy, J. M. (1977). Misconceptions of probability: An experiment with a small-group, activity-based, model building approach to introductory probability at the college level. *Educational Studies in Mathematics, 8*, 295–316. https://doi.org/10.1007/BF00385927

Shaughnessy, J. M. (1992). Research in probability and statistics: Reflections and directions. In D. A. Grouws (Ed.), *Handbook of research on mathematics teaching and learning: A project of the national council of teachers of mathematics* (pp. 465–494). Macmillan Publishing Co, Inc.

Shaughnessy, J. M. (1997). Missed opportunities in research on the teaching and learning of data and chance. In F. Biddulph & K. Carr (Eds.), *People in mathematics education: Proceedings of the twentieth annual meeting of the mathematics education research group of Australasia* (Vol. 1, pp. 6–22). The University of Waikato Printery.

Shaughnessy, J. M. (2007). Research on statistics learning and reasoning. In F. K. Lester Jr. (Ed.), *Second handbook of research on mathematics teaching and learning, 2* (pp. 957–1009). Information Age.

Shaughnessy, J. M., & Pfannkuch, M. (2002). How faithful is old faithful? Statistical thinking: A story of variation and prediction. *The Mathematics Teacher, 95*(4), 252–259. https://doi.org/10.5951/MT.95.4.0252

Shvarts, A. (2017). Eye movements in emerging conceptual understanding of rectangle area. In B. Kaur, W. K. Ho, T. L. Toh, & B. H. Choy (Eds.), *Proceedings of the 41st conference of the International Group for the Psychology of Mathematics Education* (Vol. 1, p. 268). PME. https://www.igpme.org/wp-content/uploads/2019/05/PME41-2017-Singapore.zip

Shvarts, A., & Alberto, R. A. (2021). Melting cultural artifacts back to personal actions: Embodied design for a sine graph. In M. Prasitha, N. Changsri, & N. Boonsena (Eds.), *Proceedings of the 44th conference of the International Group for the Psychology of Mathematics Education* (Vol. 4, pp. 49–56). https://pme44.kku.ac.th/home/uploads/volumn/pme44_vol4.pdf#page=61

Shvarts, A., Alberto, R. A., Bakker, A., Doorman, M., & Drijvers, P. (2021). Embodied instrumentation in learning mathematics as the genesis of a body-artifact functional system. *Educational Studies in Mathematics, 107*, 447–469. https://doi.org/10.1007/s10649-021-10053-0

Shvarts, A., Bos, R., Doorman, M., & Drijvers, P. (2022). *Reifying actions into artifacts: An embodied perspective on process-object dialectics in higher-order mathematical thinking.* [submitted]. Utrecht University.

Sill, H.-D., & Kurtzmann, G. (2019). *Didaktik der Stochastik in der Primarstufe [didactics of stochastics for primary level]*. Springer.

Simon, M. A., & Tzur, R. (2004). Explicating the role of mathematical tasks in conceptual learning: An elaboration of the hypothetical learning trajectory. *Mathematical Thinking and Learning, 6*(2), 91–104. https://doi.org/10.1207/s15327833mtl0602_2

Skovsmose, O. (1999). *Hacia una filosofía de la educación matemática crítica.* [Towards a philosophy of critical mathematics education.] (P. Valero, Trad.). Una Empresa Docente (Trabajo original publicado en 1994).

Skovsmose, O., & Borba, M. (2004). Research methodology and critical mathematics education. In P. Valero & R. Zevenbergen (Eds.), *Researching the socio-political dimensions of mathematics education: Issues of power in theory and methodology* (pp. 207–226). Springer. https://doi.org/10.1007/1-4020-7914-1_17

Snee, R. D. (1999). Discussion: Development and use of statistical thinking: A new era. *International Statistical Review, 67*(3), 255–258. https://doi.org/10.1111/j.1751-5823.1999.tb00446.x

Sorto, M. A., White, A., & Lesser, L. M. (2011). Understanding student attempts to find a line of fit. *Teaching Statistics, 33*(2), 49–52. https://doi.org/10.1111/j.1467-9639.2010.00458.x

Southerland, S. A., Sinatra, G. M., & Matthews, M. R. (2001). Belief, knowledge, and science education. *Educational Psychology Review, 13*(4), 325–351. https://doi.org/10.1023/A:1011913813847

Souza, R. R. (2018). Algorithms, future and digital rights: Some reflections. *Education for Information, 34*(3), 179–183. https://doi.org/10.3233/efi-180200

Souza, L., Lopes, C. E., & Pfannkuch, M. (2015). Collaborative professional development for statistics teaching: A case study of two middle-school mathematics teachers. *Statistics Education Research Journal, 14*(1), 112–134.

Stanic, G. M. A., & Kilpatrick, J. (2003). *A history of school mathematics* (Vol. 1 & 2). National Council of Teachers of Mathematics.

Stat Trek. Statistics dictionary. Teach yourself statistics. Accessed 6/30/2019. https://stattrek.com/statistics/dictionary.aspx?definition=margin%20of%20error

Statistics Education Research Journal (SERJ) [website]. (2022). https://iase-web.org/Publications.php?p=SERJ

Steffe, L. P., & Thompson, P. W. (2000). Teaching experiment methodology: Underlying principles and essential elements. In R. Lesh & A. E. Kelly (Eds.), *Research design in mathematics and science education* (pp. 267–307). Erlbaum.

Stein, A.-L. (2019). *Planung, durchführung und evaluation einer unterrichtsreihe zur förderung der datenkompetenz in klasse 4 unter besonderer berücksichtigung des lesens und interpretierens von ein-dimensionalen streudiagrammen* (Planning, implementation and evaluation of a series of lessons to promote data literacy in grade 4 with special consideration of reading and interpreting one-dimensional scatter plots). (Bachelor of Education), University of Paderborn.

Stern, D. (2017). Seeding the African data initiative. In *Proceedings of the IASE satellite conference "Teaching statistics in a data rich world"*.

Strauss, A., & Corbin, J. M. (1990). *Basics of qualitative research: Grounded theory procedures and techniques*. Sage Publications, Inc.

Tall, D., & Vinner, S. (1981). Concept image and concept definition in mathematics with particular reference to limits and continuity. *Educational Studies in Mathematics, 12*, 151–169. https://doi.org/10.1007/BF00305619

Tarim, K., & Tarku, H. (2022). Investigation of the questions in 8th grade mathematics textbook in terms of mathematical literacy. *International Electronic Journal of Mathematics Education, 17*(2). https://doi.org/10.29333/iejme/11819

Tarran, B. (2020). Statistical literacy for all! *Significance, 17*(1), 42–43.

Tempelaar, D. T., Schim van der Loeff, S., & Gijselaers, W. H. (2007). A structural equation model analyzing the relationship of students attitudes toward statistics, prior reasoning abilities and course performance. *Statistics Education Research Journal, 6*(2), 78–102. https://doi.org/10.52041/serj.v6i2.486

Thompson, P. W., & Liu, Y. (2005). Understandings of margin of error. In S. Wilson (Ed.), *Proceedings of the twenty-seventh annual meeting of the International Group for the Psychology of mathematics education*. Virginia Tech.

Thornton, R., & Thornton, J. (2004). Erring on the margin of error. *Southern Economic Journal, 71*(1), 130–135. https://doi.org/10.1002/j.2325-8012.2004.tb00628.x
Tietze, U.-P., Klika, M., & Wolpers, H. (2002). *Mathematikunterricht in der Sekundarstufe II. Band 3. Didaktik der Stochastik [teaching mathematics in secondary school. Volume 3. Didactics of stochastics]*. Vieweg.
TinkerPlots [Computer software]. (2022). Retrieved from https://www.tinkerplots.com/
Tintle, N., Carver, R., Chance, B., Cobb, G., Rossman, A., Roy, S., Swanson, T., & Vander Stoep, J. (2019). *Introduction to statistical investigations*. Wiley.
To, K. K., & Pang, M. F. (2019). A study of variation theory to enhance students' genre awareness and learning of genre features. *International Journal for Lesson and Learning Studies, 8*(3), 183–195. https://doi.org/10.1108/IJLLS-10-2018-0070
Toulmin, S. E. (2003). *The uses of argument* (Updated ed.). Cambridge University Press. (Original work published 1958).
Tractenberg, R. (2017). How the mastery rubric for statistical literacy can generate actionable evidence about statistical and quantitative learning outcomes. *Education Sciences, 7*(1), 3. https://doi.org/10.3390/educsci7010003
Trouche, L. (2014). Instrumentation in mathematics education. In S. Lerman (Ed.), *Encyclopedia of mathematics education*. Springer. https://doi.org/10.1007/978-94-007-4978-8_80
Tsai, C. C., Ho, H. N. J., Liang, J. C., & Lin, H. M. (2011). Scientific epistemic beliefs, conceptions of learning science and self-efficacy of learning science among high school students. *Learning and Instruction, 21*(6), 757–769. https://doi.org/10.1016/j.learninstruc.2011.05.002
Tunstall, S. L. (2018). Investigating college students' reasoning with messages of risk and causation. *Journal of Statistics Education, 26*(2), 76–86. https://doi.org/10.1080/10691898.2018.1456989
Tversky, A., & Kahneman, D. (1971). Belief in the law of small numbers. *Psychological Bulletin, 76*, 105–110. https://doi.org/10.1037/h0031322
Tversky, A., & Kahneman, D. (1973). Availability: A heuristic for judging frequency and probability. *Cognitive Psychology, 5*(2), 207–232. https://doi.org/10.1016/0010-0285(73)90033-9
Tversky, A., & Kahneman, D. (1974). Judgment under uncertainty: Heuristics and biases. *Science, 185*(4157), 1124–1131. https://doi.org/10.1126/science.185.4157.1124
Tygel, A., & Kirsch, R. (2015). Contributions of Paulo Freire for a critical data literacy. In *Proceedings of web science 2015 workshop on data literacy* (pp. 318–334).
Umugiraneza, O., Bansilal, S., & North, D. (2017). Exploring teachers' practices in teaching mathematics and statistics in KwaZulu-Natal schools. *South African Journal of Education, 37*(2). https://doi.org/10.15700/SAJE.V37N2A1306
Umugiraneza, O., Bansilal, S., & North, D. (2022). An analysis of teachers' confidence in teaching mathematics and statistics. *Statistics Education Research Journal, 21*(3). https://doi.org/10.52041/serj.v21i3.422
Used Numbers. (1989). Dale Seymour Publications.
Van Blokland, P., & Van de Giessen, C. (2020). *VUSTAT* [computer software]. Amsterdam, the Netherlands: VUSOFT. https://www.vustat.eu/apps/yesno/index.html
Van Dijke-Droogers, M. (2021). *Introducing statistical inference: Design and evaluation of a learning trajectory*. Doctoral dissertation. Utrecht University. https://www.fisme.science.uu.nl/publicaties/literatuur/2021_van_dijke_introducing_statistical_inferences.pdf
Van Dijke-Droogers, M. J. S., Drijvers, P. H. M., & Bakker, A. (2020). Repeated sampling with a black box to make informal statistical inference accessible. *Mathematical Thinking and Learning, 22*(2), 116–138. https://doi.org/10.1080/10986065.2019.1617025
Van Dijke-Droogers, M., Drijvers, P., & Bakker, A. (2021). Introducing statistical inference: Design of a theoretically and empirically based learning trajectory. *International Journal of Science and Mathematics Education, 1–24*, 1743–1766. https://doi.org/10.1007/s10763-021-10208-8
Van Dooren, W., De Bock, D., Depaepe, F., Janssens, D., & Verschaffel, L. (2003). The illusion of linearity: Expanding the evidence towards probabilistic reasoning. *Educational Studies in Mathematics, 53*(2), 113–138. https://doi.org/10.1023/A:1025516816886

Van Griethuijsen, R. A., van Eijck, M. W., Haste, H., Den Brok, P. J., Skinner, N. C., Mansour, N., et al. (2015). Global patterns in students' views of science and interest in science. *Research in Science Education, 45*(4), 581–603. https://doi.org/10.1007/s11165-014-9438-6

Vance, E. A., & Pruitt, T. (2016). *Virginia Tech's laboratory for interdisciplinary statistical analysis annual report 2015–16*. Virginia Tech. Laboratory for Interdisciplinary Statistical Analysis. http://hdl.handle.net/10919/72099. Accessed 19 Mar 2022.

Vance, E. A., Alzen, J. L., & Smith, H. S. (2022). Creating shared understanding in statistics and data science collaborations. *Journal of Statistics and Data Science Education, 30*, 1–17.

Varela, F. J., Thompson, E., & Rosch, E. (1991). *The embodied mind: Cognitive science and human experience*. MIT Press.

Verillon, P., & Rabardel, P. (1995). Cognition and artifacts: A contribution to the study of thought in relation to instrumented activity. *European Journal of Psychology of Education, 10*, 77–103. https://www.jstor.org/stable/23420087

Vygotsky, L. S. (1997). *Educational psychology*. CRC Press. https://www.taylorfrancis.com/chapters/mono/10.4324/9780429273070-9/

Walter, D. (2018). *Nutzungsweisen bei der Verwendung von Tablet-Apps* (Use patterns when using tablet apps). Springer. https://doi.org/10.1007/978-3-658-19067-5

Wang, L., & Cai, R. (2016). Classroom questioning tendencies from the perspective of big data. *Frontiers of Education in China, 11*(2), 125–164. https://doi.org/10.1007/bf03397112

Wassan, J. T. (2015). Discovering big data modelling for educational world. *Procedia-Social and Behavioral Sciences, 176*, 642–649.

Watkins, A., Bargagliotti, A., & Franklin, C. (2014). Simulation of the sampling distribution of the mean can mislead. *Journal of Statistics Education, 22*(3), 1–21. https://doi.org/10.1080/10691898.2014.11889716

Watson, J. M. (2006). *Statistical literacy at school: Growth and goals*. Routledge. https://doi.org/10.4324/9780203053898

Watson, J. M. (2013). *Statistical literacy at school: Growth and goals*. Routledge. https://doi.org/10.4324/9780203053898

Watson, J. (2016). Whither statistics education research? In *Proceedings of the 39th annual conference of the mathematics education research group of Australasia (MERGA), 3–7 July 2016* (pp. 33–58).

Watson, J., & Callingham, R. (2003). Statistical literacy: A complex hierarchical construct. *Statistics Education Research Journal, 2*(2), 3–46.

Watson, J., & English, L. D. (2016). Repeated random sampling in year 5. *Journal of Statistics Education, 24*(1), 27–37. https://doi.org/10.1080/10691898.2016.1158026

Watson, J. M., & English, L. (2017). Reaction time in grade 5: Data collection within the practice of statistics. *Statistics Education Research Journal, 16*(1), 262–293. https://doi.org/10.52041/serj.v16i1.231

Watson, J. M., & English, L. (2018). Eye color and the practice of statistics in grade 6: Comparing two groups. *The Journal of Mathematical Behavior, 49*, 35–60. https://doi.org/10.1016/j.jmathb.2017.06.006

Watson, J., Fitzallen, N., Wilson, K., & Creed, J. (2008). The representational value of HATS. *Mathematics Teaching in Middle School, 14*(1), 4–10. https://doi.org/10.5951/MTMS.14.1.0004

Weber, K., Maher, C., Powell, A., & Lee, H. S. (2008). Learning opportunities from group discussions: Warrants become the objects of debate. *Educational Studies in Mathematics, 68*, 247–261. https://doi.org/10.1007/s10649-008-9114-8

Weiland, T. (2017). Problematizing statistical literacy: An intersection of critical and statistical literacies. *Educational Studies in Mathematics, 96*, 33–47.

Weiland, T. (2019a). Critical mathematics education and statistics education: Possibilities for transforming the school mathematics curriculum. In G. Burrill & D. Ben-Zvi (Eds.), *Topics and trends in current statistics education research. International perspectives* (pp. 391–411). Springer.

Weiland, T. (2019b). The contextualized situations constructed for the use of statistics by school mathematics textbooks. *Statistics Education Research Journal, 18*(2), 18–38. https://doi.org/10.52041/serj.v18i2.13

Wessels, H., & Nieuwoudt, H. (2011). Teachers' professional development needs in data handling and probability. *Pythagoras, 32*(1). https://doi.org/10.4102/pythagoras.v32i1.10

Western Cape Education Department (WCED). (1998). Guideline document: National examination 2001 mathematics paper 1 & 2. *Higher Grade and Standard grade. Unpublished.* WCED.

Wild, C. (2006). The concept of distribution. *Statistics Education Research Journal, 5*(2), 10–26. https://doi.org/10.52041/serj.v5i2.497

Wild, C. J., & Pfannkuch, M. (1999a). Statistical thinking in empirical enquiry. *International Statistical Review, 67*(3), 223–265. https://doi.org/10.1111/j.1751-5823.1999.tb00442.x

Wild, C. J., & Pfannkuch, M. (1999b). Statistical thinking in empirical enquiry (with discussion). *International Statistical Review, 67*(3), 223–265.

Wild, C. J., Pfannkuch, M., Regan, M., & Horton, N. J. (2011a). Towards more accessible conceptions of statistical inference. *Journal of the Royal Statistical Society: Series A (Statistics in Society), 174*(2), 247–295.

Wild, C. J., Pfannkuch, M., Regan, M., & Horton, N. J. (2011c). Towards more accessible conceptions of statistical inference. *Journal of the Royal Statistical Society, 174*(2), 247–295. https://doi.org/10.1111/j.1467-985X.2010.00678.x

Wild, C. J., Utts, J. M., & Horton, N. J. (2018a). What is statistics? In D. Ben-Zvi, K. Makar, & J. Garfield (Eds.), *International handbook of research in statistics education* (pp. 5–36). Springer International Publishing. https://doi.org/10.1007/978-3-319-66195-7_1

Wild, C. J., Utts, J. M., & Horton, N. J. (2018b). What is statistics? In D. Ben-Zvi, J. Garfield, & K. Makar (Eds.), *The first handbook of research on statistics teaching and learning*. Springer. https://doi.org/10.1007/978-3-319-66195-7_

Wilson, A. D., & Golonka, S. (2013). Embodied cognition is not what you think it is. *Frontiers of Psychology, 4*, 58. https://doi.org/10.3389/fpsyg.2013.00058

Wise, A. (2019). Educating data scientists and data literate citizens for a new generation of data. *Journal of the Learning Sciences, 29*(1), 165–181. https://doi.org/10.1080/10508406.2019.1705678

Wise, A. F. (2020). Educating data scientists and data literate citizens for a new generation of data. *Journal of the Learning Sciences, 29*(1), 165–181.

Wolff, A., Gooch, D., Montaner, J. J., Rashid, U., & Kortuem, G. (2016). Creating an understanding of data literacy for a data-driven society. *The Journal of Community Informatics, 12*(3). https://doi.org/10.15353/joci.v12i3.3275

Xidong, W., & Xiaoye, L. (2016). Study of higher education reform under the background of big data. *Innovation in Regional Public Service for Sustainability, 505.* https://doi.org/10.2991/icpm-16.2016.136

Xu, X., Wang, Y., & Yu, S. (2018). Teaching performance evaluation in smart campus. *IEEE Access, 6*, 77754–77766.

Yang, Q., Li, J., & Zou, X. (2019). Big data and higher vocational and technical education: Green tourism curriculum. In *Proceedings of the 2019 international conference on big data and education* (pp. 108–112). https://doi.org/10.1145/3322134.3322149

Yawei, L., & Shiming, Z. (2019). The role and task of innovation and entrepreneurship teachers under the background of big data. In *Proceedings of the 2019 international conference on big data and education* (pp. 98–102). https://doi.org/10.1145/3322134.3322146

Ying, Y. (2019). Research on college students' information literacy based on big data. *Cluster Computing, 22*(S2), 3463–3470. https://doi.org/10.1007/s10586-018-2193-0

Yolcu, A. (2012). *An investigation of eighth grade students' statistical literacy, attitudes towards statistics and their relationship* (Unpublished master's thesis). Middle East Technical University, Ankara.

Yu, X., & Wu, S. (2015). Typical applications of big data in education. In *2015 International conference of Educational Innovation Through Technology (EITT)* (pp. 103–106). IEEE. https://doi.org/10.1109/EITT.2015.29

Zapata-Cardona, L. (2016a). ¿Estamos promoviendo el pensamiento estadístico en la enseñanza? [Are we promoting statistical thinking in teaching?] *Segundo Encuentro Colombiano de Educación Estocástica (2 ECEE)*. Bogotá, Colombia.

Zapata-Cardona, L. (2016b). Enseñanza de la estadística desde una perspectiva crítica [Teaching statistics from a critical perspective]. *Yupana, 10*, 30–41.

Zapata-Cardona, L. (2018). Students' construction and use of statistical models: A socio–critical perspective. *ZDM, 50*(7), 1213–1222. https://doi.org/10.1007/s11858-018-0967-8

Zapata-Cardona, L., & González-Gómez, D. (2017). Imágenes de los profesores sobre la estadística y su enseñanza [Images of teachers on statistics and its teaching]. *Educación Matemática, 29*(1), 61–89.

Zapata-Cardona, L., & Marrugo, L. M. (2019). Critical citizenship in Colombian statistics textbooks. In G. Burrill & D. Ben-Zvi (Eds.), *Topics and trends in current statistics education research. International perspectives* (pp. 373–389). Springer.

Zeelenberg, K., & Braaksma, B. (2017). Big data in official statistics. In T. Prodromou (Ed.), *Data visualisation and statistical literacy for open and big data* (pp. 274–296). IGI Global.

Zeide, E. (2017). The structural consequences of big data-driven education. *Big Data, 5*(2), 164–172. https://doi.org/10.1089/big.2016.0061

Zieffler, A., Garfield, J., Delmas, R., & Reading, C. (2008a). A framework to support research on informal inferential reasoning. *Statistics Education Research Journal, 7*(2), 40–58. https://doi.org/10.52041/serj.v7i2.469

Zieffler, A., Garfield, J., Alt, S., Dupuis, D., Holleque, K., & Chang, B. (2008b). What does research suggest about the teaching and learning of introductory statistics at the college level? A review of the literature. *Journal of Statistics Education, 16*(2). https://doi.org/10.1080/10691898.2008.11889566

GPSR Compliance
The European Union's (EU) General Product Safety Regulation (GPSR) is a set of rules that requires consumer products to be safe and our obligations to ensure this.

If you have any concerns about our products, you can contact us on

ProductSafety@springernature.com

In case Publisher is established outside the EU, the EU authorized representative is:

Springer Nature Customer Service Center GmbH
Europaplatz 3
69115 Heidelberg, Germany

www.ingramcontent.com/pod-product-compliance
Ingram Content Group UK Ltd.
Pitfield, Milton Keynes, MK11 3LW, UK
UKHW020232290926
15470UKWH00002B/496
* 9 7 8 3 0 3 1 2 9 4 5 8 7 *